AF573878

Cognitive Technologies

Titles in this series now included in the Thomson Reuters Book Citation Index and Scopus!

The Cognitive Technologies (CT) series is committed to the timely publishing of high-quality manuscripts that promote the development of cognitive technologies and systems on the basis of artificial intelligence, image processing and understanding, natural language processing, machine learning and human-computer interaction.

It brings together the latest developments in all areas of this multidisciplinary topic, ranging from theories and algorithms to various important applications. The intended readership includes research students and researchers in computer science, computer engineering, cognitive science, electrical engineering, data science and related fields seeking a convenient way to track the latest findings on the foundations, methodologies and key applications of cognitive technologies.

The series provides a publishing and communication platform for all cognitive technologies topics, including but not limited to these most recent examples:

- Interactive machine learning, interactive deep learning, machine teaching
- Explainability (XAI), transparency, robustness of AI and trustworthy AI
- Knowledge representation, automated reasoning, multiagent systems
- Common sense modelling, context-based interpretation, hybrid cognitive technologies
- Human-centered design, socio-technical systems, human-robot interaction, cognitive robotics
- Learning with small datasets, never-ending learning, metacognition and introspection
- Intelligent decision support systems, prediction systems and warning systems
- Special transfer topics such as CT for computational sustainability, CT in business applications and CT in mobile robotic systems

The series includes monographs, introductory and advanced textbooks, state-of-the-art collections, and handbooks. In addition, it supports publishing in Open Access mode.

Massih-Reza Amini

Advanced Supervised and Semi-supervised Learning

Theory and Algorithms

Springer

Massih-Reza Amini
Grenoble Alpes University
Saint Martin d'Hères, France

ISSN 1611-2482 ISSN 2197-6635 (electronic)
Cognitive Technologies
ISBN 978-3-031-99927-7 ISBN 978-3-031-99928-4 (eBook)
https://doi.org/10.1007/978-3-031-99928-4

This is a revised and improved translation from the French language edition: "Apprentissage machine: de la théorie à la pratique" by Massih-Reza Amini, © Author 2015. Published by Editions Eyrolles. All Rights Reserved.

This Springer imprint is published by the registered company Springer Nature Switzerland AG
The registered company address is: Gewerbestrasse 11, 6330 Cham, Switzerland

To my mother and family, whose unwavering belief in me has been a constant source of strength and inspiration.

Preface

Machine learning is one of the leading areas of artificial intelligence. It concerns the study and development of quantitative algorithms that allow a computer to perform tasks without being programmed to do them. Learning in this context is about recognizing complex patterns and making intelligent decisions. Given all the existing inputs, the difficulty in accomplishing this task lies in the fact that the set of possible decisions is usually very complex to provide. To overcome this difficulty, machine learning algorithms have been designed with the aim of acquiring knowledge about the problem to be treated based on a set of limited data from this problem.

Studied Concepts

To illustrate this principle, let us consider the empirical risk minimization framework in supervised learning which we will treat in this textbook. According to this framework, the decision to be made on a given input is made based on the output of a prediction function which is inferred using a set of labeled examples (or training set), where each such example is a pair consisting of the vector representation of an observation in a given vector space, and a response associated with the example (also called desired output). After the training phase, the function returned by the algorithm must be able to predict the response associated with new observations. The underlying assumption in this case is that the examples are, in general, representative of the prediction problem on which the function will be applied. In practice, an error function measures the deviation between the model's prediction on an example and its desired output. From a given training set, the learning algorithm then chooses a function, resulting from a set of functions defined beforehand, which achieves the lowest average error on the examples of the training set.

This empirical error is generally not representative of the performance of the algorithm on new examples. It is then necessary to have a second set of labeled

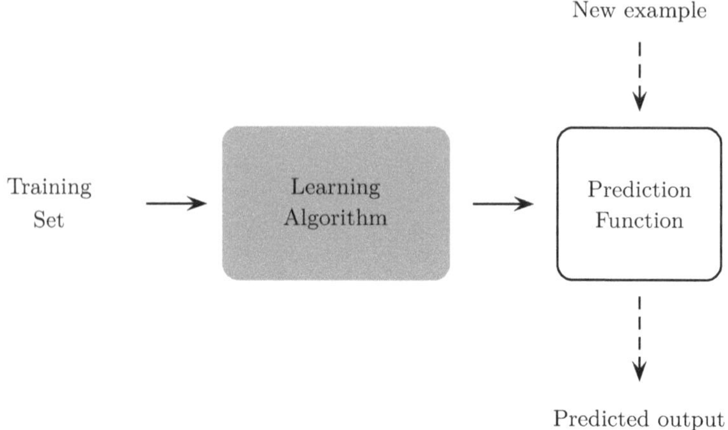

Fig. 1 Illustration of the two phases of the empirical risk minimization framework in a supervised learning problem. In the training phase (schematized by the solid lines), a function minimizing the empirical error on a training set is found among a class of predefined functions. In the test phase (schematized by the dotted lines), the outputs of new examples are predicted by the prediction function that has been found.

examples, or a test set, to which the algorithm did not have access during its training phase in order to estimate its average error. The learning algorithm is expected to find a function with good performance on new examples and not the one that will be able to perfectly reproduce the desired outputs associated with the training examples (see Fig. 1). The learnability guarantees of the empirical risk minimization process have been studied in the machine learning theory largely initiated by Vapnik [177]. They depend on the size of the training base and the complexity of the class of functions where the prediction function is sought.

Historically, the two main tasks of the supervised learning framework were classification and regression. These tasks are similar unlike the desired output space of the examples. In the case of classification, the output space is discrete whereas in regression this space is real.

In the late 1990s, driven by advancements in new technologies, particularly those associated with the growth of the Internet, new learning paradigms began to emerge. Among these approaches is the concept of learning with partially labeled data, commonly referred to as semi-supervised learning. This concept was born out of the realization that constituting labeled training datasets require significant effort and resources, making labeled data acquisition costly, whereas unlabeled data are readily available and contain valuable information pertaining to the specific problem at hand. Consequently, numerous research efforts were initiated with the aim of leveraging a limited amount of labeled data in conjunction with a substantial pool of unlabeled data to develop predictive models.

The other framework that has generated much work in the learning community since the 2000s concerns the development of ranking models. This framework initially formalized the problems of Information Retrieval and was subsequently extended to other more general problems.

For many years, learning algorithms designed within these frameworks have proven their effectiveness across a diverse range of applications, encompassing domains such as speech and handwriting recognition, computer vision, structural prediction, protein analysis, recommendation systems, document classification, search engines, and more.

Organization of the Textbook

This textbook presents the scientific foundations of supervised learning theory, the most widespread algorithms developed according to this framework as well as the two learning frameworks mentioned above, at a level accessible to master's students and engineering students. Each chapter presents a specific supervised or semi-supervised learning topic and concludes with a series of carefully designed exercises to deepen understanding and promote student's active learning process. In addition, the textbook introduces a curriculum structure suitable for both target groups, making it an ideal resource for advanced undergraduate and graduate courses in machine learning.

It is organized into eight main chapters and an appendix; our concern has been to provide a coherent presentation linking the theory to the algorithms developed in this field.

The sequence of ideas presented in each of the chapters is as follows:

- In Chap. 1, we describe the fundamental concepts of the statistical learning theory of Vapnik [177]. We expose the notion of consistency of the empirical risk minimization principle according to which most supervised learning algorithms have been developed. The study of this consistency will lead us to the presentation of the second fundamental principle in supervised learning, which is the structural risk minimization, opening the field to the development of new models in machine learning. In particular, we present in this chapter the notion of a bound on the generalization error by describing the assumptions and the tools necessary to obtain it.
- In Chap. 2, we will present bounds on the generalization error which can be estimated on a training set used to learn a model. These bounds are based on a notion of data-dependent class complexity, called Rademacher complexity. With this notion it is also possible to easily derive bounds on the generalization error for multi-class classification problems. We expose multi-class approaches based on binary classification, called combined approaches, and derive a bound on the generalization error of linear classifiers in this case.
- In Chap. 3 we present the basic optimization algorithms for the minimization of a convex surrogate loss function called *direction of descent algorithms*. In particular, we present the necessary conditions to verify a descent direction

algorithm to converge to the minimizer of a convex objective function and we describe some simple and efficient variants of this algorithm.
- Chap. 4 presents the main models based on formal neurons, more commonly called artificial neural networks or *Deep Learning*. These models were precursors to the development of quantitative methods in Artificial Intelligence, and particular care has been taken to present these models in their historical context.
- In Chap. 5, we present the Support Vector Machines (SVM) which are derived from the structural risk minimization principal. These models have become very popular thanks to their theoretical justifications. In particular, we will see how to use the kernel trick to embed the input space into a higher dimensional space in which the learning problem becomes simpler to solve and we present the extension of these to the multi-class case.
- Chap. 6 presents the AdaBoost algorithm. This algorithm combines several base classifiers, called weak learners, to build a final classifier, called strong learner, which performs better than each of these base classifiers. In particular, we will make the link between this algorithm and the empirical risk minimization principle stated by Vapnik [177]. We also expose the extension of this algorithm to the multi-class case.
- In Chap. 7 we formally describe the framework for learning to rank functions by focusing on two particular forms of ranking called ranking of alternatives and ranking of instances. We then expose some algorithms developed following classical approaches to learning-to-rank functions. We end this chapter by showing the reduction of some ranking problems to the binary classification of pairs of observations. This reduction opens the way to learning classifiers with interdependent examples that we analyze with the result of Janson [88].
- We then expose the semi-supervised learning framework in Chap. 8. We start this chapter by presenting the EM and CEM algorithms developed in the framework of unsupervised learning, detailing some special cases leading to well-known unsupervised models such as the K-means algorithm. We then present the basic assumptions in semi-supervised learning by detailing the three generative, discriminative, and graph-based approaches developed according to this framework.

How to Use This Textbook

This textbook provides a clear and accessible path through the foundations and advances of supervised and semi-supervised learning. To facilitate its use as a curriculum resource, we propose the following reading scheme:

- *Core Chapters:* Chaps. 1, 2, and 3 are foundational and should be read in sequence. They introduce the main theoretical tools and optimization methods essential for understanding modern machine learning.

- *Application Chapters:* Chaps. 4, 5, and 6 can be read in any order after the core chapters. Each focuses on a major family of algorithms and their theoretical underpinnings.
- *Advanced Topics:* Chaps. 7 (Learning-to-Rank) and 8 (Semi-Supervised Learning) build on the earlier material but can be approached independently, depending on the reader's interests or course requirements.
- *Appendix:* The appendix provides mathematical background and additional proofs; it is recommended for readers seeking a deeper or more rigorous understanding.

For instructors, this structure allows the textbook to be used flexibly in semester-long or modular courses, with clear indications of which chapters are essential and which can be treated as electives or advanced topics.

What Is New in This Textbook?

This textbook brings several original and distinguishing features:

- *Integration of Supervised and Semi-Supervised Learning:* Unlike most texts, this textbook gives equal weight to both supervised and semi-supervised paradigms, reflecting the latest research and practical needs.
- *Modern Theory Accessible to Students:* The textbook presents advanced theoretical concepts (such as Rademacher complexity, generalization bounds, and structural risk minimization) at a level accessible to master's and engineering students, with step-by-step explanations and intuitive illustrations.
- *Curriculum-Ready with Exercises:* Each chapter concludes with a carefully designed set of exercises, designed to reinforce key ideas and support self-study or classroom use. These exercises range from theoretical questions to conceptual ones, making the textbook suitable for a wide range of learners.
- *Bridging Theory and Practice:* Throughout, we emphasize the connection between theoretical guarantees and practical algorithm design, helping readers understand not just how algorithms work, but why they work.
- *Flexible Reading Paths:* By clearly indicating prerequisites and dependencies between chapters, the textbook supports both linear and modular reading, making it adaptable to different course structures and learning goals.

Saint Martin d'Hères, France Massih-Reza Amini

Declarations

Competing Interests The author has no competing interests to declare that are relevant to the content of this manuscript.

Contents

Chapter 1
Fundamentals of Supervised Learning

In this chapter, we present the theory of machine learning according to the framework of Vapnik [177] which will be used as a basis in our description of the learning algorithms described in the following chapters. More specifically, we present the notion of consistency which guarantees the learnability of a prediction function. The definitions and basic assumptions of this theory, as well as the principle of empirical risk minimization, are described in Sect. 1.1. The study of the consistency of this principle, presented in Sect. 1.2, leads us to the second principle of structural risk minimization, which states that learning is a compromise between a small empirical error and a large capacity of the class of prediction functions.

A learning algorithm builds a prediction function using a finite set of examples, known as the training set [21, 49, 61, 151]. According to the supervised framework, each example is a pair generally made up of the vector representation of an observation and its associated desired output.

The goal of learning is to induce a function that predicts the outputs associated with new observations by committing the lowest possible generalization error. This output is usually a real value or a class label, as we will see below. The underlying assumption here is that the phenomenon which has generated the data is stationary, i.e. the examples in the training set, on which the prediction function is learned, are somehow representative of the general problem that the we want to solve. We will return to this assumption in the next section.

In practice, among an existing class of functions, the learning algorithm chooses the function which achieves the lowest average prediction error (or empirical error) on the training set. The error function quantifies the disagreement between the output prediction given by the function that one wishes to learn for an example of the training set and its associated desired output. The aim here, is not to induce a function that gives exactly the desired outputs of the examples of the training set (or to overfit), but to find, as we have just evoked it, the function that will have good generalization performance.

M.-R. Amini, *Advanced Supervised and Semi-supervised Learning*,
Cognitive Technologies, https://doi.org/10.1007/978-3-031-99928-4_1

In logic, this process of finding a general rule from a finite set of examples is called "induction" [63, chapter 7, pp.161–176].[1] In machine learning, the inductive framework has been set up following the Empirical Risk Minimization (ERM) principle and its statistical properties have been studied in the theory developed by [177]. The striking result of this theory is an upper bound on the generalization error of the learned function which is expressed as a function of the empirical error of the latter on the training set and the complexity of the class of functions that is used.

This complexity translates the capacity of the class of functions to solve the prediction problem and the greater the flexibility in assigning diverse output labels to examples, the larger this capacity becomes. In other words, the greater the capacity, the lower the empirical risk would be and the less guaranteed we have to achieve the main learning objective, which is to have a low generalization error. This bound thus exhibits the trade-off that exists between the empirical error and the capacity of the class of functions, and shows a way to minimize the bound on the generalization error by minimizing the empirical error while controlling the capacity of the class of functions.

This principle is called structural risk minimization (SRM), which, along with the ERM principle, forms the basis of a large number of learning algorithms. Moreover, they can explain the foundational principles of algorithms designed prior to the establishment of the theory of [177]. The rest of this chapter is devoted to the more formal presentation of these different concepts according to the framework of the binary classification, which constituted the initial framework for the development of this theory.

1.1 Empirical Risk Minimization Principle

In this section, we will present the empirical risk minimization principle by first fixing the notations which will be used thereafter.

1.1.1 Assumption and Definitions

We assume that the examples have a representation in a fixed d-dimensional vector space, called the "input space", $\mathcal{X} \subseteq \mathbb{R}^d$. The desired outputs of the examples are assumed to be part of an output set $\mathcal{Y} \subset \mathbb{R}$. Until the early 2000s, there were two major frameworks of supervised learning problems; classification and regression. In

[1] The opposite reasoning called deduction is based on axioms and produces specific rules (which are always true) as consequences of the axioms.

classification, the output set $\mathcal{Y}$ is discrete and the prediction function $f : \mathcal{X} \rightarrow \mathcal{Y}$ is called a classifier. When $\mathcal{Y}$ is continuous, f is called a regression function.

In Chap. 7, we will present the framework for ranking functions that has been developed recently in the machine learning and information retrieval communities. A pair $(\mathbf{x}, y) \in \mathcal{X} \times \mathcal{Y}$ thus designates a labeled example and $S = (\mathbf{x}_i, y_i)_{i=1}^m \in (\mathcal{X} \times \mathcal{Y})^m$ denotes a set of training examples.

In the particular case of the binary classification that we consider in this chapter, we denote the output space by $\mathcal{Y} = \{-1, +1\}$ and an example $(\mathbf{x}, +1)$ (respectively $(\mathbf{x}, -1)$) is called a positive (respectively negative) example. For instance, consider the problem of classifying emails, consisting in labeling them according to two classes: hams and spams. We will represent the emails by vectors in a given vector space and we will designate one of the classes (for example the class of hams) by the class label $+1$ and the other class by the class label -1.

The fundamental assumption of machine learning theory is that all examples are generated *independently and identically* according to a fixed, but unknown, probability distribution, denoted $\mathcal{D}$. The "identically" distributed hypothesis ensures that the phenomenon is stationary, while the independently distributed hypothesis states that each individual example contributes maximum information to solve the prediction problem. According to this hypothesis, the examples $(\mathbf{x}_i, y_i)$ of any training set S and test set are assumed to be identically and independently distributed (i.i.d.) according to $\mathcal{D}$. That is, each set is a sample of i.i.d. examples according to $\mathcal{D}$.

This assumption thus characterizes the notion of representativeness of a training and test sets with respect to the prediction problem, i.e. the training examples as well as test examples and their desired output are assumed to be generated from the same source of information.

Another basic concept in learning is the notion of loss, also called risk or error. For a given prediction function f, the disagreement between the desired output y of an example $\mathbf{x}$ and the prediction $f(\mathbf{x})$ is measured using an instantaneous loss function defined by:

$$\ell : \mathcal{Y} \times \mathcal{Y} \rightarrow \mathbb{R}^+.$$

In general, this function is a distance on the output set $\mathcal{Y}$ and it measures the difference between the desired and predicted outputs for a given example. In regression, the usual instantaneous loss functions are ℓ_1 and ℓ_2 norms of the difference between the desired and the predicted outputs of a given example. In binary classification, the commonly considered instantaneous loss is the 0/1 loss, which for an example $(\mathbf{x}, y)$ and a prediction function f is defined by:

$$\ell(f(\mathbf{x}), y) = \mathbb{1}_{f(\mathbf{x}) \neq y},$$

where $\mathbb{1}_\pi$ equals 1 if the predicate π is true and 0 otherwise. In practice, in the case of binary classification, the learned function $h : \mathcal{X} \rightarrow \mathbb{R}$ is a real-valued function and the associated classifier $f : \mathcal{X} \rightarrow \{-1, +1\}$ is defined by taking the sign function on the output of h.

In this case, the instantaneous error equivalent to the 0/1 loss, defined for the function h is:

$$\ell_0 : \mathbb{R} \times \mathcal{Y} \to \mathbb{R}^+$$
$$(h(\mathbf{x}), y) \mapsto \mathbb{1}_{y \times h(\mathbf{x}) \leqslant 0}.$$

From an instantaneous loss and the i.i.d. generation of the examples according to the distribution $\mathcal{D}$, we can define the generalization error of a learned function $f \in \mathcal{F}$ as:

$$\mathcal{L}(f) = \mathbb{E}_{(\mathbf{x},y)\sim\mathcal{D}}[\ell(f(\mathbf{x}), y)] = \int_{\mathcal{X}\times\mathcal{Y}} \ell(f(\mathbf{x}), y) d\mathcal{D}(\mathbf{x}, y), \tag{1.1}$$

where $\mathbb{E}_{(\mathbf{x},y)\sim\mathcal{D}}[X(\mathbf{x}, y)]$ is the expectation of the random variable $X(\mathbf{x}, y)$ when $(\mathbf{x}, y)$ follows the probability distribution $\mathcal{D}$. As $\mathcal{D}$ is unknown, this generalization error cannot be estimated exactly, and to measure the performance of a function f, one often uses a set of examples S of size m on which we can estimate the empirical error of f defined by:

$$\hat{\mathcal{L}}_m(f, S) = \frac{1}{m} \sum_{i=1}^{m} \ell(f(\mathbf{x}_i), y_i). \tag{1.2}$$

Thus, to solve a classification problem for which we have a training set S, it is natural to choose a class of functions $\mathcal{F}$ and look for the classifier f_S which minimizes the empirical error on S (since this error is an unbiased estimator of the generalization error of f_S that cannot be measured).

1.1.2 Statement of the Principle

The foundation of the earliest machine learning models lies in this learning approach known as the empirical risk minimization principle.

The essential question that emerges is whether, by adhering to the ERM principle, *it is possible to consistently generate a prediction function that generalizes well from a finite set of observations?* The answer to this question is obviously no. To demonstrate this, consider the following toy problem of binary classification.

Example 1.1 (Overfitting [23]) Suppose the input dimension is $d = 1$. Let us take the space of observations the interval $[a, b] \subset \mathbb{R}$ where a and b are real numbers such that $a < b$ and the output space is $\{-1, +1\}$. Furthermore, assume that the distribution $\mathcal{D}$ generating the example pairs $(\mathbf{x}, y)$ is a uniform distribution over $[a, b] \times \{-1\}$. In other words, the examples are chosen randomly over the interval $[a, b]$ and, for each observation, the desired output is -1.

Now consider a learning algorithm that minimizes the empirical risk, by choosing a function from the class of functions $\mathcal{F} = \{f : [a, b] \to \{-1, +1\}\}$ as follows; given a training set $S = \{(\mathbf{x}_1, y_1), \ldots, (\mathbf{x}_m, y_m)\}$ the algorithm produces the prediction function f_S such that:

$$f_S(\mathbf{x}) = \begin{cases} -1, & \text{if } \mathbf{x} \in \{\mathbf{x}_1, \ldots, \mathbf{x}_m\}; \\ +1, & \text{otherwise.} \end{cases}$$

In this case, the classifier produced by the learning algorithm has an empirical risk equal to 0. However, as the classifier makes an error on the entire infinite set $[a, b]$ except for the examples of a finite training set, of measure zero, its generalization error is always equal to 1.

1.2 Consistency of the ERM Principle

The question underlying the earlier question is: *in which case the ERM principle is likely to generate a general learning rule?* The answer to this question lies in a statistical notion called consistency.

1.2.1 Definition

The concept of consistency indicates the two conditions that a learning algorithm must satisfy, namely *(a)* the algorithm must return a function whose empirical error reflects its generalization error when the size of the training set tends towards infinity and, *(b)* in the asymptotic case, the algorithm must make it possible to find a function which minimizes the generalization error in the class of functions considered $\mathcal{F}$. Formally:

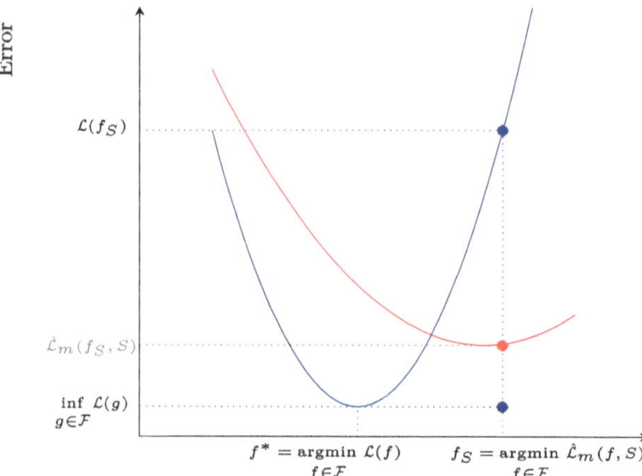

Fig. 1.1 Schematic description of the notion of consistency. The abscissa axis represents the class of functions $\mathcal{F}$ and the empirical (red) and generalization (blue) losses are shown as a function of $f \in \mathcal{F}$. The ERM principle consists of finding the function f_S in the class $\mathcal{F}$ which minimizes the empirical error on a training set S. This principle is consistent if in probability $\hat{\mathcal{L}}_m(f_S, S)$ converges to $\mathcal{L}(f_S)$ and $\inf_{g\in\mathcal{F}} \mathcal{L}(g)$

Definition 1.1 (Consistency)

(a) $\forall\epsilon > 0, \lim_{m\to\infty} \mathbb{P}(|\hat{\mathcal{L}}_m(f_S, S) - \mathcal{L}(f_S)| > \epsilon) = 0$, noted, $\hat{\mathcal{L}}_m(f_S, S) \xrightarrow{\mathbb{P}} \mathcal{L}(f_S)$.

(b) $\hat{\mathcal{L}}_m(f_S, S) \xrightarrow{\mathbb{P}} \inf_{g\in\mathcal{F}} \mathcal{L}(g)$.

These two conditions thus imply the convergence in probability of the empirical error $\hat{\mathcal{L}}_m(f_S, S)$ of the prediction function found by the learning algorithm on the training set S, f_S, towards its generalization error $\mathcal{L}(f_S)$ and $\inf_{g\in\mathcal{F}} \mathcal{L}(g)$ (Fig. 1.1).

A natural way to analyze the condition *(a)* of consistency, expressing the concept of generalization, is to use the following inequality:

$$|\mathcal{L}(f_S) - \hat{\mathcal{L}}_m(f_S, S)| \leqslant \sup_{g\in\mathcal{F}} |\mathcal{L}(g) - \hat{\mathcal{L}}_m(g, S)|. \tag{1.3}$$

We see clearly from this inequality that a sufficient condition for generalizing is that asymptotically, the empirical error of the prediction function, including the difference in absolute value between this error and its generalization error among all the other functions of a given class of functions $\mathcal{F}$ is the greatest, tends towards the generalization error of the function, i.e.:

$$\sup_{g\in\mathcal{F}} |\mathcal{L}(g) - \hat{\mathcal{L}}_m(g, S)| \xrightarrow{\mathbb{P}} 0. \tag{1.4}$$

This sufficient condition for generalization is a worst case consideration and, according to (1.3), it implies a bilateral uniform convergence for all functions of the class $\mathcal{F}$. Furthermore, the condition (1.4) does not depend on the algorithm considered but only on the class of functions $\mathcal{F}$. Thus, a necessary condition for the ERM principle to be consistent is that the class of functions considered is restricted (see the example on overfitting in the previous section).

1.2.2 *Worst Case Study*

The fundamental result of the learning theory [177, Theorem 2.1, p.38] exhibits another relation concerning the consistency of the ERM principle in the form of a unique unilateral uniform convergence in the worst case.

The ERM principle is consistent if and only if:

$$\forall\epsilon > 0, \lim_{m\to\infty} \mathbb{P}\left(\sup_{f\in\mathcal{F}}\left[\mathcal{L}(f) - \hat{\mathcal{L}}_m(f, S)\right] > \epsilon\right) = 0. \tag{1.5}$$

This condition states that the ERM principle is consistent if and only if, unilateral uniform convergence is ensured for the function in the class of functions $\mathcal{F}$ for which the difference between its generalization error and its empirical error is the largest. This phenomenon has led to its characterization as the "worst case study". The direct implication of this result is a uniform bound on the generalization error of any prediction function $f \in \mathcal{F}$ learned on a training set S of size m and which is of the following form:

$$\forall\delta \in]0, 1], \mathbb{P}\left(\forall f \in \mathcal{F}, (\mathcal{L}(f) - \hat{\mathcal{L}}_m(f, S)) \leqslant \mathfrak{C}(|\mathcal{F}|, m, \delta)\right) \geqslant 1 - \delta, \tag{1.6}$$

where $\mathfrak{C}$ is a term which depends on the size of the function class, $|\mathcal{F}|$, the size of the training set, m, and the desired precision $\delta \in]0, 1]$. Various methods have been explored to quantify the size of a class of functions, with the metric employed for this purpose typically referred to as the "complexity" or "capacity". In this chapter and the next one, we will introduce two of these measures, namely the VC dimension and the Rademacher complexity. These measures lead to distinct forms of generalization bounds and also give rise to the structural risk minimization principle.

1.3 Structural Risk Minimization Principle

Before presenting a generalization bound estimated on the training set which was used to find the prediction function, we will first consider the estimation of the generalization error of a prediction function on a test set [98]. The goal here is to show that it is possible to estimate an upper bound on the generalization error using a test set and that the empirical error on this set converges, in probability, towards the generalization error when the number of examples in the test set tends towards infinity and this independently of the capacity of the class of functions considered.

1.3.1 Estimation of Generalization Error on a Test Set

Recall that the examples of a test set are generated i.i.d. following the same probability distribution $\mathcal{D}$ which was used to generate the examples of a training set. Consider f_S a function learned on a training set, S. Let $T = \{(\mathbf{x}_i, y_i); i \in \{1, \ldots, n\}\}$ be a test set of size n. As the examples of this set do not intervene in the learning phase, the function f_S does not depend on the values of the instantaneous errors of the examples $(\mathbf{x}_i, y_i)$ of this set, and the random variables $(f_S(\mathbf{x}_i), y_i) \mapsto \ell(f_S(\mathbf{x}_i), y_i)$ can be considered as independent copies of the same random variable, i.e.:

$$\begin{aligned}\mathbb{E}_{T\sim\mathcal{D}^n}[\hat{\mathcal{L}}_n(f_S, T)] &= \mathbb{E}_{T\sim\mathcal{D}^n}\left[\frac{1}{n}\sum_{i=1}^{n}\ell(f_S(\mathbf{x}_i), y_i)\right] \\ &= \frac{1}{n}\sum_{i=1}^{n}\mathbb{E}_{(\mathbf{x},y)\sim\mathcal{D}}[\ell(f_S(\mathbf{x}), y)] = \mathcal{L}(f_S).\end{aligned}$$

Thus, the empirical error of f_S on the test base $\hat{\mathcal{L}}_(f_S, T)$ is an unbiased estimator of its generalization error.

Furthermore, for each example $(\mathbf{x}_i, y_i)$, let X_i denote the random variable defined by $\frac{1}{n}\ell(f_S(\mathbf{x}_i), y_i)$. As the random variables $X_i, i \in \{1, \ldots, n\}$ are independent and they have values in $\{0, \frac{1}{n}\}$, noting that $\hat{\mathcal{L}}_n(f_S, T) = \sum_{i=1}^{n} X_i$ and $\mathcal{L}(f_S) = \mathbb{E}\left(\sum_{i=1}^{n} X_i\right)$, we have from the Hoeffding inequality [81] (see Appendix A):

$$\forall\epsilon > 0, \mathbb{P}\left(\left[\mathcal{L}(f_S) - \hat{\mathcal{L}}_n(f_S, T)\right] > \epsilon\right) \leqslant e^{-2n\epsilon^2}. \tag{1.7}$$

To better understand this result, let us solve the equation $e^{-2n\epsilon^2} = \delta$ as a function of ϵ, i.e. $\epsilon = \sqrt{\frac{\ln 1/\delta}{2n}}$, and consider the opposite event:

$$\forall \delta \in]0, 1], \mathbb{P}\left(\mathcal{L}(f_S) \leqslant \hat{\mathcal{L}}_n(f_S, T) + \sqrt{\frac{\ln 1/\delta}{2n}}\right) \geqslant 1 - \delta. \tag{1.8}$$

For a low δ, we have according to (1.8) and with a high probability:

$$\mathcal{L}(f_S) \leqslant \hat{\mathcal{L}}_n(f_S, T) + \sqrt{\frac{\ln 1/\delta}{2n}},$$

which holds for all possible test sets of size n. Based on this result, we now possess an upper bound on the generalization error of a function learned from a training set, which can be computed on any test set. In scenarios where n is adequately large, this bound provides a reliable approximation of the actual generalization error.

Example 1.2 (Estimation of the Generalization Error on a Test Set [98]) Suppose that the empirical error of a prediction function f_S on a test set T of size $n = 1000$ is $\hat{\mathcal{L}}_n(f_S, T) = 0.23$. For $\delta = 0.01$, i.e. $\sqrt{\frac{\ln(1/\delta)}{2n}} \approx 0.047$, the generalization error of the function f_S is thus upper-bounded by 0.277 with a probability of at least $1 - \delta = 0.99$.

1.3.2 Uniform Bound on the Generalization Error

For a given prediction function, we know from the previous result how to bound its generalization error, using a test set on which the function parameters have not been found. In the context of studying the consistency of the ERM principle, we would now like to establish a uniform bound on the generalization error of a learned function as a function of its empirical error on a training set. We cannot answer this question using the previous development. This is mainly due to the fact that, when the function f_S has been found on the training set $S = ((\mathbf{x}_i, y_i))_{1\leqslant i\leqslant m}$, the random variables $X_i = \frac{1}{m}\ell(f_S(\mathbf{x}_i), y_i); i \in \{1, \ldots, m\}$, which are involved in estimating the empirical error of the function f_S on S, are all dependent on each other. Indeed, if we change one example in the training set, the found function f_S also changes, as do the weighted instantaneous errors of all the other examples. As a result, since the random variables X_i can no longer be considered independently distributed, we are no longer able to apply Hoeffding's inequality.

In the following, we will expose a uniform bound on the generalization error following the framework of Vapnik [177]. In the next section, we present another more recent framework, developed in the early 2000s, showing the link with the work of Vapnik [177]. For the uniform bound, our starting point is to upper-bound the probability:

$$\mathbb{P}\left(\sup_{f\in\mathcal{F}}\left|\mathcal{L}(f)-\hat{\mathcal{L}}_m(f,S)\right|>\epsilon\right). \tag{1.9}$$

Given a training set S of size m, if we consider the following set:

$$\mathfrak{F}(\mathcal{F},S)=\left\{\left((\mathbf{x}_1,f(\mathbf{x}_1)),\ldots,(\mathbf{x}_m,f(\mathbf{x}_m))\right)\mid f\in\mathcal{F}\right\}, \tag{1.10}$$

the size of this set corresponds to the number of possible ways in which the functions of $\mathcal{F}$ can label examples $(\mathbf{x}_1,\ldots,\mathbf{x}_m)$. Since these functions have only two possible outputs (-1 or $+1$), the size of $\mathfrak{F}(\mathcal{F},S)$ is finite, bounded by 2^m, whatever the class of $\mathcal{F}$ functions considered. Thus, a learning algorithm minimizing the empirical risk on a learning set S chooses the function among $|\mathfrak{F}(\mathcal{F},S)|$ functions of $\mathcal{F}$ that performs the labeling of S examples resulting in the smallest error. In this way, only a finite number of functions are involved in calculating the empirical error in the expression of (1.9).

However, for a second set S' different from the first one, the set $\mathfrak{F}(\mathcal{F},S')$ will be different from $\mathfrak{F}(\mathcal{F},S)$ and it is impossible to apply the bound obtained for finite sets. The solution proposed by Vapnik and Chervonenkis is an elegant way of solving this problem. It consists in replacing the true error $\mathcal{L}(f)$ in the expression (1.9) by the empirical error of f on another sample of the same size as S, called *ghost sample*, and is formally stated as follows.

Lemma 1.3 (Symmetrization [177]) *Let $\mathcal{F}$ be a class of functions (which can be infinite) and S and S' be two training sets of the same size m. For any real $\epsilon>0$, such that $m\epsilon^2\geqslant 2$ we then have:*

$$\begin{aligned}&\mathbb{P}\left(\sup_{f\in\mathcal{F}}\left|\mathcal{L}(f)-\hat{\mathcal{L}}_m(f,S)\right|>\epsilon\right)\\&\quad\leqslant 2\mathbb{P}\left(\sup_{f\in\mathcal{F}}\left|\hat{\mathcal{L}}_m(f,S')-\hat{\mathcal{L}}_m(f,S)\right|>\epsilon/2\right).\end{aligned} \tag{1.11}$$

Proof Let $\epsilon > 0$ and $f_S^* \in \mathfrak{F}(\mathcal{F}, S)$ be the function that realizes the supremum $\sup_{f\in\mathcal{F}} \left|\mathcal{L}(f) - \hat{\mathcal{L}}_m(f, S)\right|$. From the remark above, f_S^* depends on the sample S, and we have:

$$
\begin{aligned}
\mathbb{1}_{|\mathcal{L}(f_S^*)-\hat{\mathcal{L}}_m(f_S^*,S)|>\epsilon}\mathbb{1}_{|\mathcal{L}(f_S^*)-\hat{\mathcal{L}}_m(f_S^*,S')|<\epsilon/2} \\
&= \mathbb{1}_{|\mathcal{L}(f_S^*)-\hat{\mathcal{L}}_m(f_S^*,S)|>\epsilon \wedge |\hat{\mathcal{L}}_m(f_S^*,S')-\mathcal{L}(f_S^*)|\geqslant -\epsilon/2} \\
&\leqslant \mathbb{1}_{|\hat{\mathcal{L}}_m(f_S^*,S')-\hat{\mathcal{L}}_m(f_S^*,S)|>\epsilon/2}.
\end{aligned}
$$

Taking the expectation over the sample S' in the previous inequality gives:

$$
\begin{aligned}
\mathbb{1}_{|\mathcal{L}(f_S^*)-\hat{\mathcal{L}}_m(f_S^*,S)|>\epsilon}\mathbb{E}_{S'\sim\mathcal{D}^m}[\mathbb{1}_{|\mathcal{L}(f_S^*)-\hat{\mathcal{L}}_m(f_S^*,S')|<\epsilon/2}] \\
&\leqslant \mathbb{E}_{S'\sim\mathcal{D}^m}[\mathbb{1}_{|\hat{\mathcal{L}}_m(f_S^*,S')-\hat{\mathcal{L}}_m(f_S^*,S)|>\epsilon/2}].
\end{aligned}
$$

Thus:

$$
\begin{aligned}
\mathbb{1}_{|\mathcal{L}(f_S^*)-\hat{\mathcal{L}}_m(f_S^*,S)|>\epsilon}\mathbb{P}(|\mathcal{L}(f_S^*) - \hat{\mathcal{L}}_m(f_S^*, S')| < \epsilon/2) \\
&\leqslant \mathbb{E}_{S'\sim\mathcal{D}^m}[\mathbb{1}_{|\hat{\mathcal{L}}_m(f_S^*,S')-\hat{\mathcal{L}}_m(f_S^*,S)|>\epsilon/2}]. \qquad (1.12)
\end{aligned}
$$

For each example $(\mathbf{x}'_i, y'_i) \in S'$ let us denote by X_i the random variable $\frac{1}{m}\ell(f_S^*(\mathbf{x}'_i), y'_i)$. Since f_S^* is independent of the sample S', the random variables $X_i, i \in \{1, \dots, m\}$ are independent. The variance of the random variable $\hat{\mathcal{L}}_m(f_S^*, S')$, $\mathbb{V}(\hat{\mathcal{L}}_m(f_S^*, S'))$, is thus equal to:

$$
\mathbb{V}(\hat{\mathcal{L}}_m(f_S^*, S')) = \frac{1}{m}\mathbb{V}(\ell(f_S^*(\mathbf{x}'), y')),
$$

and, according to Chebyshev's inequality (Appendix A), we have:

$$
\mathbb{P}(|\mathcal{L}(f_S^*) - \hat{\mathcal{L}}_m(f_S^*, S')| \geqslant \epsilon/2) \leqslant \frac{4\mathbb{V}(\ell(f_S^*(\mathbf{x}'), y'))}{m\epsilon^2} \leqslant \frac{1}{m\epsilon^2}. \qquad (1.13)
$$

The last inequality is due to the fact that $\ell(f_S^*(\mathbf{x}'), y'))$ is a random variable taking its values in $[0, 1]$ and its variance is less than $1/4$. By taking the opposite event in (1.13), from (1.12) we obain:

$$
\left(1 - \frac{1}{m\epsilon^2}\right)\mathbb{1}_{|\mathcal{L}(f_S^*)-\hat{\mathcal{L}}_m(f_S^*,S)|>\epsilon} \leqslant \mathbb{E}_{S'\sim\mathcal{D}^m}[\mathbb{1}_{|\hat{\mathcal{L}}_m(f_S^*,S')-\hat{\mathcal{L}}_m(f_S^*,S)|>\epsilon/2}].
$$

The result follows by taking the expectation on the sample S and noting that $m\epsilon^2 \geqslant 2$, i.e. $\frac{1}{2} \leqslant \left(1 - \frac{1}{m\epsilon^2}\right)$. □

We note that the expectation in the left-hand side of the inequality (1.11) follows the distribution of an i.i.d. sample of size m, while the expectation in the right-hand side follows the distribution of an i.i.d. sample of size $2m$.

The extension of the generalization bound for an infinite class of functions, $\mathcal{F}$, is done by studying the largest deviation between the empirical risks of the functions of $\mathcal{F}$ on any two training sets S and S' of the same size. Indeed, the important quantity involved in the previous result is the maximum number of possible labels for two sets of the same size, m, denoted $\mathfrak{G}(\mathcal{F}, 2m)$, where:

$$\mathfrak{G}(\mathcal{F}, m) = \max_{S \in \mathcal{X}^m} |\mathfrak{F}(\mathcal{F}, S)|. \tag{1.14}$$

$\mathfrak{G}(\mathcal{F}, m)$ is called the growth function and measures the maximum number of possible labelings of a sequence of m points of $\mathcal{X}$ by the function class, $\mathcal{F}$. $\mathfrak{G}(\mathcal{F}, m)$ can thus be seen as a measure of the size of the $\mathcal{F}$ function class. In the forthcoming theorem, we will establish a generalization bound that provides an upper bound on the generalization error of any function within a class of functions. This bound is determined by the empirical loss of $f \in \mathcal{F}$ on a training set S of size m and a complexity measure associated with the growth function $\mathfrak{G}(\mathcal{F}, m)$. The proof of this theorem relies on the use of Rademacher variables, defined as:

Definition 1.2 Rademacher variables, $(\sigma_1, \ldots, \sigma_m)$ are discrete random variables that are symmetric and have a mean of zero, that is for all $i \in \{1, \ldots, m\}$:

- $\mathbb{P}(\sigma_i = 1) = P(\sigma_i = -1) = 0.5$,
- $\mathbb{E}(\sigma_i) = 0$.

These variables are involved in the expression of another complexity measure of a class of functions that we will present in the next chapter.

Theorem 1.4 (Vapnik and Chervonenkis's Theorem [177], Chapter 3) *For any infinite class of functions $\mathcal{F}$, let $\delta \in]0, 1]$ and S be a training set of size m generated i.i.d. according to a probability distribution $\mathcal{D}$; the following inequality holds with probability at least $1 - \delta$:*

$$\forall f \in \mathcal{F}, \mathcal{L}(f) \leqslant \hat{\mathcal{L}}_m(f, S) + \sqrt{\frac{32 \ln(\mathfrak{G}(\mathcal{F}, m)) + 32 \ln(\frac{8}{\delta})}{m}}. \tag{1.15}$$

Proof Let ϵ be a positive real. According to the symmetrization Lemma (1.3), we have:

$$\mathbb{P}\left(\sup_{f\in\mathcal{F}}\left|\mathcal{L}(f)-\hat{\mathcal{L}}_m(f,S)\right|>\epsilon\right)\leqslant 2\mathbb{P}\left(\sup_{f\in\mathcal{F}}\left|\hat{\mathcal{L}}_m(f,S')-\hat{\mathcal{L}}_m(f,S)\right|>\epsilon/2\right). \tag{1.16}$$

Consider the expression on the right side of (1.16):

$$\mathbb{P}\left(\sup_{f\in\mathcal{F}}\left|\hat{\mathcal{L}}_m(f,S')-\hat{\mathcal{L}}_m(f,S)\right|>\epsilon/2\right)$$
$$=\mathbb{P}\left(\sup_{f\in\mathcal{F}}\frac{1}{m}\left|\sum_{i=1}^{m}\mathbb{1}_{f(\mathbf{x}_i')\neq y_i'}-\mathbb{1}_{f(\mathbf{x}_i)\neq y_i}\right|>\epsilon/2\right).$$

Both random variables $\mathbb{1}_{f(\mathbf{x}_i')\neq y_i'}$ and $\mathbb{1}_{f(\mathbf{x}_i)\neq y_i}$ exhibit identical distributions, implying that $\mathbb{1}_{f(\mathbf{x}_i')\neq y_i'}-\mathbb{1}_{f(\mathbf{x}_i)\neq y_i}$ possesses a mean of zero and follows a symmetric distribution. Consequently, if we randomly interchange the signs using m Rademacher variables $(\sigma_1,\ldots,\sigma_m)$, it does not alter the probability, hence:

$$\mathbb{P}\left(\sup_{f\in\mathcal{F}}\frac{1}{m}\left|\sum_{i=1}^{m}\mathbb{1}_{f(\mathbf{x}_i')\neq y_i'}-\mathbb{1}_{f(\mathbf{x}_i)\neq y_i}\right|>\epsilon/2\right)$$
$$=\mathbb{P}\left(\sup_{f\in\mathcal{F}}\frac{1}{m}\left|\sum_{i=1}^{m}\sigma_i(\mathbb{1}_{f(\mathbf{x}_i')\neq y_i'}-\mathbb{1}_{f(\mathbf{x}_i)\neq y_i})\right|>\epsilon/2\right)$$
$$\leqslant\mathbb{P}\left(\sup_{f\in\mathcal{F}}\frac{1}{m}\left|\sum_{i=1}^{m}\sigma_i\mathbb{1}_{f(\mathbf{x}_i')\neq y_i'}\right|>\epsilon/4\vee\sup_{f\in\mathcal{F}}\frac{1}{m}\left|\sum_{i=1}^{m}\sigma_i\mathbb{1}_{f(\mathbf{x}_i)\neq y_i}\right|>\epsilon/4\right).$$

From (1.16) and the union bound we have:

$$\mathbb{P}\left(\sup_{f\in\mathcal{F}}\left|\mathcal{L}(f)-\hat{\mathcal{L}}_m(f,S)\right|>\epsilon\right)\leqslant 4\mathbb{P}\left(\sup_{f\in\mathcal{F}}\frac{1}{m}\left|\sum_{i=1}^{m}\sigma_i\mathbb{1}_{f(\mathbf{x}_i)\neq y_i}\right|>\epsilon/4\right). \tag{1.17}$$

Now let $\mathfrak{F}(\mathcal{F}, S)$ be the smallest subset of $\mathcal{F}$ which provides all the different predictions on the training set $S = \{(\mathbf{x}_i, y_i); i \in \{1, \ldots, m\}\}$ (1.10). The cardinal of this set is less than the growth function, e.g. $|\mathfrak{F}(\mathcal{F}, S)| \leqslant \mathfrak{G}(\mathcal{F}, m)$, and we have:

$$\mathbb{P}\left(\sup_{f\in\mathcal{F}} \frac{1}{m}\left|\sum_{i=1}^{m}\sigma_i \mathbb{1}_{f(\mathbf{x}_i)\neq y_i}\right| > \epsilon/4\right) = \mathbb{P}\left(\max_{f\in\mathfrak{F}(\mathcal{F},S)} \frac{1}{m}\left|\sum_{i=1}^{m}\sigma_i \mathbb{1}_{f(\mathbf{x}_i)\neq y_i}\right| > \epsilon/4\right)$$
$$= \mathbb{P}\left(\bigcup_{f\in\mathfrak{F}(\mathcal{F},S)} \frac{1}{m}\left|\sum_{i=1}^{m}\sigma_i \mathbb{1}_{f(\mathbf{x}_i)\neq y_i}\right| > \epsilon/4\right).$$

From the union bound, we get:

$$\mathbb{P}\left(\sup_{f\in\mathcal{F}} \frac{1}{m}\left|\sum_{i=1}^{m}\sigma_i \mathbb{1}_{f(\mathbf{x}_i)\neq y_i}\right| > \epsilon/4\right) \leqslant \sum_{f\in\mathfrak{F}(\mathcal{F},S)} \mathbb{P}\left(\frac{1}{m}\left|\sum_{i=1}^{m}\sigma_i \mathbb{1}_{f(\mathbf{x}_i)\neq y_i}\right| > \epsilon/4\right)$$
$$\leqslant |\mathfrak{F}(\mathcal{F}, S)| \sup_{f\in\mathfrak{F}(\mathcal{F},S)} \mathbb{P}\left(\frac{1}{m}\left|\sum_{i=1}^{m}\sigma_i \mathbb{1}_{f(\mathbf{x}_i)\neq y_i}\right| > \epsilon/4\right)$$
$$\leqslant \mathfrak{G}(\mathcal{F}, m) \sup_{f\in\mathcal{F}} \mathbb{P}\left(\frac{1}{m}\left|\sum_{i=1}^{m}\sigma_i \mathbb{1}_{f(\mathbf{x}_i)\neq y_i}\right| > \epsilon/4\right). \tag{1.18}$$

Note that the expression in the absolute value of the right side of (1.17) pertains to a sum of i.i.d. random variables with zero mean. The terms inside the sum are bounded between -1 and $+1$, and according to the Hoeffding inequality (Appendix A), we have:

$$\forall\epsilon > 0, \mathbb{P}\left(\frac{1}{m}\left|\sum_{i=1}^{m}\sigma_i \mathbb{1}_{f(\mathbf{x}_i)\neq y_i}\right| > \epsilon/4\right) = \mathbb{P}\left(\left|\sum_{i=1}^{m}\sigma_i \mathbb{1}_{f(\mathbf{x}_i)\neq y_i}\right| > \frac{m\epsilon}{4}\right)$$
$$\leqslant 2e^{-\frac{m\epsilon^2}{32}}. \tag{1.19}$$

Hence, from Eqs. (1.17), (1.18), and (1.19) we get:

$$\mathbb{P}\left(\sup_{f\in\mathcal{F}}\left|\mathcal{L}(f) - \hat{\mathcal{L}}_m(f, S)\right| > \epsilon\right) \leqslant 8\mathfrak{G}(\mathcal{F}, m)e^{-m\epsilon^2/32}.$$

The result follows by solving $8\mathfrak{G}(\mathcal{F}, m)e^{-m\epsilon^2/32} = \delta$ for ϵ. □

An important result of this theorem is that the ERM principle is consistent in the case where $\sqrt{\frac{\ln(\mathfrak{G}(\mathcal{F},m))}{m}}$ tends towards 0 when m tends to infinity. Moreover, since the distribution $\mathcal{D}$ of observations is not involved in the definition of the growth function, the preceding analysis is valid whatever the probability distribution, $\mathcal{D}$.

Thus, a sufficient condition for the ERM principle to be consistent, for all probability distributions $\mathcal{D}$ and an infinite class of functions, is:

$$\lim_{m\to\infty}\sqrt{\frac{\ln(\mathfrak{G}(\mathcal{F},m))}{m}}=0.$$

However, $\mathfrak{G}(\mathcal{F},m)$ is not measurable, and the only certainty we have is that it is bounded by 2^m. Furthermore, if the growth function were to reach this bound this would mean that there exists a sample of size m such that the function class $\mathcal{F}$ can generate all possible labelings on this sample, so we say that the sample is shattered by $\mathcal{F}$. Based on this observation, Vapnik and Chervonenkis proposed an auxiliary quantity, called the VC dimension, to study the growth function and which is defined as follows [177].

Definition 1.3 (VC Dimension, [177]) Let $\mathcal{F}=\{f:\mathcal{X}\to\{-1,+1\}\}$ be a class of discrete-valued functions. The VC dimension of $\mathcal{F}$ is the largest integer $vc_{\mathcal{F}}$ verifying $\mathfrak{G}(\mathcal{F},vc_{\mathcal{F}})=2^{vc_{\mathcal{F}}}$. In other words, $vc_{\mathcal{F}}$ is the greatest number of points that the function class manages to shatter. If no such integer exists, then the VC dimension of $\mathcal{F}$ is considered to be infinite.

Figure 1.2 illustrates the estimation of the VC dimension of a class of linear functions in the plane. From the previous definition, we can see that the larger the

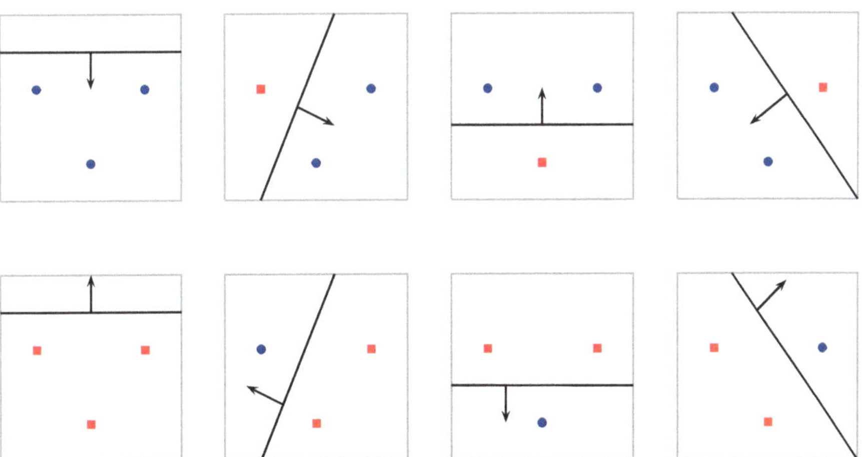

Fig. 1.2 Shattering of points in a $d=2$-dimensional plane by a class of linear functions. Each linear classifier divides the plane into two subspaces, with a normal vector pointing to the subspace containing examples belonging to class $+1$ (represented by blue circles). The maximum number of points in the plane that can be shattered by the class of linear functions, or the VC dimension of this function class, is in this case equal to 3

VC dimension, $vc_{\mathcal{F}}$, of a function class, the higher the growth function $\mathfrak{G}(\mathcal{F}, m)$ of this class, and this for any $m \geqslant vc_{\mathcal{F}}$. An important property proved by Sauer [147] and Shelah [155], is that the dimension VC of a class of functions $\mathcal{F}$ is a measure of the capacity of $\mathcal{F}$, and it is exhibited in the following lemma.

Lemma 1.5 (Growth Function Bound for Finite VC Dimension [147, 155]) *Let $\mathcal{F}$ be a class of functions with values in $\{-1, +1\}$ and with finite VC dimension, $vc_{\mathcal{F}}$.*

For any natural number m, the growth function $\mathfrak{G}(\mathcal{F}, m)$ is bounded by:

$$\mathfrak{G}(\mathcal{F}, m) \leqslant \sum_{i=0}^{vc_{\mathcal{F}}} \binom{m}{i}, \tag{1.20}$$

and, for all $m \geqslant vc_{\mathcal{F}}$:

$$\mathfrak{G}(\mathcal{F}, m) \leqslant \left(\frac{m}{vc_{\mathcal{F}}}\right)^{vc_{\mathcal{F}}} e^{vc_{\mathcal{F}}}. \tag{1.21}$$

Proof There are various proofs of this lemma [25, 29, 119, 147, 155], including the one based on induction with respect to $m + vc_{\mathcal{F}}$ that we will present in the following. Note first that the inequality (1.20) is true for $vc_{\mathcal{F}} = 0$ and $m = 0$. Indeed:

- If $vc_{\mathcal{F}} = 0$, this means that the function class does not manage to shatter any points, always producing the same labeling, i.e. $\mathfrak{G}(\mathcal{F}, m) = 1 = \binom{m}{0}$.
- If $m = 0$, this means we are dealing with the trivial labeling of the empty set, e.g. $\mathfrak{G}(\mathcal{F}, 0) = 1 = \sum_{i=0}^{vc_{\mathcal{F}}} \binom{0}{i}$.

Now assume that the inequality (1.20) is true for any $m' + vc'_{\mathcal{F}} < m + vc_{\mathcal{F}}$. Given a set $S = \{\mathbf{x}_1, \ldots, \mathbf{x}_m\}$ and a class of functions $\mathcal{F}$ with dimension VC, $vc_{\mathcal{F}}$, let us show that $|\mathfrak{F}(\mathcal{F}, S)| \leqslant \sum_{i=0}^{vc_{\mathcal{F}}} \binom{m}{i}$.

Consider two subsets of classes $\mathcal{F}_1$ and $\mathcal{F}_2$, of $\mathcal{F}$, defined on the set $S' = S \backslash \{\mathbf{x}_m\}$ of size $m - 1$. Let us construct the class $\mathcal{F}_1$, adding to it the set of $\mathcal{F}$ functions such that the prediction vectors of these functions on S' are all different, and let us posit the class $\mathcal{F}_2 = \mathcal{F} \setminus \mathcal{F}_1$. So, if two functions of $\mathcal{F}$ have the same prediction vectors on S' and their only difference lies in their predictions on the single example $\mathbf{x}_m$, one of these functions will be put in $\mathcal{F}_1$ and the other in $\mathcal{F}_2$.

Figure 1.3 illustrates this construction for a toy problem. The pairs of functions (f_1, f_2) and (f_4, f_5) have the same prediction vectors on $S' = S \setminus \{\mathbf{x}_5\}$, and the sets $\mathcal{F}_1$ and $\mathcal{F}_2$ will each contain one of the functions of these pairs.

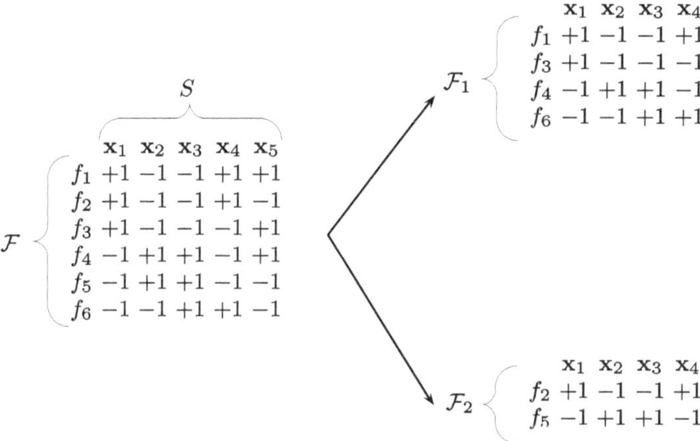

Fig. 1.3 Construction of the sets $\mathcal{F}_1$ and $\mathcal{F}_2$ from the function class $\mathcal{F}$ for the proof of the Sauer's lemma on a toy example

We now note that if a set is shattered by the class $\mathcal{F}_1$, it will also be shattered by the class $\mathcal{F}$ since $\mathcal{F}_1$ contains all non-redundant functions on S' of $\mathcal{F}$, so:

$$\text{VC dimension}(\mathcal{F}_1) \leqslant \text{VC dimension}(\mathcal{F}) = vc_{\mathcal{F}}.$$

Furthermore, if a set S' is shattered by $\mathcal{F}_2$, the set $S' \cup \{\mathbf{x}_m\}$ will also be shattered by $\mathcal{F}$ since, for any function in $\mathcal{F}_2$, $\mathcal{F}$ also contains the other function whose output on $\mathbf{x}_m$ differs from the first.

Thus, VC dimension$(\mathcal{F}) \geqslant$ VC dimension$(\mathcal{F}_2) + 1$, so:

$$\text{VC dimension}(\mathcal{F}_2) \leqslant vc_{\mathcal{F}}.$$

From the induction hypothesis, we have:

$$|\mathcal{F}_1| \leqslant \mathfrak{G}(\mathcal{F}_1, m-1) \leqslant \sum_{i=0}^{vc_{\mathcal{F}}} \binom{m-1}{i},$$

$$|\mathcal{F}_2| \leqslant \mathfrak{G}(\mathcal{F}_2, m-1) \leqslant \sum_{i=0}^{vc_{\mathcal{F}}-1} \binom{m-1}{i}.$$

We can now employ the inductive hypothesis on both $\mathcal{F}_1$ and $\mathcal{F}_2$. The reasoning ends after a change of variable and the use of Pascal's formula:

$$\begin{aligned}
\mathfrak{G}(\mathcal{F}, m) &= \mathfrak{G}(\mathcal{F}_1, m-1) + \mathfrak{G}(\mathcal{F}_2, m-1) \\
&\leqslant \sum_{i=0}^{vc_{\mathcal{F}}} \binom{m-1}{i} + \sum_{i=0}^{vc_{\mathcal{F}}-1} \binom{m-1}{i} \\
&\leqslant \sum_{i=0}^{vc_{\mathcal{F}}} \binom{m-1}{i} + \sum_{i=0}^{vc_{\mathcal{F}}} \binom{m-1}{i-1} \\
&\leqslant \sum_{i=0}^{vc_{\mathcal{F}}} \binom{m}{i}.
\end{aligned}$$

To demonstrate the inequality (1.21), we will use Newton's binomial formula. Thus, according to the inequality (1.20) and in the case where $\frac{vc_{\mathcal{F}}}{m} \leqslant 1$, we have:

$$\begin{aligned}
\left(\frac{vc_{\mathcal{F}}}{m}\right)^{vc_{\mathcal{F}}} \mathfrak{G}(\mathcal{F}, m) &\leqslant \left(\frac{vc_{\mathcal{F}}}{m}\right)^{vc_{\mathcal{F}}} \sum_{i=0}^{vc_{\mathcal{F}}} \binom{m}{i} \\
&\leqslant \sum_{i=0}^{vc_{\mathcal{F}}} \left(\frac{vc_{\mathcal{F}}}{m}\right)^{i} \binom{m}{i}.
\end{aligned}$$

Multiplying the term on the right by $1^{m-i} = 1$ and using the binomial formula, we get:

$$\begin{aligned}
\left(\frac{vc_{\mathcal{F}}}{m}\right)^{vc_{\mathcal{F}}} \mathfrak{G}(\mathcal{F}, m) &\leqslant \sum_{i=0}^{vc_{\mathcal{F}}} \binom{m}{i} \left(\frac{vc_{\mathcal{F}}}{m}\right)^{i} 1^{m-i} \\
&= \left(1 + \frac{vc_{\mathcal{F}}}{m}\right)^{m}.
\end{aligned}$$

Finally, from the inequality $\forall z \in \mathbb{R}, (1-z) \leqslant e^{-z}$, we get:

$$\mathfrak{G}(\mathcal{F}, m) \leqslant \left(\frac{m}{vc_{\mathcal{F}}}\right)^{vc_{\mathcal{F}}} \left(1 + \frac{vc_{\mathcal{F}}}{m}\right)^{m} \leqslant \left(\frac{m}{vc_{\mathcal{F}}}\right)^{vc_{\mathcal{F}}} e^{vc_{\mathcal{F}}}.$$

□

Note that this lemma is better known as Sauer's lemma but it was first stated, and in a slightly different form, in [177]. From the previous result, we can see that the values taken by the growth function associated with the class of functions $\mathcal{F}$ will depend on the existence or non-existence of the VC dimension of $\mathcal{F}$:

$$\forall m, \mathfrak{G}(\mathcal{F}, m) = \begin{cases} O(m^{vc_{\mathcal{F}}}) & \text{if } vc_{\mathcal{F}} \text{ is finite,} \\ 2^m & \text{if } vc_{\mathcal{F}} \text{ is infinite.} \end{cases}$$

Moreover, in the case where the VC dimension, $vc_{\mathcal{F}}$, of a class of functions $\mathcal{F}$ is finite and there are enough training examples such that $m \geqslant vc_{\mathcal{F}}$, the evolution of the growth function becomes polynomial as a function of m, i.e. $\ln \mathfrak{G}(\mathcal{F}, m) \leqslant vc_{\mathcal{F}} \ln \frac{em}{vc_{\mathcal{F}}}$ (1.21). This result exhibits a new expression for the generalization bound of (1.15) measurable for a known value of $vc_{\mathcal{F}}$ and any fixed training set.

Corollary 1.6 (Generalization Bound with VC Dimension) *Let $\mathcal{X} \subseteq \mathbb{R}^d$ be a vector space, $\mathcal{Y} = \{-1, +1\}$ an output space and $\mathcal{F}$ a class of functions with values in $\mathcal{Y}$ and VC dimension, $vc_{\mathcal{F}}$. Suppose that the pairs of examples $(\mathbf{x}, y) \in \mathcal{X} \times \mathcal{Y}$ are generated i.i.d. according to a probability distribution $\mathcal{D}$. For any $\delta \in]0, 1]$, we have for any function $f \in \mathcal{F}$ and any set $S \in (\mathcal{X} \times \mathcal{Y})^m$ of size $m \geqslant vc_{\mathcal{F}}$ generated i.i.d. according to the same probability distribution, the following inequality which holds with probability at least equal to $1 - \delta$:*

$$\mathcal{L}(f) \leqslant \hat{\mathcal{L}}_m(f, S) + \sqrt{\frac{32 vc_{\mathcal{F}} \ln \frac{me}{vc_{\mathcal{F}}} + 32 \ln(\frac{8}{\delta})}{m}}. \tag{1.22}$$

As $\lim_{m \to \infty} \frac{32 vc_{\mathcal{F}} \ln \frac{em}{vc_{\mathcal{F}}} + 32 \ln \frac{8}{\delta}}{m} = 0$, we can deduce from the previous result a sufficient condition on the consistency of the ERM principle, which can be stated as follows:

For a given class of binary functions $\mathcal{F}$, if the VC dimension of $\mathcal{F}$ is finite, then, the ERM principle is consistent for all $\mathcal{D}$ distributions generating the examples.

Furthermore, Vapnik and Chervonenkis demonstrated that for the ERM principle to be consistent for all $\mathcal{D}$-distributions, it is also necessary for the VC dimension of the function class under consideration to be finite [177]. Thus, for any probability distribution, the ERM principle is consistent if and only if the VC dimension of the function class under consideration is finite.

1.3.3 Statement of the Principle

Our analysis reveals that as the capacity of a function class increases, its ability to learn from data improves, thereby reducing the empirical error of a function within that class when evaluated on a training set. This observation underscores the trade-off between the capacity of a function class and its performance on training data, a key concept in machine learning. However, it is crucial to note that a larger capacity does not inherently guarantee better generalization performance, as it may also heighten the risk of overfitting the training data. Overfitting happens when a model captures not just the underlying data patterns but also the noise, leading to poor performance on unseen data. This balance, depicted in Fig. 1.4, is known as structural risk minimization. Structural risk minimization seeks to harmonize the model's complexity (its capacity) with its performance on both training and new, unseen data, thereby enhancing the model's generalization capabilities. This approach is essential for ensuring that the model remains accurate on familiar data while also being robust and reliable when applied to new data.

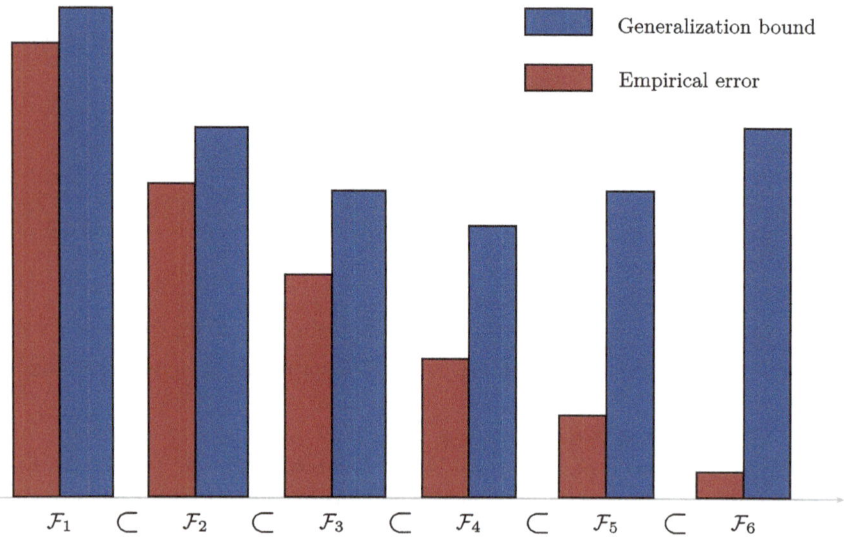

Fig. 1.4 Illustration of the structural risk minimization principle. The x-axis shows a hierarchy of nested function classes with increasing capacity, from left to right. The greater the capacity of a function class, the smaller the empirical error of a function of this class on a training set, and the worse the bound on its generalization error. The structural risk minimization principle consists in choosing the function from the function class for which we have the best estimate of its generalization bound (i.e. $\mathcal{F}_4$ in this example)

To Sum Up

1. *The ERM principle involves selecting a function from a hypothesis class that minimizes the empirical risk, which is the average loss measured on the training data. It aims to approximate the true risk, which is the expected loss over the entire data distribution.*
2. *The worst-case study in learning theory focuses on scenarios where the empirical risk minimizer performs poorly on unseen data. This approach emphasizes the importance of considering the worst possible outcomes to ensure robustness and generalization of learning algorithms, leading to the development of bounds on generalization error.*
3. *The generalization bound quantifies the difference between the empirical risk and the true risk. This bound is typically expressed in terms of the VC dimension and the number of training samples, providing a probabilistic guarantee that the empirical risk minimizer will perform well on unseen data, thereby ensuring generalization.*
4. *The VC dimension is a measure of the capacity of a hypothesis class, defined as the largest number of points that can be shattered (i.e., correctly classified in all possible ways) by the hypothesis class. It is a key concept in understanding the complexity and generalization ability of learning models.*
5. *The growth function, which describes the maximum number of ways a hypothesis set can classify points in a dataset, is crucial in understanding the capacity of a class of functions. It is directly related to the VC dimension and helps in analyzing the generalization ability of learning algorithms.*
6. *The empirical risk minimization (ERM) principle is consistent for all $\mathcal{D}$ distributions generating the data, if and only if the VC dimension of the class of functions is finite.*
7. *The study of the consistency of the ERM principle led to the structural risk minimization (SRM), the second fundamental principle in machine learning.*
8. *Learning is a compromise between low empirical error and high capacity of the class of functions.*

1.4 Exercises

1. How does the VC theory explain the convergence of a learning algorithm's performance as the training size increases?
2. Linear classifiers in $\mathbb{R}^d$.

   (a) Prove that the VC dimension of a class of linear functions in an input space of dimension d is $d+1$.
   (b) Compare the generalization bounds for $d = 10$, and $d = 100$ with the size of the training set $m = 5000$. How does dimensionality affect the bound?

3. Consider the following class of threshold functions:

$$\mathcal{F} = \{f : \mathbb{R} \to \{-1, +1\}, f(x) = \text{sign}(x - \theta), \theta \in \mathbb{R}\}.$$

   (a) Show that its VC dimension is 1.
   (b) Compute the generalization bound of a function in this class that has an empirical error of 0.25 on a training set of size $m = 200$, and that holds with a probability of at least 0.95.

4. Growth function and VC dimension.

   (a) What is an upper bound of a growth function for a class of functions with VC dimension $vc_{\mathcal{F}} = 3$ on a training set of size $m = 10$?
   (b) Derive a generalization bound that holds with a probability of at least 0.99 for a function from this class which has an empirical error of 0.12 over a training set of size $m = 1000$.

5. Given a class of functions with VC dimension $vc_{\mathcal{F}} = 50$.

   (a) What is the smallest training set size needed to ensure that the approximation error of the generalization error, when compared to the empirical error for a function within this class, is less than 0.1, with a probability of at least 0.95?
   (b) How does this sample size change if this approximation error is halved?

6. Suppose that the input space and the output space are respectively, $\mathcal{X} = [0, 1]$ and $\mathcal{Y} = \{-1, +1\}$.

   (a) Estimate the VC dimension of the following class of functions:

$$\mathcal{F}_1 = \{f : \mathcal{X} \to \mathcal{Y}, f(\mathbf{x}) = 2\mathbb{1}_{\alpha_1 < \mathbf{x} < \alpha_2} - 1, 0 \leqslant \alpha_1 < \alpha_2 \leqslant 1\},$$

$$\mathcal{F}_2 = \{f : \mathcal{X} \to \mathcal{Y}, f(\mathbf{x}) = \sum_{i=0}^{d} (2\mathbb{1}_{\alpha_{2i} \leqslant \mathbf{x} < \alpha_{2i+1}} - 1), 0 \leqslant \alpha_0 < \ldots < \alpha_{2d+1} \leqslant 1\},$$

where $\mathbb{1}_\pi$ is the indicator function: $\mathbb{1}_\pi = 1$ if the predicate π is true and 0 otherwise.

(b) Compute the approximation errors of the generalization error, when compared to the empirical error, for functions from $\mathcal{F}_1$ and $\mathcal{F}_2$ on a training set of size $m = 10{,}000$ that hold with a probability of at least 0.95.

(c) How do these bounds compare to that of linear classifiers in 2D space for the same training set size?

7. Consider the hypothesis class $\mathcal{H}_k$ consisting of indicator functions of unions of k intervals on the real line:

$$\mathcal{H}_k = \Big\{h_{a,b} : \mathbb{R} \to \{0, 1\} \mid h_{a,b}(x) = \mathbb{1}_{\left(x \in \bigcup_{i=1}^k [a_i, b_i]\right)},$$
$$a_i \leqslant b_i, \quad a_i, b_i \in \mathbb{R}\Big\},$$

where $\mathbb{1}_\pi = 1$ if the predicate π is true and 0 otherwise.

(a) Show that the VC dimension of $\mathcal{H}_1$, the class of single intervals on the real line, is 2.

(b) For general $k \geqslant 1$, prove that the VC dimension of $\mathcal{H}_k$ is $2k$.
Hint: Construct a set of $2k$ points on the real line that can be shattered by $\mathcal{H}_k$. Show that no set of $2k+1$ points can be shattered by $\mathcal{H}_k$.

(c) Provide an intuitive explanation for why the VC dimension grows linearly with k.

(d) Discuss how this result extends to unions of other geometric shapes, such as rectangles in $\mathbb{R}^2$.

8. Consider the hypothesis class $\mathcal{B}_d$ of all closed balls in $\mathbb{R}^d$, i.e., sets of the form

$$B(\mathbf{x}_0, r) = \{\mathbf{x} \in \mathbb{R}^d : \|\mathbf{x} - \mathbf{x}_0\|_2 \leqslant r\}$$

for some center $\mathbf{x}_0 \in \mathbb{R}^d$ and radius $r \geqslant 0$.

(a) Show that there exists a set of $d+1$ points in $\mathbb{R}^d$ that can be shattered by $\mathcal{B}_d$.
Hint: Consider the origin and the standard basis vectors.

(b) Prove that the VC dimension of $\mathcal{B}_d$ is at most $d+2$.

(c) For the special case $n = 2$, argue why no set of 4 points in the plane can be shattered by disks (closed balls in $\mathbb{R}^2$), using geometric intuition and the fact that any 4 points in the plane must have some point inside the convex hull of the others, which prevents certain labelings from being realized by disks.

(d) Conclude that the VC dimension of the set of all closed balls in $\mathbb{R}^d$ satisfies

$$d + 1 \leqslant vc_{\mathcal{B}_d} \leqslant d + 2.$$

(e) Discuss the implications of this VC dimension bound on the sample complexity of learning classifiers based on closed balls in $\mathbb{R}^n$.

Hints: The condition $\mathbf{x} \in B(\mathbf{x}_0, r)$ *is equivalent to* $\|\mathbf{x}\|^2 - 2\langle \mathbf{x}_0, \mathbf{x}\rangle + \|\mathbf{x}_0\|^2 - r^2 \leqslant 0$*. Define the feature mapping* $\Phi : \mathbb{R}^d \to \mathbb{R}^{d+1}$ *by* $\Phi(\mathbf{x}) = (\|\mathbf{x}\|^2, x_1, \ldots, x_d)$*. Membership in a ball corresponds then to a linear inequality in* $\mathbb{R}^{d+1}$*:* $\langle \boldsymbol{w}, \Phi(x)\rangle + w_0 \leqslant 0$*, for some* $\boldsymbol{w} \in \mathbb{R}^{d+1}$*,* $w_0 \in \mathbb{R}$*. The VC dimension of halfspaces in* $\mathbb{R}^{d+1}$ *is* $d + 2$*, which provides an upper bound on* $vc_{\mathcal{B}_d}$*.*

9. Consider the class of sinusoidal classifiers on the interval [0, 1], defined as:

$$\mathcal{F} = \{f : [0, 1] \to \{-1, +1\},\ f(x) = \text{sign}(\sin(\gamma x + \Phi)),\ \gamma \in \mathbb{R},\ \Phi \in [0, 2\pi]\}.$$

(a) Estimate the VC dimension of the class $\mathcal{F}$.
(b) Compute the approximation errors of the generalization error, when compared to the empirical error, for functions from $\mathcal{F}$ on a training set of size $m = 10{,}000$ that hold with a probability of at least 0.95.
(c) Discuss how the complexity of this classifier, as measured by its VC dimension, affects its generalization performance compared to linear classifiers in 2D space.

10. We are comparing two models for binary classification to understand how the VC dimension can guide model selection.

- **Model A**: Linear classifier with VC dimension $vc_{\mathcal{F}} = d + 1$, where d is the input dimension.
- **Model B**: Polynomial kernel SVM (Chap. 5) with VC dimension $vc_{\mathcal{F}} = \binom{d+p}{p}$, where p is the polynomial degree.

(a) For $d = 10$, $p = 3$, compute the approximation error of the generalization error, when compared to the empirical error, of these two models on a training set of size $m = 1000$ that holds with a probability of at least 0.95.
(b) Which model has a tighter bound?
(c) Same questions for a training set of size $m = 100{,}000$.
(d) How does these observations guide model selection for small and large datasets?

11. Consider a finite class of functions $\mathcal{F}$ of size $|\mathcal{F}|$, and a training set of m examples.

(a) Derive the generalization error bound for any hypothesis $\mathcal{F}$ with probability at least $1 - \delta$, using the following steps:

– Show that for a single $f \in \mathcal{F}$:

$$\mathbb{P}(\left|\mathcal{L}(f) - \hat{\mathcal{L}}_m(f, S)\right| > \epsilon/2) \leqslant 2e^{-2\epsilon^2 m}$$

- Considering the worst-case study (Sect. 1.2.2), show that:

$$\mathbb{P}\left(\sup_{f\in\mathcal{F}}\left|\mathcal{L}(f)-\hat{\mathcal{L}}_m(f,S)\right|>\epsilon\right)=\mathbb{P}\left(\bigcup_{f\in\mathcal{F}}\left|\mathcal{L}(f)-\hat{\mathcal{L}}_m(f,S)\right|>\epsilon\right)$$

- Apply the union bound to extend the bound found in (a) to all hypotheses in $\mathcal{F}$.
- Set the right-hand side equal to δ and solve for ϵ.

(b) Using the derived bound, calculate the minimum number of training examples needed to ensure that with probability at least 0.95, the generalization error of any hypothesis is within 0.1 of its empirical error, given $|\mathcal{F}| = 1000$.

12. Consider a set of m points in $\mathbb{R}^n$. Select any one of these points as the origin. Prove that the m points can be shattered by oriented hyperplanes if and only if the position vectors of the remaining points are linearly independent.
Hint: Consider the convex hulls of any two disjoint subsets of the m points, with one subset containing the origin. If the position vectors of the remaining $m-1$ *points are linearly independent, the convex hulls of these subsets will not intersect, allowing them to be separated by an oriented hyperplane. Conversely, if the position vectors are not linearly independent, the convex hulls may overlap, making it impossible to separate the points with hyperplanes, thus preventing the points from being shattered.*

Chapter 2
Data-Dependent Generalization Bounds

The generalization bounds based on the VC dimension presented in the previous chapter are defined on concepts that only apply to binary classification. Moreover, these bounds are independent of the training data. In this sense, this independence guarantees that they hold for any training set of a given classification problem; but as they are based on a worst-case study, this independence means that they generally do not give a good estimate of the upper bound of the generalization error. In this chapter, we introduce a new notion of data-dependent complexity called Rademacher complexity (Sect. 2.1), which allows us to obtain data-dependent generalization bounds. In Sect. 2.4, we present different approaches for multi-class classification based on binary classification techniques, so-called combined approaches, and derive a generalization bound for this case that can be estimated on training data.

2.1 Rademacher Complexity

The growth function and VC dimension are two quantities that measure the capacity of a class of functions independently of the probability distribution generating the data, which is unknown. However, the growth function is virtually impossible to estimate in most cases, and the VC dimension is a fairly broad measure in general.

There are other quantities that lead to more refined estimates of the capacity of a function class, and which are moreover dependent on the training data. Among the existing quantities, we will focus in this section on the Rademacher complexity introduced by Koltchinskii [95, 96] and which has opened up a new avenue of active research in learning theory.

Empirical Rademacher complexity estimates the richness of a function class $\mathcal{F}$ by measuring the degree to which it can adjust to random noise on a data set $S = \{(\mathbf{x}_1, y_1), \ldots, (\mathbf{x}_m, y_m)\}$ of size m generated i.i.d. according to a probability

M.-R. Amini, *Advanced Supervised and Semi-supervised Learning*,
Cognitive Technologies, https://doi.org/10.1007/978-3-031-99928-4_2

distribution $\mathcal{D}$. This complexity is estimated via the Rademacher variables $\boldsymbol{\sigma} = (\sigma_1, \ldots, \sigma_m)^\top$ (Chap. 1, Definition 1.2) and is defined as:

Definition 2.1 (Empirical Rademacher Complexity)

$$\hat{\mathfrak{R}}_S(\mathcal{F}) = \frac{2}{m}\mathbb{E}_{\sigma}\left[\sup_{f\in\mathcal{F}}\sum_{i=1}^{m}\sigma_i f(\mathbf{x}_i) \mid \mathbf{x}_1, \ldots, \mathbf{x}_m\right], \tag{2.1}$$

where $\mathbb{E}_\sigma$ is a conditional expectation with respect to the data and taken over the Rademacher variables $\boldsymbol{\sigma} = (\sigma_1, \ldots, \sigma_m)^\top$ taken as random noises.

As for all $i \in \{1, \ldots, m\}$ both the random noise σ_i and the prediction function $f(\mathbf{x}_i)$ take values in $\{-1, +1\}$; the better a function fits random noise, the higher will be $\sum_{i=1}^{m}\sigma_i f(\mathbf{x}_i)$. In other words, this sum measures the correlation between the vector of predictions and the vector of random noises over S.

The supremum $\sup_{f\in\mathcal{F}}\sum_{i=1}^{m}\sigma_i f(\mathbf{x}_i)$ thus estimates the extent to which the function class $\mathcal{F}$ is correlated with the random noise $\boldsymbol{\sigma}$ on the set S, and the empirical Rademacher complexity is a mean measure of this estimate. This complexity then describes the richness of the function class $\mathcal{F}$; the greater the capacity of this class, the greater the chance of finding a function f_S that will correlate with the noise on S.

In addition, we define the Rademacher complexity of the function class $\mathcal{F}$ independently of a given set by taking the expectation of the empirical Rademacher complexity of this class over all generated sets i.i.d. of the same size, m:

Definition 2.2 (Rademacher Complexity)

$$\mathfrak{R}_m(\mathcal{F}) = \mathbb{E}_{S\sim\mathcal{D}^m}\hat{\mathfrak{R}}_S(\mathcal{F}) = \frac{2}{m}\mathbb{E}_{S\sigma}\left[\sup_{f\in\mathcal{F}}\sum_{i=1}^{m}\sigma_i f(\mathbf{x}_i)\right]. \tag{2.2}$$

2.2 Link Between Rademacher Complexity and VC Dimension

There is a link between the Rademacher complexity, $\mathfrak{R}_m(\mathcal{F})$ of a class of functions $\mathcal{F}$ and the associated growth function $\mathfrak{G}(\mathcal{F}, m)$ (and hence the dimension of VC in the case where the upper-bound in (1.21) holds). This connection follows from the result presented in [111, lemma 5.2] and known as Massart's lemma, which states:

Lemma 2.1 (Finite Set Rademacher Complexity Bound [111]) *Let $\mathcal{A}$ be a finite subset of $\mathbb{R}^m$ and $\boldsymbol{\sigma} = (\sigma_1, \dots, \sigma_m)^\top$ be m independent Rademacher random variables. We then have:*

$$\mathbb{E}_{\boldsymbol{\sigma}}\left[\sup_{a\in\mathcal{A}}\sum_{i=1}^{m}\sigma_i a_i\right] \leqslant r\sqrt{2\ln|\mathcal{A}|}, \tag{2.3}$$

where $r = \sup_{a\in\mathcal{A}} ||a||$.

Proof From the convexity of the exponential function and Jensen's inequality, for any real $\lambda > 0$ we have:

$$\exp\left(\lambda\mathbb{E}_{\boldsymbol{\sigma}}\left[\sup_{a\in\mathcal{A}}\sum_{i=1}^{m}\sigma_i a_i\right]\right) \leqslant \mathbb{E}_{\boldsymbol{\sigma}}\left[\exp\left(\lambda\sup_{a\in\mathcal{A}}\sum_{i=1}^{m}\sigma_i a_i\right)\right],$$

with $\mathbb{E}_{\boldsymbol{\sigma}}\left[\exp\left(\lambda\sup_{a\in\mathcal{A}}\sum_{i=1}^{m}\sigma_i a_i\right)\right] = \mathbb{E}_{\boldsymbol{\sigma}}\left[\sup_{a\in\mathcal{A}}\exp\left(\lambda\sum_{i=1}^{m}\sigma_i a_i\right)\right].$

Furthermore,

$$\mathbb{E}_{\boldsymbol{\sigma}}\left[\sup_{a\in\mathcal{A}}\exp\left(\lambda\sum_{i=1}^{m}\sigma_i a_i\right)\right] \leqslant \mathbb{E}_{\boldsymbol{\sigma}}\left[\sum_{a\in\mathcal{A}}\exp\left(\lambda\sum_{i=1}^{m}\sigma_i a_i\right)\right],$$

where $\mathbb{E}_{\boldsymbol{\sigma}}\left[\sum_{a\in\mathcal{A}}\exp\left(\lambda\sum_{i=1}^{m}\sigma_i a_i\right)\right] = \sum_{a\in\mathcal{A}}\mathbb{E}_{\boldsymbol{\sigma}}\prod_{i=1}^{m}\left[\exp(\lambda\sigma_i a_i)\right].$

As the random variables $\sigma_1, \dots, \sigma_m$ are independent and by definition $\forall i$, $\mathbb{P}(\sigma_i = -1) = \mathbb{P}(\sigma_i = +1) = 1/2$, we get:

$$\exp\left(\lambda\mathbb{E}_{\boldsymbol{\sigma}}\left[\sup_{a\in\mathcal{A}}\sum_{i=1}^{m}\sigma_i a_i\right]\right) \leqslant \sum_{a\in\mathcal{A}}\prod_{i=1}^{m}\mathbb{E}_{\sigma_i}\left[\exp(\lambda\sigma_i a_i)\right] = \sum_{a\in\mathcal{A}}\prod_{i=1}^{m}\frac{e^{-\lambda a_i}+e^{\lambda a_i}}{2}.$$

Using the inequality $\forall z \in \mathbb{R}$, $\frac{e^{-z}+e^{z}}{2} \leqslant e^{z^2/2}$, we have:

$$\exp\left(\lambda\mathbb{E}_{\boldsymbol{\sigma}}\left[\sup_{a\in\mathcal{A}}\sum_{i=1}^{m}\sigma_i a_i\right]\right) \leqslant \sum_{a\in\mathcal{A}}\prod_{i=1}^{m}e^{\lambda^2 a_i^2/2} = \sum_{a\in\mathcal{A}}e^{\lambda^2||a||^2/2} \leqslant |\mathcal{A}|e^{\lambda^2 r^2/2}.$$

So, using the logarithm and dividing by λ:

$$\mathbb{E}_\sigma\left[\sup_{a\in\mathcal{A}}\sum_{i=1}^m \sigma_i a_i\right] \leqslant \frac{\ln|\mathcal{A}|}{\lambda} + \frac{\lambda r^2}{2}.$$

As this inequality holds for any $\lambda > 0$, it is also true for $\lambda^* = \frac{\sqrt{2\ln|\mathcal{A}|}}{r}$ which minimizes the right-hand term of the inequality. The inequality (2.3) is then deduced using the latter value for λ^*. □

With this result, we can bound the Rademacher complexity of a class of functions with respect to the associated growth function.

Corollary 2.2 (Link with the VC Dimension) *Let $\mathcal{F}$ be a class of functions with values in $\{-1,+1\}$, with finite dimension VC, $vc_{\mathcal{F}}$, and $\mathfrak{G}(\mathcal{F},m)$ the associated growth function. For any non-zero natural number m, the Rademacher complexity $\mathfrak{R}_m(\mathcal{F})$ is bounded by:*

$$\mathfrak{R}_m(\mathcal{F}) \leqslant \sqrt{\frac{8\ln(2\mathfrak{G}(\mathcal{F},m))}{m}}, \tag{2.4}$$

and, in the case where $m \geqslant vc_{\mathcal{F}}$, we have:

$$\mathfrak{R}_m(\mathcal{F}) \leqslant \sqrt{8\left(\frac{\ln 2}{m} + \frac{vc_{\mathcal{F}}}{m}\ln\frac{em}{vc_{\mathcal{F}}}\right)}. \tag{2.5}$$

Proof For a fixed $S = \{(\mathbf{x}_1, y_1), \ldots, (\mathbf{x}_m, y_m)\}$ data set of size m, since the functions $f \in \mathcal{F}$ have values in $\{-1,+1\}$, the norm of the vectors $(f(\mathbf{x}_1), \ldots, f(\mathbf{x}_m))^\top$ for any $f \in \mathcal{F}$ is bounded by $\sqrt{m}$. According to the inequality (2.3), the definition (1.10) and the definition of the growth function (1.14), we get:

$$\mathfrak{R}_m(\mathcal{F}) = \frac{2}{m}\mathbb{E}_S\left[\mathbb{E}_\sigma\left[\sup_{f\in\mathfrak{F}(\mathcal{F},S)}\sum_{i=1}^m \sigma_i f(\mathbf{x}_i)\right]\right] \leqslant \frac{2}{m}\mathbb{E}_S\left[\sqrt{2m\ln(2\mathfrak{G}(\mathcal{F},m))}\right],$$

where $\frac{2}{m}\mathbb{E}_S\left[\sqrt{2m\ln(2\mathfrak{G}(\mathcal{F},m))}\right] = \sqrt{\frac{8\ln(2\mathfrak{G}(\mathcal{F},m))}{m}}$.

The inequality (2.5) follows from the previous result and the second part of Sauer's lemma (1.5, Chap. 1). □

2.2.1 Generalization Bounds Using Rademacher Complexity

In this section, we give an outline of how to obtain a generalization bound in the case of i.i.d. data classification using the theory developed around Rademacher complexity ([13, 164]—Chapter 4):

Theorem 2.3 (Data-Dependent Generalization Bound) *Let $\mathcal{X} \in \mathbb{R}^d$ be a vector space and $\mathcal{Y} = \{-1, +1\}$ an output space. Suppose that the pairs of examples $(\mathbf{x}, y) \in \mathcal{X} \times \mathcal{Y}$ are generated i.i.d. according to a probability distribution $\mathcal{D}$. Let $\mathcal{F}$ be a class of functions with values in $\mathcal{Y}$ and $\ell : \mathcal{Y} \times \mathcal{Y} \to [0, 1]$ a given loss function. For any $\delta \in]0, 1]$, we have that for any function $f \in \mathcal{F}$ and any set $S \in (\mathcal{X} \times \mathcal{Y})^m$ of size m generated i.i.d. following the same probability distribution, the following inequality which holds with probability at least $1 - \delta$:*

$$\mathcal{L}(f) \leqslant \hat{\mathcal{L}}_m(f, S) + \mathfrak{R}_m(\ell \circ \mathcal{F}) + \sqrt{\frac{\ln \frac{1}{\delta}}{2m}}, \tag{2.6}$$

and, also with a probability of at least $1 - \delta$ we have:

$$\mathcal{L}(f) \leqslant \hat{\mathcal{L}}_m(f, S) + \hat{\mathfrak{R}}_S(\ell \circ \mathcal{F}) + 3\sqrt{\frac{\ln \frac{2}{\delta}}{2m}}. \tag{2.7}$$

where $\ell \circ \mathcal{F} = \{(\mathbf{x}, y) \mapsto \ell(f(\mathbf{x}), y) \mid f \in \mathcal{F}\}$.

The main interest of this theorem lies in the second bound, which involves the empirical Rademacher complexity of the class $\ell \circ \mathcal{F}$. We will see in the following and also in Chap. 5 (Theorem 5.4, Eq. (5.37)) that it is, indeed, possible to easily estimate this empirical Rademacher complexity for certain classes of functions and thus have a computable generalization bound on any training set. In what follows, we present the three major steps involved in obtaining these bounds.

2.2.1.1 Linking the Supremum of $\mathcal{F}$ de $\mathcal{L}(f) - \hat{\mathcal{L}}_m(f, S)$ to Its Expectation

As in the case of the development presented in the previous section, here we are looking for a generalization bound that is true in a uniform way for all functions f of a given class $\mathcal{F}$ and all learning bases S, noting that:

$$\forall f \in \mathcal{F}, \forall S, \mathcal{L}(f) - \hat{\mathcal{L}}_m(f, S) \leqslant \sup_{f \in \mathcal{F}} [\mathcal{L}(f) - \hat{\mathcal{L}}_m(f, S)]. \tag{2.8}$$

The study of this bound is carried out by linking the supremum appearing, in the right-hand term of the preceding inequality, with its expectation, by means of a powerful tool developed for empirical processes [17, 50, 65, 178] by McDiarmid [114] and known as the bounded difference inequality.

Theorem 2.4 (The Bounded Difference Inequality [114]) *Let $I \subset \mathbb{R}$ be a real interval and $(X_1, \ldots, X_m)$, m be independent random variables with values in I^m. Let $\Phi : I^m \to \mathbb{R}$ be such that: $\forall i \in \{1, \ldots, m\}, \exists a_i \in \mathbb{R}$. The following inequality is true whatever $(x_1, \ldots, x_m) \in I^m$ and $\forall x' \in I$:*

$$|\Phi(x_1, \ldots, x_{i-1}, x_i, x_{i+1}, \ldots, x_m) - \Phi(x_1, \ldots, x_{i-1}, x', x_{i+1}, \ldots, x_m)| \leqslant a_i.$$

We have then:

$$\forall \epsilon > 0, \mathbb{P}(\Phi(x_1, \ldots, x_m) - \mathbb{E}[\Phi] > \epsilon) \leqslant e^{\frac{-2\epsilon^2}{\sum_{i=1}^m a_i^2}}.$$

Hence, considering the following function:

$$\Phi : S \mapsto \sup_{f \in \mathcal{F}} [\mathcal{L}(f) - \hat{\mathcal{L}}_m(f, S)],$$

we can see that for two samples S and S^i of size m containing exactly the same examples apart from the pair $(x_i, y_i) \in S$, instead of which S^i contains the pair (x', y') sampled according to the same $\mathcal{D}$ distribution, the difference $|\Phi(S) - \Phi(S^i)|$ is bounded by $a_i = 1/m$ since the cost function ℓ (involved in $\mathcal{L}$ and $\hat{\mathcal{L}}_m$) has a value in $[0, 1]$. We then place ourselves in the case of application of the bounded difference inequality for the function Φ under consideration and $a_i = 1/m, \forall i$ and obtain the result:

$$\forall \epsilon > 0, \mathbb{P}\left(\sup_{f \in \mathcal{F}} [\mathcal{L}(f) - \hat{\mathcal{L}}_m(f, S)] - \mathbb{E}_S \sup_{f \in \mathcal{F}} [\mathcal{L}(f) - \hat{\mathcal{L}}_m(f, S)] > \epsilon\right) \leqslant e^{-2m\epsilon^2}.$$

2.2.1.2 Bounding $\mathbb{E}_S \sup_{f \in \mathcal{F}} [\mathcal{L}(f) - \hat{\mathcal{L}}_m(f, S)]$ with Respect to $\mathfrak{R}_m(\ell \circ \mathcal{F})$

This step is a symmetrization step and it consists in introducing into $\mathbb{E}_S \sup_{f \in \mathcal{F}} [\mathcal{L}(f) - \hat{\mathcal{L}}_m(f, S)]$ a second set S', also sampled along $\mathcal{D}^m$ and which plays a symmetrical role with respect to S. This step is the most technical in obtaining the generalization bound, and allows us to find (2.6).

The upper-bounding of $\mathbb{E}_S \sup_{f\in\mathcal{F}}[\mathcal{L}(f) - \hat{\mathcal{L}}_m(f,S)]$ begins by noting that the empirical error of a function $f \in \mathcal{F}$ on a given virtual sample S' is an unbiased estimator of its generalization error, i.e. $\mathcal{L}(f) = \mathbb{E}_{S'}\hat{\mathcal{L}}_m(f,S')$ and $\hat{\mathcal{L}}_m(f,S) = \mathbb{E}_{S'}\hat{\mathcal{L}}_m(f,S)$. Thus, we have:

$$\begin{aligned}\mathbb{E}_S \sup_{f\in\mathcal{F}}(\mathcal{L}(f) - \hat{\mathcal{L}}_m(f,S)) &= \mathbb{E}_S \sup_{f\in\mathcal{F}}[\mathbb{E}_{S'}(\hat{\mathcal{L}}_m(f,S') - \hat{\mathcal{L}}_m(f,S))]\\ &\leqslant \mathbb{E}_S\mathbb{E}_{S'} \sup_{f\in\mathcal{F}}[\hat{\mathcal{L}}_m(f,S') - \hat{\mathcal{L}}_m(f,S)].\end{aligned}$$

The previous inequality is due to the fact that the supremum of an expectation is less than the expectation of the supremum.

The second point in this step is to introduce the Rademacher variables into the supremum:

$$\mathbb{E}_S\mathbb{E}_{S'} \sup_{f\in\mathcal{F}}\left[\frac{1}{m}\sum_{i=1}^m(\ell(f(\mathbf{x}_i'),y_i') - \ell(f(\mathbf{x}_i),y_i))\right]. \tag{2.9}$$

For a fixed i, this introduction has the following effect: $\sigma_i = 1$ changes nothing, but $\sigma_i = -1$ inverts the two examples $(\mathbf{x}_i', y_i')$ and $(\mathbf{x}_i, y_i)$. So when we take the expectations on S and S', the introduction changes nothing. This being true for all i, we have, taking the expectation on $\boldsymbol{\sigma} = (\sigma_1,\dots,\sigma_m)$:

$$\begin{aligned}&\mathbb{E}_S\mathbb{E}_{S'} \sup_{f\in\mathcal{F}}[\hat{\mathcal{L}}_m(f,S') - \hat{\mathcal{L}}_m(f,S)] =\\ &\qquad\mathbb{E}_S\mathbb{E}_{S'}\mathbb{E}_{\boldsymbol{\sigma}} \sup_{f\in\mathcal{F}}\left[\frac{1}{m}\sum_{i=1}^m \sigma_i(\ell(f(\mathbf{x}_i'),y_i') - \ell(f(\mathbf{x}_i),y_i))\right].\end{aligned} \tag{2.10}$$

Applying the triangle inequality on $\sup = ||.||_\infty$ we get:

$$\begin{aligned}&\mathbb{E}_S\mathbb{E}_{S'}\mathbb{E}_{\boldsymbol{\sigma}} \sup_{f\in\mathcal{F}}\left[\frac{1}{m}\sum_{i=1}^m \sigma_i(\ell(f(\mathbf{x}_i'),y_i') - \ell(f(\mathbf{x}_i),y_i))\right] \leqslant\\ &\mathbb{E}_S\mathbb{E}_{S'}\mathbb{E}_{\sigma} \sup_{f\in\mathcal{F}}\frac{1}{m}\sum_{i=1}^m \sigma_i\ell(f(\mathbf{x}_i'),y_i') + \mathbb{E}_S\mathbb{E}_{S'}\mathbb{E}_{\boldsymbol{\sigma}} \sup_{f\in\mathcal{F}}\frac{1}{m}\sum_{i=1}^m(-\sigma_i)\ell(f(\mathbf{x}_i'),y_i').\end{aligned}$$

Finally, since $\forall i, \sigma_i$ and $-\sigma_i$ have the same distribution, we have:

$$\mathbb{E}_S\mathbb{E}_{S'} \sup_{f\in\mathcal{F}}[\hat{\mathcal{L}}_m(f,S') - \hat{\mathcal{L}}_m(f,S)] \leqslant \underbrace{2\mathbb{E}_S\mathbb{E}_{\boldsymbol{\sigma}} \sup_{f\in\mathcal{F}}\frac{1}{m}\sum_{i=1}^m \sigma_i\ell(f(\mathbf{x}_i),y_i)}_{=\mathfrak{R}_m(\ell\circ\mathcal{F})}. \tag{2.11}$$

Summing up the results obtained so far, we get:

1. $\forall f \in \mathcal{F}, \forall S, \mathcal{L}(f) - \hat{\mathcal{L}}_m(f, S) \leqslant \sup_{f \in \mathcal{F}}[\mathcal{L}(f) - \hat{\mathcal{L}}_m(f, S)]$.
2. $\forall \epsilon > 0, \mathbb{P}\left(\sup_{f \in \mathcal{F}}[\mathcal{L}(f) - \hat{\mathcal{L}}_m(f, S)] - \mathbb{E}_S \sup_{f \in \mathcal{F}}[\mathcal{L}(f) - \hat{\mathcal{L}}_m(f, S)] > \epsilon\right)$ $\leqslant e^{-2m\epsilon^2}$.
3. $\mathbb{E}_S \sup_{f \in \mathcal{F}} \left[\mathcal{L}(f) - \hat{\mathcal{L}}_m(f, S)\right] \leqslant \mathfrak{R}_m(\ell \circ \mathcal{F})$.

And so the first point of Theorem (2.3) is obtained by solving the equation $e^{-2m\epsilon^2} = \delta$ with respect to ϵ.

2.2.1.3 Bounding $\mathfrak{R}_m(\ell \circ \mathcal{F})$ with Respect to $\hat{\mathfrak{R}}_S(\ell \circ \mathcal{F})$

This step is carried out by applying the bounded difference inequality to the function $\Phi : S \mapsto \hat{\mathfrak{R}}_S(\ell \circ \mathcal{F})$. For a given S and any $i \in \{1, \dots, m\}$ let us consider the set S^i obtained by replacing the $(\mathbf{x}_i, y_i) \in S$ example with a $(\mathbf{x}', y')$ example sampled i.i.d. following the same $\mathcal{D}$ distribution generating the examples in S. In this case, the absolute value of the difference $|\Phi(S) - \Phi(S^i)|$ will be bounded by $a_i = 2/m$, and we have according to bounded difference inequality (Theorem 2.4).

$$\forall \epsilon > 0, \mathbb{P}(\mathfrak{R}_m(\ell \circ \mathcal{F}) > \hat{\mathfrak{R}}_S(\ell \circ \mathcal{F}) + \epsilon) \leqslant e^{-m\epsilon^2/2}. \tag{2.12}$$

So for $\delta/2 = e^{-m\epsilon^2/2}$, we have with probability at least equal to $1 - \delta/2$:

$$\mathfrak{R}_m(\ell \circ \mathcal{F}) \leqslant \hat{\mathfrak{R}}_S(\ell \circ \mathcal{F}) + 2\sqrt{\frac{\ln \frac{2}{\delta}}{2m}}.$$

According to the first point (2.6) of Theorem 2.3 that we just proved, we also have with probability at least $1 - \delta/2$:

$$\forall f \in \mathcal{F}, \forall S, \mathcal{L}(f) \leqslant \hat{\mathcal{L}}_m(f, S) + \mathfrak{R}_m(\ell \circ \mathcal{F}) + \sqrt{\frac{\ln \frac{2}{\delta}}{2m}}.$$

The second point (2.7) of Theorem (2.3) is then obtained by combining the two previous results with the union bound.

2.2.2 *Rademacher Complexity Properties*

In the following, we will present two data-dependent bounds that are directly derived from Theorem (2.3), the previous developments, and properties 4 and 5 of the

Rademacher complexity, as outlined in the subsequent theorem. Note that property 5 applies to Lipschitz functions, defined as:

Definition 2.3 (Lipschitz Function) A function $\mathfrak{h}_L : \mathbb{R} \to \mathbb{R}$ such that:

$$\forall (x, y) \in \mathbb{R}^2, |\mathfrak{h}_L(x) - \mathfrak{h}_L(y)| \leqslant L|x - y|, \tag{2.13}$$

where $L \in \mathbb{R}_+^\star$ is a constant independent of x and y, is called a Lipschitz function.

Theorem 2.5 (Rademacher Complexity Properties) *Let $\mathcal{F}_1, \dots, \mathcal{F}_\ell$ and $\mathcal{G}$ be classes of real functions. We then have:*

1. *For all $a \in \mathbb{R}$, $\hat{\mathfrak{R}}_S(a\mathcal{F}) = a\hat{\mathfrak{R}}_S(\mathcal{F})$;*
2. *If $\mathcal{F} \subseteq \mathcal{G}$, then $\hat{\mathfrak{R}}_S(\mathcal{F}) \leqslant \hat{\mathfrak{R}}_S(\mathcal{G})$;*
3. $\hat{\mathfrak{R}}_S\left(\sum_{i=1}^{\ell} \mathcal{F}_i\right) \leqslant \sum_{i=1}^{\ell} \hat{\mathfrak{R}}_S(\mathcal{F}_i)$;
4. *Let $conv(\mathcal{F})$ be the set of convex combinations (or the convex hull) of $\mathcal{F}$, defined by:*

$$conv(\mathcal{F}) = \left\{ \sum_{t=1}^{T} \lambda_t f_t \mid \forall t \in \{1, \dots, T\}, f_t \in \mathcal{F}, \lambda_t \geqslant 0 \wedge \sum_{t=1}^{T} \lambda_t = 1 \right\}. \tag{2.14}$$

We then have $\hat{\mathfrak{R}}_S(conv(\mathcal{F})) = \hat{\mathfrak{R}}_S(\mathcal{F})$;

5. *If $\mathfrak{h}_L : \mathbb{R} \to \mathbb{R}$ be a Lipschitz function with constant $L > 0$, then:*

$$\hat{\mathfrak{R}}_S(\mathfrak{h}_L \circ \mathcal{F}) \leqslant L\hat{\mathfrak{R}}_S(\mathcal{F}); \tag{2.15}$$

6. *For all functions g, we have* $\hat{\mathfrak{R}}_S(\mathcal{F} + g) \leqslant \hat{\mathfrak{R}}_S(\mathcal{F}) + \frac{2}{m}\left(\sum_{i=1}^{m} g(\mathbf{x}_i)^2\right)^{1/2}$.

Proof Properties 1, 2 and 3 are derived from the definition of empirical Rademacher complexity (2.1).

The proof of property 4 is the following. From (2.1) we have:

$$\begin{aligned}\hat{\mathfrak{R}}_S(conv(\mathcal{F})) &= \frac{2}{m}\mathbb{E}_\sigma\left[\sup_{f_1,\dots,f_T\in\mathcal{F},\lambda_1,\dots,\lambda_T,||\boldsymbol{\lambda}||_1\leqslant 1}\sum_{i=1}^m\sigma_i\sum_{t=1}^T\lambda_t f_t(\mathbf{x}_i)\right]\\ &= \frac{2}{m}\mathbb{E}_\sigma\left[\sup_{f_1,\dots,f_T\in\mathcal{F}}\ \sup_{\lambda_1,\dots,\lambda_T,||\boldsymbol{\lambda}||_1\leqslant 1}\sum_{t=1}^T\lambda_t\left(\sum_{i=1}^m\sigma_i f_t(\mathbf{x}_i)\right)\right].\end{aligned}$$

Since all the weights of the combination $\lambda_t, \forall t$ are positive and $\sum_{t=1}^T\lambda_t = 1$ we have $\forall(z_1,\dots,z_T)\in\mathbb{R}^T$:

$$\sup_{\lambda_1,\dots,\lambda_T,||\boldsymbol{\lambda}||_1\leqslant 1}\sum_{t=1}^T\lambda_t z_t = \max_{t\in\{1,\dots,T\}} z_t.$$

That is, the supremum is reached by putting all the weights (of sum equal to 1) on the largest term of the combination, which gives:

$$\begin{aligned}\hat{\mathfrak{R}}_S(conv(\mathcal{F})) &= \frac{2}{m}\mathbb{E}_\sigma\left[\sup_{f_1,\dots,f_T\in\mathcal{F}}\ \sup_{\lambda_1,\dots,\lambda_T,||\boldsymbol{\lambda}||_1\leqslant 1}\sum_{t=1}^T\lambda_t\left(\sum_{i=1}^m\sigma_i f_t(\mathbf{x}_i)\right)\right]\\ &= \frac{2}{m}\mathbb{E}_\sigma\left[\sup_{f_1,\dots,f_T\in\mathcal{F}}\ \max_{t\in\{1,\dots,T\}}\sum_{i=1}^m\sigma_i f_t(\mathbf{x}_i)\right]\\ &= \underbrace{\frac{2}{m}\mathbb{E}_\sigma\left[\sup_{h\in\mathcal{F}}\sum_{i=1}^m\sigma_i h(\mathbf{x}_i)\right]}_{=\hat{\mathfrak{R}}_S(\mathcal{F})}. \end{aligned} \tag{2.16}$$

Property 5 is better known as Talagrand's lemma and it is the basis for the derivation of margin-based generalization bounds [8]. For a fixed training set S, we have:

$$\begin{aligned}\hat{\mathfrak{R}}_S(\mathfrak{h}_L\circ\mathcal{F}) = \frac{2}{m}\mathbb{E}_\sigma\left[\sup_{f\in\mathcal{F}}\sum_{i=1}^m\sigma_i(\mathfrak{h}_L\circ f)(\mathbf{x}_i)\right] =\\ \frac{2}{m}\mathbb{E}_{\sigma_1,\dots,\sigma_{m-1}}\left[\mathbb{E}_{\sigma_m}\left[\sup_{f\in\mathcal{F}}\Delta_{m-1}(f)+\sigma_m(\mathfrak{h}_L\circ f)(\mathbf{x}_m)\right]\right],\end{aligned}$$

where $\Delta_{m-1}(f) = \sum_{i=1}^{m-1} \sigma_i (\mathfrak{h}_L \circ f)(\mathbf{x}_i)$, by supposing that the suprema can be attained and let $f_1, f_2 \in \mathcal{F}^2$ be such that:

$$\Delta_{m-1}(f_1) + (\mathfrak{h}_L \circ f_1)(\mathbf{x}_m) = \sup_{f \in \mathcal{F}} [\Delta_{m-1}(f) + (\mathfrak{h}_L \circ f)(\mathbf{x}_m)],$$
$$\Delta_{m-1}(f_2) - (\mathfrak{h}_L \circ f_2)(\mathbf{x}_m) = \sup_{f \in \mathcal{F}} [\Delta_{m-1}(f) - (\mathfrak{h}_L \circ f)(\mathbf{x}_m)].$$

By Definition (1.2) we obtain:

$$\mathbb{E}_{\sigma_m}\left[\sup_{f \in \mathcal{F}} \Delta_{m-1}(f) + \sigma_m (\mathfrak{h}_L \circ f)(\mathbf{x}_m)\right] = \frac{1}{2}[\Delta_{m-1}(f_1) + (\mathfrak{h}_L \circ f_1)(\mathbf{x}_m)] + \frac{1}{2}[\Delta_{m-1}(f_2) - (\mathfrak{h}_L \circ f_2)(\mathbf{x}_m)].$$

If $s = \operatorname{sgn}(f_1(\mathbf{x}_m) - f_2(\mathbf{x}_m))$, then from the L-Lipschitzness of $\mathfrak{h}_L$ the above equation writes:

$$\begin{aligned}\mathbb{E}_{\sigma_m}\left[\sup_{f \in \mathcal{F}} \Delta_{m-1}(f) + \sigma_m (\mathfrak{h}_L \circ f)(\mathbf{x}_m)\right] &\leqslant \frac{1}{2}[\Delta_{m-1}(f_1) + \Delta_{m-1}(f_2) + sL(f_1(\mathbf{x}_m) - f_2(\mathbf{x}_m))] \\ &\leqslant \frac{1}{2}[\Delta_{m-1}(f_1) + sLf_1(\mathbf{x}_m)] + \frac{1}{2}[\Delta_{m-1}(f_2) - sLf_2(\mathbf{x}_m))].\end{aligned}$$

By the definition of the expectation over σ_m in the last inequality we get:

$$\mathbb{E}_{\sigma_m}\left[\sup_{f \in \mathcal{F}} \Delta_{m-1}(f) + \sigma_m (\mathfrak{h}_L \circ f)(\mathbf{x}_m)\right] \leqslant \mathbb{E}_{\sigma_m}\left[\sup_{f \in \mathcal{F}} \Delta_{m-1}(f) + \sigma_m L f(\mathbf{x}_m)\right].$$

Applying the same approach to all other $\sigma_i, i \neq m$ values establishes the lemma.

Property 6 is proved by following the same development used for the Rademacher complexity upper-bound presented in the previous section. By definition and according to the triangle inequality:

$$\begin{aligned}\forall g, \hat{\mathfrak{R}}_S(\mathcal{F} + g) &\leqslant \mathbb{E}_{\boldsymbol{\sigma}}\left[\sup_{f \in \mathcal{F}} \left|\frac{2}{m}\sum_{i=1}^{m} \sigma_i (f(\mathbf{x}_i) + g(\mathbf{x}_i))\right|\right] \\ &\leqslant \frac{2}{m}\mathbb{E}_{\boldsymbol{\sigma}}\left[\sup_{f \in \mathcal{F}} \left|\sum_{i=1}^{m} \sigma_i f(\mathbf{x}_i)\right|\right] + \frac{2}{m}\mathbb{E}_{\boldsymbol{\sigma}}\left|\sum_{i=1}^{m} \sigma_i g(\mathbf{x}_i)\right|.\end{aligned}$$

From the equality, $\forall z \in \mathbb{R}$, $|z| = \sqrt{z^2}$, we have:

$$\forall g, \hat{\mathfrak{R}}_S(\mathcal{F} + g) \leqslant \hat{\mathfrak{R}}_S(\mathcal{F}) + \frac{2}{m}\mathbb{E}_\sigma \left[\left(\sum_{i=1}^{m} \sigma_i g(\mathbf{x}_i) \right)^2 \right]^{1/2}$$

$$\leqslant \hat{\mathfrak{R}}_S(\mathcal{F}) + \frac{2}{m}\mathbb{E}_\sigma \left[\sum_{i=1}^{m} \sum_{j=1}^{m} \sigma_i \sigma_j g(\mathbf{x}_i) g(\mathbf{x}_j) \right]^{1/2}.$$

Using Jensen's inequality, the concavity of the square root function and the definition of Rademacher variables, the result is:

$$\forall g, \hat{\mathfrak{R}}_S(\mathcal{F} + g) \leqslant \hat{\mathfrak{R}}_S(\mathcal{F}) + \frac{2}{m} \left(\sum_{i=1}^{m} g(\mathbf{x}_i)^2 \right)^{1/2}.$$

□

2.3 Practical Considerations

In this section, we present how the principle of structural risk minimization is implemented in practice, and how empirical error minimization is considered in learning algorithms.

2.3.1 Regularization

We have seen that learning a model is a compromise between a low empirical error and a high function class capacity (Chap. 1, Sect. 1.3.3). In practice, this compromise is implemented by adding a term, called *regularization*, to the empirical error, which penalizes the algorithm for choosing a complex model:

$$\min_h \hat{\mathcal{L}}(h, S) + \lambda Reg(h), \tag{2.17}$$

where $Reg(h)$ is the regularization term that reflects the complexity of the prediction function h, and its value increases as the complexity of h increases; $\lambda \in \mathbb{R}_+$ is the regularization parameter that controls the importance of the term $Reg(h)$.

In concrete terms, the notions of complexity used in the literature depend on the number of parameters in the prediction model and their values, and they generally express a norm on the parameters. The most commonly used norms are

---

Input :

- A training set S divided into K disjoint subsets $(S_k)_{1\leqslant k\leqslant K}$; $S = \bigsqcup_{k=1}^{K} S_k$;
- A predefined set of hyperparameter values Λ;

for $\lambda \in \Lambda$ **do**
  $\pi_\lambda \leftarrow 0$;
  for $k = 1 \ldots K$ **do**
    Train h_k on $S \setminus S_k$;
    $perf_k \leftarrow$ performance of h_k on S_k;
  $\pi_\lambda = \frac{1}{K}\sum_{k=1}^{K} perf_k$;

Output : $\lambda^* = argmin_{\lambda\in\Lambda}\, \pi_\lambda$.

Algorithme 1: K-fold cross-validation

the ℓ_2 norm and the ℓ_1 norm defined for a prediction function h of n parameters $\boldsymbol{w} = (w_1, \ldots, w_n)$ as:

$$\ell_1(\boldsymbol{w}) = \sum_{j=1}^{n} |w_j|, \quad (\ell_1 - \text{norm}),$$

$$\ell_2(\boldsymbol{w}) = \sum_{j=1}^{n} w_j^2, \quad (\ell_2 - \text{norm}).$$

Typically, the parameter λ is not learned on the training set and is considered as a hyperparameter passed as input to the learning algorithm.

Different strategies have been proposed for defining the optimal value of a hyperparameter [94], and *K-fold cross-validation* is probably the most widely used technique in the literature.

This technique consists in dividing the training set into K disjoint subsets. For a given hyperparameter value chosen from a set of predefined values, the model is trained K times on $K - 1$ disjoint subsets of the initial training set, and its performance is estimated on the last subset (called the validation set) not used to train the model each time.

For this hyperparameter value, a final performance is estimated as the average performance of the K performances obtained on the different validation sets. In the end, the hyperparameter value that led to the best average performance value is chosen to train the model on the entire training set. This procedure is summarized in Algorithm 1.

There exists another cross-validation approach called *leave-K-out cross-validation* (LKO) where K samples are left out during each iteration and the model is trained on the remaining data. This process is repeated such that every possible combination of K samples is used as a validation set at least once. This approach can be computationally expensive, especially when the number of training samples is large.

2.3.2 Minimization of a Convex Bound of the Classification Error

Geometrically, learning a two-class classification model $h : \mathcal{X} \to \mathbb{R}$ results in the search for a decision frontier in the input space $\{\mathbf{x}|h(\mathbf{x}) = 0\}$ which separates the representation space into two subspaces: $\{\mathbf{x}|h(\mathbf{x}) > 0\}$ and $\{\mathbf{x}|h(\mathbf{x}) < 0\}$ corresponding to the subspaces that respectively contain examples of classes $+1$ and -1. Once the function h is fixed, the associated classifier is then defined as $\forall \mathbf{x}, f(\mathbf{x}) = \text{sign}(h(\mathbf{x}))$:

$$\forall z \in \mathbb{R}, \text{sgn}(z) = \begin{cases} +1 & \text{if } z > 0, \\ 0 & \text{if } z = 0, \\ -1 & \text{if } z < 0. \end{cases}$$

We recall that the instantaneous loss of the h classification function for an example $(\mathbf{x}, y)$ is in this case:

$$\forall (\mathbf{x}, y); \ell(h(\mathbf{x}), y) = \mathbb{1}_{y \times h(\mathbf{x}) \leqslant 0}. \tag{2.18}$$

The empirical error h on a training set $S = (\mathbf{x}_i, y_i)_{i=1}^m \in (\mathbb{R}^d \times \{-1, +1\})^m$ is also written:

$$\hat{\mathcal{L}}_m(h, S) = \frac{1}{m} \sum_{i=1}^{m} \mathbb{1}_{y_i \times h(\mathbf{x}_i) \leqslant 0}. \tag{2.19}$$

This error is not differentiable, and direct optimization of the empirical 0/1 loss estimated on a training set is impossible. In practice, classification functions are trained by minimizing a convex, continuous and differentiable upper-bound of the empirical 0/1 loss (2.19).

Numerous works have investigated the validity of this procedure, and in particular established that the empirical error and its majorant defined with a convex, continuous, differentiable loss function taking the value 1 in 0, admit the same minimizer [12].

Among the existing bounds on classification error, the following four loss functions have been particularly studied in the state of the art because of their mathematical properties, and they are used in classical models namely, *AdaBoost*, *logistic regression*, *support vector machines* and *Adaline* that we will present in the next chapter:

$$\begin{aligned}
\ell_e(h(\mathbf{x}), y) &= e^{-y \times h(\mathbf{x})}, && \textbf{(Exponentialloss)}, \\
\ell_l(h(\mathbf{x}), y) &= \ln(1 + \exp(-y \times h(\mathbf{x}))), && \textbf{(Logisticloss)}, \\
\ell_\varsigma(h(\mathbf{x}), y) &= \max(0, 1 - y \times h(\mathbf{x})), && \textbf{(Hingeloss)}, \\
\ell_q(h(\mathbf{x}), y) &= (1 - y \times h)^2, && \textbf{(Leastsquareloss)}.
\end{aligned}$$

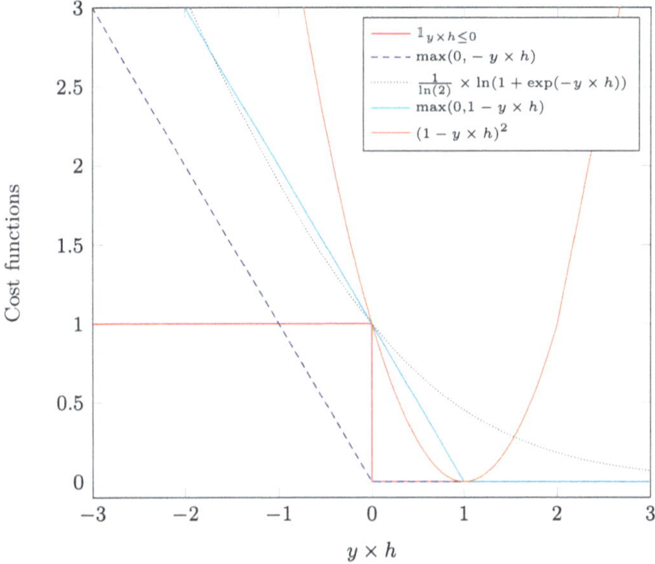

Fig. 2.1 Four loss functions for a two-class classification problem as a function of the product $y \times h$, where h is the learned function. The loss functions are the 0/1 loss: $\mathbb{1}_{y \times h \leqslant 0}$, the exponential loss: $e^{-y \times h}$, the logistic loss: $\frac{1}{\ln(2)} \times \ln(1 + \exp(-y \times h))$ and the hinge loss: $\max(0, 1 - y \times h)$, and the least square loss: $(1 - y \times h)^2$. Positive (negative) values of $y \times h$ imply good (bad) classification

Figure 2.1 shows the graphs of these four loss functions and the 0/1 loss as a function of the product $y \times h$.

The minimization of multivariate, continuous and differentiable functions has been widely studied in the past [24, 129] and most of the conventional optimization approaches proposed in this framework have been adapted and used in machine learning. In Chap. 3, we present some of these unconstrained convex optimization techniques, which belong to the family of descent-direction algorithms and which we will use in estimating the parameters of the learning models described in various chapters of this textbook.

2.4 Multi-Class Case

The binary classification considered so far has provided an ideal framework for the development of learning theory. However, real applications in classification are mostly large-scale, multi-class problems where the number of classes to be processed is very large. This is the case, for example, of the document classification problem, which has long been a favorite field for the development of multi-class models. In this section, we will present the formal framework of this type of classification, as well as some approaches that are reductions of the multi-class problem to binary classification.

In multi-class classification problems, the two cases of study are the mono-label case, where each example belongs to just one class, and the opposite case, often referred to as the multi-label case. In the mono-label case, the output space is a set of class labels $\mathcal{Y} = \{1, \dots, K\}$ and the two usual ways of representing the output of an example $\mathbf{x} \in \mathcal{X}$ is either to assign it its class label $y \in \mathcal{Y}$, or to associate it with a class indicator vector $\mathbf{y} \in \{0, 1\}^K$, referred to as one-hot encoding, which is a binary vector whose components are all equal to 0 except for the component corresponding to the example's class, which is 1:

$$\forall(\mathbf{x}, y) \in \mathcal{X} \times \mathcal{Y}, y = k \Leftrightarrow \mathbf{y}^\top = \left(\underbrace{y_1, \dots, y_{k-1}}_{\text{all equal to 0}}, \underbrace{y_k}_{=1}, \underbrace{y_{k+1}, \dots, y_K}_{\text{all equal to 0}} \right). \tag{2.20}$$

In the multi-label case, the output space is $\mathcal{Y} = \{-1, +1\}^K$ and each example $\mathbf{x} \in \mathcal{X}$ can belong to several classes, in which case the output associated with the example is a vector $\mathbf{y} \in \mathcal{Y}$ whose components corresponding to the classes of the example are equal to $+1$ and the other components are worth -1.

2.4.1 Classification Errors

As in the binary case, we assume that the pairs of examples $(\mathbf{x}, y) \in \{\mathcal{X} \to \mathcal{Y}\}$ are i.i.d. following a fixed but unknown probability distribution $\mathcal{D}$. The learning goal is to find the classification function in a given function space $\mathcal{F} = \{f : \mathcal{X} \to \mathcal{Y}\}$ that has the lowest generalization error:

$$\mathcal{L}(f) = \mathbb{E}_{(\mathbf{x}, y)\sim\mathcal{D}}[\ell(f(\mathbf{x}), y)], \tag{2.21}$$

where $f(\mathbf{x}) = (f_1(\mathbf{x}), \dots, f_K(\mathbf{x})) \in \mathcal{Y}$ is the predicted output vector for example $\mathbf{x}$ in the multi-label case or the class index in the mono-label case , and $\ell : \mathcal{Y} \times \mathcal{Y} \to \mathbb{R}_+$ is an instantaneous classification error. In the multi-label case, this error is often based on the Hamming distance [73], counting the number of different components between the predicted $f(\mathbf{x})$ and actual $\mathbf{y}$ class vectors for an example $\mathbf{x}$:

$$\ell(f(\mathbf{x}), \mathbf{y}) = \frac{1}{2} \sum_{k=1}^{K} (1 - \operatorname{sgn}(y_k f_k(\mathbf{x}))), \tag{2.22}$$

In the mono-label case, the instantaneous error is defined simply as:

$$\ell(f(\mathbf{x}), y) = \mathbb{1}_{f(\mathbf{x}) \neq y}. \tag{2.23}$$

As in the binary case, the classification function is found according to the principle of empirical risk minimization, using a training set $S \in (\mathcal{X} \times \mathcal{Y})^m$.

$$f^* = \underset{f \in \mathcal{F}}{\operatorname{argmin}}\, \hat{\mathcal{L}}_m(f, S) = \underset{f \in \mathcal{F}}{\operatorname{argmin}}\, \frac{1}{m} \sum_{i=1}^{m} \ell(f(\mathbf{x}_i), y_i). \tag{2.24}$$

In practice, we learn a function $\boldsymbol{h}$ defined as:

$$\begin{aligned} \boldsymbol{h} : \mathbb{R}^d &\rightarrow \mathbb{R}^K \\ \mathbf{x} &\mapsto (h(\mathbf{x}, 1), \ldots, h(\mathbf{x}, K)), \end{aligned}$$

where $h \in \mathbb{R}^{\mathcal{X} \times \mathcal{Y}}$, by minimizing a convex bound derivable from the empirical error as introduced in the previous section. The prediction function f for an example $\mathbf{x}$ is then obtained by thresholding the outputs of $h(\mathbf{x}, k), k \in \{1, \ldots, K\}$ for the multi-label case or by taking the class index giving the largest predicted output value following h for the mono-label case:

$$\forall \mathbf{x},\ f_{\boldsymbol{h}}(\mathbf{x}) = \underset{k \in \{1,\ldots,K\}}{\operatorname{argmax}}\ h(\mathbf{x}, k). \tag{2.25}$$

In the following sections, we will outline the most common reduction-based strategies for learning $h \in \mathbb{R}^{\mathcal{X} \times \mathcal{Y}}$ functions.

2.4.2 *Reduction of the Multi-Class Problem to Binary Classification*

The first two approaches, called one versus all and one versus one, learn a classifier according to different strategies and combine the outputs of these classifiers to define a single multi-class predictor. We will then show that these two strategies are special cases of a more general framework called *error-correcting codes*.

2.4.2.1 One Versus All

The one vs all strategy involves training K binary classifiers $f_k : \mathcal{X} \rightarrow \{-1, +1\}; k \in \{1, \ldots, K\}$ to identify each of the K existing classes from all the others. Starting from a training set $S = \{(\mathbf{x}_i, y_i), i \in \{1, \ldots, m\} \in (\mathcal{X} \times \mathcal{Y})^m$, we build each classifier $f_k, k \in \{1, \ldots, K\}$ on an induced training set $S_k \in (\mathcal{X} \times \{-1, +1\})^m$ by labeling S examples belonging to class k by $+1$ and those belonging to all other classes by -1, and we learn a function $h_k : \mathcal{X} \rightarrow \mathbb{R}$ whose outputs are used to induce the binary function f_k. The multi-class decision function $f : \mathcal{X} \rightarrow \{-1, +1\}$ is then obtained according to the following rule:

$$\forall \mathbf{x} \in \mathcal{X}, \, f(\mathbf{x}) = \underset{k \in \{1,\dots,K\}}{\operatorname{argmax}} \; h_k(\mathbf{x}). \tag{2.26}$$

The one versus all strategy is summarized in Algorithm 2.

This strategy, although simple, has several drawbacks. Firstly, the amplitude of the classifiers' output values is generally not in the same range, whereas the rule (2.26) assumes the opposite. Secondly, even if the distribution of classes is balanced in the initial training set, the proportion of the positive class in the induced sets will be lower the greater the number of classes K, which could pose a problem for learning $h_k : \mathcal{X} \rightarrow \mathbb{R}$ functions.

2.4.2.2 One Versus One

Another strategy, known as the one versus one approach, involves learning a binary classifier $f_{yy'} : \mathcal{X} \rightarrow \{-1, +1\}$ to discriminate the elements of all pairs of classes $(y, y') \in \mathcal{Y}^2, y \neq y'$.

This classifier is obtained by learning a classification function $h_{yy'} : \mathcal{X} \rightarrow \mathbb{R}$ for each pair of classes $(y, y') \in \mathcal{Y}^2$, $y < y'$ by building a new training set containing only examples of the two classes and labeling those of class y with the desired output $+1$ and those of class y' with the output -1. There are thus $\binom{K}{2} = \frac{K(K-1)}{2}$ classifiers to learn and they are combined to obtain the binary classifiers with the following rule:

$$\forall \mathbf{x} \in \mathcal{X}, \forall (y, y') \in \mathcal{Y}^2, y \neq y', \, f_{yy'}(\mathbf{x}) = \begin{cases} \operatorname{sgn}\left(h_{yy'}(\mathbf{x})\right), & \text{if } y < y', \\ -f_{y'y}(\mathbf{x}), & \text{if } y' < y. \end{cases} \tag{2.27}$$

Input :

- A training set $S = \{\mathbf{x}_1, \mathbf{y}_1), \dots, (\mathbf{x}_m, \mathbf{y}_m)\} \in (\mathcal{X} \times \mathcal{Y})^m$;

for $k = 1 \dots K$ **do**
- $\tilde{S} \leftarrow \emptyset$;
- **for** $i = 1 \dots m$ **do**
  - **if** $y_i = k$ **then**
    - $\tilde{S} \leftarrow (\mathbf{x}_i, +1)$;
  - **else**
    - $\tilde{S} \leftarrow (\mathbf{x}_i, -1)$;
- Train a classifier $h_k : \mathcal{X} \rightarrow \mathbb{R}$ on $\tilde{S}$;

Output : $\forall \mathbf{x} \in \mathcal{X}, \, f(\mathbf{x}) = \operatorname{argmax}_{k \in \{1,\dots,K\}} h_k(\mathbf{x})$.

Algorithme 2: One versus all strategy for multi-class classification

The final multi-class classification function $f : \mathcal{X} \rightarrow \mathcal{Y}$ is constructed by aggregating the outputs of multiple binary classifiers through a majority voting mechanism. When classifying a new example $\mathbf{x}$, each binary classifier casts a vote for either class y or class y'.

The class that receives the highest number of votes across all relevant binary classifiers is assigned to the example $\mathbf{x}$. In other words, the prediction is made based on the following decision rule:

$$\forall \mathbf{x} \in \mathcal{X}, f(\mathbf{x}) = \underset{y' \in \mathcal{Y}, y' \neq y}{\text{argmax}} \left|\{y \mid f_{yy'}(\mathbf{x}) = +1\}\right|. \tag{2.28}$$

This strategy does not suffer from the problem of class imbalance, mentioned for the one-versus-all approach, but in the case where the majority of classes contain few examples, which is the case for large-scale multi-class classification problems, the $h_{.}$ classifiers will be trained on few examples, which may affect their performance.

But the major drawback of this strategy is the number of classifiers, which is squared with the number of classes to be learned, and which can quickly become a handicap for large-scale problems.

Input :

- A training set $S = \{(\mathbf{x}_1, \mathbf{y}_1), \ldots, (\mathbf{x}_m, \mathbf{y}_m)\} \in (\mathcal{X} \times \mathcal{Y})^m$;

```
for k = 1 ... K − 1 do
    for ℓ = k + 1 ... K do
        S̃ ← ∅;
        for i = 1 ... m do
            if y_i = k then
                S̃ ← (x_i, +1);
            else if y_i = ℓ then
                S̃ ← (x_i, −1);
        Train a classifier h_kℓ : X → ℝ on S̃;
```

Output : $\forall \mathbf{x} \in \mathcal{X}, f(\mathbf{x}) = \underset{y' \in \mathcal{Y}, y' \neq y}{\text{argmax}} \left|\{y \mid f_{yy'}(\mathbf{x}) = +1\}\right|$ // ▷ (2.28)

where,

$$\forall \mathbf{x} \in \mathcal{X}, \forall (y, y') \in \mathcal{Y}^2, y \neq y', f_{yy'}(\mathbf{x}) = \begin{cases} \text{sgn}\left(h_{yy'}(\mathbf{x})\right), & \text{if } y < y', \\ -f_{y'y}(\mathbf{x}), & \text{if } y' < y. \end{cases}$$

Algorithme 3: One versus one approach for multi-class classification

Table 2.1 An example of error-correction output codes for multi-class classification. The code matrix is depicted, where each row corresponds to a code word of length $n = 5$ assigned to a class $k \in \{1, \dots, 6\}$

		Codes				
Classes	1	−1	−1	+1	+1	−1
	2	+1	+1	−1	+1	−1
	3	−1	+1	−1	−1	−1
	4	−1	+1	+1	−1	+1
	5	+1	−1	+1	+1	+1
	6	−1	−1	−1	+1	+1

2.4.2.3 Error Correcting Output Codes

A broader approach to transforming multi-class into binary classification involves leveraging error-correction output codes (ECOC). This method entails associating each class k by a fixed-length bit sequence n, $\mathfrak{D}_k$, called code word, which reflects the intrinsic characteristics of the class. Collectively, these vectors form a matrix, as illustrated in Table 2.1.

With the resulting code matrix $\mathfrak{D} \in \{-1, +1\}^{K\times n}$, n binary classifiers are learned by creating n training sets $\tilde{S}_j$ from the initial training set S. Each of the $\tilde{S}_j$ sets is created by considering the corresponding j column of the $\mathfrak{D}$ matrix and relabeling the examples with their corresponding input labels in that column.

$$\forall (\mathbf{x}, y) \in \mathcal{X} \times \mathcal{Y}, \forall j \in \{1, \dots, n\}, \text{ the associated code is } (\mathbf{x}, \mathfrak{D}_y(j)). \tag{2.29}$$

This process is illustrated in Algorithm 4.

Input :

- A training set $S = \{(\mathbf{x}_1, \mathbf{y}_1), \dots, (\mathbf{x}_m, \mathbf{y}_m)\} \in (\mathcal{X} \times \mathcal{Y})^m$;
- The class coding matrix $\mathfrak{D} \in \{-1, +1\}^{K\times n}$;

for $j = 1 \dots n$ **do**
  $\tilde{S}_j \leftarrow \emptyset$;
  for $i = 1 \dots m$ **do**
    $\tilde{S}_j \leftarrow (\mathbf{x}_i, \mathfrak{D}_{y_i}(j))$; // ▷ (2.29);
  Train the classifier $h_j : \mathcal{X} \to \mathbb{R}$ on $\tilde{S}_j$;

Output : $\forall \mathbf{x} \in \mathcal{X}, \mathbf{h}(\mathbf{x}) = (h_1(\mathbf{x}), \dots, h_n(\mathbf{x}))^\top$.

Algorithme 4: Error-correcting output code for multi-class classification

The second step, called prediction, consists of assigning each example to the class whose associated code $\mathfrak{D}_.$ is closest to the outputs of the binary classifiers learned from the Hamming distance:

$$\forall \mathbf{x} \in \mathcal{X}, f(\mathbf{x}) = \underset{k\in\{1,\dots,K\}}{\operatorname{argmin}} \frac{1}{2}\sum_{j=1}^{n}(1 - \operatorname{sgn}(\mathfrak{D}_k(j)h_j(\mathbf{x}))). \tag{2.30}$$

Allwin et al. [2] have extended matrix coding to ternary codes $\mathfrak{D} \in \{-1, 0, +1\}^{K \times n}$ to disregard examples with class 0 output after coding, when learning binary functions. With this coding scheme, both the one-vs-all and one-vs-one strategies can be viewed as special cases of error-correcting output codes. This perspective allows us to understand these strategies within a broader framework of coding theory, where each class is represented by a unique codeword.

In the one-versus-all strategy, the code matrix is structured as a square matrix of size $K \times K$. Each row of this matrix corresponds to a binary classifier that distinguishes one class from all the others. The diagonal elements of this matrix are set to $+1$, indicating that the classifier is trained to recognize the specific class associated with that row. All other elements in the row are set to -1, signifying that the classifier treats the remaining classes as a single negative class.

The code matrix for the one-vs-one case, meanwhile, is a $K \times n$ matrix, with $n = \frac{K(K-1)}{2}$, where each column is associated with a pair of class $(k, \ell), k < \ell$ with all its elements equal to 0 except those corresponding to row k, which is equal to $+1$, and row ℓ, which is worth -1.

2.4.3 Generalization Error Bound

The definition of the classification function $f_{\boldsymbol{h}}$ based on a quantization of the outputs of the function $h \in \mathbb{R}^{\mathcal{X} \times \mathcal{Y}}$ allows us to introduce a notion of confidence in the predictions $\boldsymbol{h}(\mathbf{x}) = (h(\mathbf{x}, 1), \ldots, h(\mathbf{x}, K))$ for an example $\mathbf{x}$ playing the role of the margin in the mono-label case:

$$\forall (\mathbf{x}, y), \mu_{\boldsymbol{h}}(\mathbf{x}, y) = h(\mathbf{x}, y) - \max_{y' \neq y} h(\mathbf{x}, y'). \tag{2.31}$$

In this case, the instantaneous loss (2.23) of the classification function $f_{\boldsymbol{h}}$ defined in (2.25) is:

$$\ell(f_{\boldsymbol{h}}(\mathbf{x}), y) = \mathbb{1}_{\mu_{\boldsymbol{h}}(\mathbf{x}, y) \leqslant 0}. \tag{2.32}$$

The generalization error of $f_{\boldsymbol{h}}$, $\mathcal{L}(f_{\boldsymbol{h}}) = \mathbb{E}_{(\mathbf{x}, y) \sim \mathcal{D}}[\mathbb{1}_{\mu_{\boldsymbol{h}}(\mathbf{x}, y) \leqslant 0}]$, can be bounded by introducing a ρ-margin loss function, which dominates the classification loss for a specified $\rho > 0$.

The ρ-margin loss function is defined as follows and it is illustrated in Fig. 2.2.

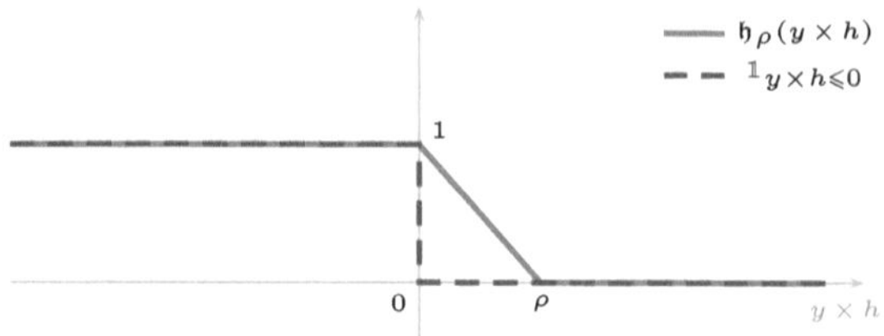

Fig. 2.2 Margin-based loss (solid grey), dominating the classification loss (dashed black) and defined with the margin parameter ρ

Definition 2.4 (ρ-Margin Loss Function) For a given $\rho > 0$, the ρ-margin loss function $\ell_\rho : \mathbb{R}_+ \times \mathbb{R} \to \mathbb{R}_+$ is defined for all $y \in \mathbb{R}_+$ and $h \in \mathbb{R}$ by $\ell_\rho(y, h) = \mathfrak{h}_\rho(yh)$ with:

$$\mathfrak{h}_\rho(z) = \begin{cases} 1, & \text{if } z \leqslant 0; \\ 1 - z/\rho, & \text{if } 0 < z \leqslant \rho; \\ 0, & \text{otherwise.} \end{cases} \quad (2.33)$$

As a result of the preceding analysis, we can now introduce the concept of margin-based empirical error specifically for the classification function $f_{\boldsymbol{h}}$, as defined in (2.25). The margin-based empirical error takes into account not only whether the predicted label is correct but also the confidence or margin by which the correct label is chosen over other possible labels and can be defined as follows:

$$\hat{\mathcal{L}}_{m,\rho}(f_{\boldsymbol{h}}, S) = \frac{1}{m}\sum_{i=1}^{m} \mathfrak{h}_\rho(\mu_{\boldsymbol{h}}(\mathbf{x}_i, y_i)). \quad (2.34)$$

With these definitions, it is possible to easily extend the results obtained for the binary case, in order to bound the margin-based generalization error for the multi-class mono-label case for classes of linear functions with bounded norm, defined as:

$$\mathcal{H}_B = \{h : (\mathbf{x}, y) \in \mathcal{X} \times \mathcal{Y} \mapsto \langle \mathbf{x}, \boldsymbol{w}_y \rangle \mid \boldsymbol{W} = (\boldsymbol{w}_1, \ldots, \boldsymbol{w}_K), \|\boldsymbol{W}\|_{\mathbb{H}} \leqslant B\}, \quad (2.35)$$

where $B \in \mathbb{R}_+^*$ is a positive real and $\|\boldsymbol{W}\|_{\mathbb{H}} = \left(\sum_y \|w_y\|^2\right)^{1/2}$. This result is stated in the following theorem:

Theorem 2.6 (Data-Generalization Bound for Multi-Class Classification [70, 71]) *Let $S = ((\mathbf{x}_i, y_i))_{i=1}^m \in (\mathcal{X} \times \mathcal{Y})^m$ be a training set of size m identically and independently distributed according to a probability distribution $\mathcal{D}$, where $\mathcal{X} \subseteq \mathbb{R}^d$ and $\mathcal{Y} = \{1, \dots, K\}$. Suppose there is a real $R > 0$ such that $\forall \mathbf{x} \in \mathcal{X}$, $\|\mathbf{x}\|^2 \leqslant R^2$ and let $\rho > 0$. For any $\delta \in (0, 1)$ and any function $h \in \mathcal{H}_B$ (2.35), the generalization error of the function $f_{\boldsymbol{h}}$ (2.25) is then upper-bounded with a probability of at least $1 - \delta$:*

$$\mathcal{L}(f_{\boldsymbol{h}}) \leqslant \hat{\mathcal{L}}_{m,\rho}(f_{\boldsymbol{h}}, S) + \frac{8BR}{\rho\sqrt{m}}K(K-1) + 3\sqrt{\frac{\ln(2/\delta)}{2m}}. \tag{2.36}$$

Proof Define $\Delta = \{(\mathbf{x}, y) \mapsto \mu_h(\mathbf{x}, y) \mid h \in \mathcal{H}_B\}$, and let $\mathcal{G}_\rho$ be the class of $\mathcal{X} \times \mathcal{Y}$ functions in $[0, 1]$ defined by:

$$\mathcal{G}_\rho = \mathfrak{h}_\rho \circ \Delta = \{\mathfrak{h}_\rho \circ \mu_h, \mu_h \in \Delta\}.$$

The loss function $\mathfrak{h}_\rho$ has values in $[0, 1]$, according to Theorem 2.3, we have, for any $\delta \in (0, 1)$ and any function $\mu_h \in \Delta$, the following inequality which holds with probability at least $1 - \delta$:

$$\mathbb{E}_{(\mathbf{x},y)\sim\mathcal{D}}[\mathfrak{h}_\rho(\mu_h(\mathbf{x}, y))] \leqslant \frac{1}{m}\sum_{i=1}^m(\mathfrak{h}_\rho(\mu_h(\mathbf{x}_i, y_i))) + \hat{\mathfrak{R}}_S(\mathfrak{h}_\rho \circ \Delta) + 3\sqrt{\frac{\ln(2/\delta)}{2m}},$$

where $\hat{\mathfrak{R}}_S(\mathfrak{h}_\rho \circ \Delta)$ is the empirical Rademacher complexity of the class of functions $\mathcal{G}_\rho$.

The function $z \mapsto \mathfrak{h}_\rho(z)$ is $1/\rho$-lipschitzian (Definition 2.3) then according to Talagrand's lemma (Theorem 2.5, property 5) we have:

$$\hat{\mathfrak{R}}_S(\mathfrak{h}_\rho \circ \Delta) \leqslant \frac{1}{\rho}\hat{\mathfrak{R}}_S(\Delta), \tag{2.37}$$

with:

$$\begin{aligned}
\hat{\mathfrak{R}}_S(\Delta) &= \frac{2}{m}\mathbb{E}_\sigma\left[\sup_{g\in\Delta}\sum_{i=1}^m \sigma_i g(\mathbf{x}_i, y_i)\right] \\
&= \frac{2}{m}\mathbb{E}_\sigma\left[\sup_{h\in\mathcal{H}_B}\sum_{i=1}^m \sigma_i\left(h(\mathbf{x}_i, y_i) - \max_{y'\neq y_i} h(\mathbf{x}_i, y')\right)\right].
\end{aligned}$$

Let $\pi : \mathbb{H}_B \times \mathcal{X} \times \mathcal{Y} \to \mathcal{Y} \times \mathcal{Y}$ be defined as:

$$\pi(h, \mathbf{x}, y) = (k, \ell) \Rightarrow (y = k) \wedge (\ell \neq y) \wedge h(x, \ell) = \max_{y'\neq y} h(\mathbf{x}, y').$$

By construction, we have then:

$$\hat{\mathfrak{R}}_S(\Delta) \leqslant \frac{2}{m}\mathbb{E}_\sigma\left[\sup_{h\in\mathcal{H}_B}\sum_{(k,\ell)\in\mathcal{Y}^2,k\neq\ell}\left|\sum_{i:\pi(h,\mathbf{x},y)=(k,\ell)}\sigma_i\left(h(\mathbf{x}_i,k)-h(\mathbf{x}_i,\ell)\right)\right|\right].$$

By definition, $h(\mathbf{x}_i,k) - h(\mathbf{x}_i,\ell) = \langle \boldsymbol{w}_k - \boldsymbol{w}_\ell, \mathbf{x}_i\rangle$. From the Cauchy-Schwarz inequality, we have:

$$\begin{aligned}
\hat{\mathfrak{R}}_S(\Delta) &\leqslant \frac{2}{m}\mathbb{E}_\sigma\left[\sup_{\|\boldsymbol{W}\|_{\mathbb{H}}\leqslant B}\sum_{(k,\ell)\in\mathcal{Y}^2,k\neq\ell}\left|\left\langle \boldsymbol{w}_k-\boldsymbol{w}_\ell,\sum_{i:\pi(h,\mathbf{x},y)=(k,\ell)}\sigma_i\mathbf{x}_i\right\rangle\right|\right]\\
&\leqslant \frac{2}{m}\mathbb{E}_\sigma\left[\sup_{\|\boldsymbol{W}\|\leqslant B}\sum_{(k,\ell)\in\mathcal{Y}^2,k\neq\ell}\|\boldsymbol{w}_k-\boldsymbol{w}_\ell\|_{\mathbb{H}}\left\|\sum_{i:\pi(h,\mathbf{x},y)=(k,\ell)}\sigma_i\mathbf{x}_i\right\|\right]\\
&\leqslant \frac{4B}{m}\sum_{(k,\ell)\in\mathcal{Y}^2,k\neq\ell}\mathbb{E}_\sigma\left[\left\|\sum_{i:\pi(h,\mathbf{x},y)=(k,\ell)}\sigma_i\mathbf{x}_i\right\|\right].
\end{aligned}$$

As $\forall i,j\in\{u \mid \pi(h,\mathbf{x}_u,y_u)=(k,\ell)\}^2$, $\mathbb{E}_\sigma[\sigma_i\sigma_j]=0$, from Jensen' inequality, we get:

$$\hat{\mathfrak{R}}_S(\Delta) \leqslant \frac{4B}{m}\sum_{(k,\ell)\in\mathcal{Y}^2,k\neq\ell}\left(\mathbb{E}_\sigma\left[\left\|\sum_{i:\pi(h,\mathbf{x},y)=(k,\ell)}\sigma_i\mathbf{x}_i\right\|^2\right]\right)^{1/2}.$$

Hence, from the Jensen inequality we have:

$$\begin{aligned}
\hat{\mathfrak{R}}_S(\Delta) &\leqslant \frac{4B}{m}\sum_{(k,\ell)\in\mathcal{Y}^2,k\neq\ell}\left(\sum_{i:\pi(h,\mathbf{x},y)=(k,\ell)}\|\mathbf{x}_i\|^2\right)^{1/2}\\
&\leqslant \frac{4B}{m}\sum_{(k,\ell)\in\mathcal{Y}^2,k\neq\ell}\left(mR^2\right)^{1/2}\\
&= \frac{4BR}{\sqrt{m}}K(K-1).
\end{aligned}$$

The result comes from the inequality (2.37) and by noting that the function $(\mathbf{x},y)\mapsto\mu_h(\mathbf{x},y)$ is upper-bounded by $(\mathbf{x},y)\mapsto\mathfrak{h}_\rho(\mu_h(\mathbf{x},y))$. □

A variant of the previous theorem is stated in [119, Chapter 8]. In addition, this important result has made it possible to derive single-label multi-class algorithms and, in particular, to extend support vector machine to this case. We also note that, in this case, the ρ margin still plays a central role: the more confident a classifier is in its predictions, the greater the value of ρ and the tighter the bound on its generalization error.

In the multi-label case, there are also theoretical studies to bound the generalization error, but these are mostly based on the VC dimension and the reduction of the multi-class problem to binary classification.

Other multi-class approaches, known as *uncombined approaches*, will be presented in subsequent chapters. These approaches seek to directly predict the classes of an observation without being combined with binary techniques.

To Sum Up

1. *Rademacher complexity is a measure of the richness of a class of functions with respect to a given dataset. It quantifies the ability of the function class to fit random noise.*
2. *Despite their differences, Rademacher complexity and VC dimension are closely related, and bounds on Rademacher complexity can be derived from VC dimension and vice versa.*
3. *The empirical Rademacher complexity is computed directly from the training data, and it provides an estimate of how well a class of functions can fit random noise in the given dataset.*
4. *By leveraging the empirical Rademacher complexity, it is possible to derive generalization bounds that can be estimated from the training set. These bounds are generally tighter than those derived using the VC dimension.*
5. *Regularization is a crucial technique in implementing the Structural Risk Minimization (SRM) principle. It involves adding a penalty term to the loss function to prevent overfitting by controlling the complexity of the model.*
6. *Minimizing the 0/1 loss directly is often computationally infeasible due to its non-convex nature. Instead, the common practice is to minimize a convex surrogate loss, such as hinge loss or logistic loss, which provides a tractable upper bound on the classification error.*
7. *In multi-class classification, two main approaches are used: uncombined and combined.*
8. *The uncombined approach for multi-class classification reduces the multi-class problem to multiple binary classification problems, such as one-vs-all or one-vs-one, and then combines the results.*
9. *The combined approach for multi-class classification treats the multi-class problem as a single optimization problem, often using a joint loss function that considers all classes simultaneously.*
10. *Data-dependent generalization bounds can be extended to multi-class settings. This extension involves adapting the concepts of Rademacher complexity to handle multiple classes.*

2.5 Exercises

1. Let $\mathcal{F}$ be a hypothesis class and S a training set of size m.
   (a) Explain the role of Rademacher variables σ_i in the empirical Rademacher complexity $\hat{\mathfrak{R}}_S(\mathcal{F})$.
   (b) What does it mean for $\mathcal{F}$ to shatter S? Prove that if $\mathcal{F}$ shatters S, then the empirical Rademacher complexity $\hat{\mathfrak{R}}_S(\mathcal{F})$ equals 1.
   (c) Prove that if $\mathcal{F}$ contains a single function, then $\hat{\mathfrak{R}}_S(\mathcal{F}) = 0$.
   (d) Compare the empirical Rademacher complexity in the two scenarios (shattering and single function) and discuss how these extremes illustrate the concept of model complexity.

2. Show that the Rademacher complexity is invariant to linear transformations of the function class. For $\mathcal{G} = \{af + b \mid f \in \mathcal{F}\}$, prove

$$\hat{\mathfrak{R}}_S(\mathcal{G}) = |a|\hat{\mathfrak{R}}_S(\mathcal{F}).$$

3. Let $\mathcal{F}$ be a class of functions mapping from an input space $\mathcal{X}$ to $\{-1, +1\}$. Suppose we observe a training set $S = \{(\mathbf{x}_i, y_i)\}_{i=1}^m$, where the labels y_i are noisy. Specifically, assume that each label y_i is flipped with probability $\eta \in [0, 1/2]$, i.e., $\mathbb{P}(y_i' \neq y_i) = \eta$, where y_i' is the observed label.
   (a) Define the empirical Rademacher complexity $\hat{\mathfrak{R}}_S(\mathcal{F})$ for the original (unnoisy) labels.
   (b) Let $\hat{\mathfrak{R}}_S^\eta(\mathcal{F})$ be the empirical Rademacher complexity computed using the noisy labels y_i'. Show how $\hat{\mathfrak{R}}_S^\eta(\mathcal{F})$ relates to $\hat{\mathfrak{R}}_S(\mathcal{F})$. Specifically, prove that:

$$\hat{\mathfrak{R}}_S^\eta(\mathcal{F}) \leqslant (1 - 2\eta)\hat{\mathfrak{R}}_S(\mathcal{F}).$$

   (c) Discuss the implications of noisy labels on the generalization error bound derived using Rademacher complexity. How does the noise affect the bound?

4. Let $\mathcal{X}$, be the input space and $\mathcal{F} = \{f : \mathcal{X} \to \{-1, +1\}\}$ a class of functions. If $\mathcal{F}$ is finite, show that the empirical Rademacher complexity, $\hat{\mathfrak{R}}_S(\mathcal{F})$, on a training set S of size m is bounded by:

$$\hat{\mathfrak{R}}_S(\mathcal{F}) \leqslant \sqrt{\frac{8 \ln |\mathcal{F}|}{m}}.$$

Hint: Follow the proof of Lemma 2.1.

5. For a finite binary class of functions $\mathcal{F}$ with VC dimension $vc_{\mathcal{F}}$, and a training set of size m, with $m > vc_{\mathcal{F}}$ show:

$$\hat{\mathfrak{R}}_S(\mathcal{F}) \leqslant C\sqrt{\frac{vc_{\mathcal{F}}}{m}},$$

where $C > 0$ is a constant.
Hint: From the previous result, use Lemma 1.5.

6. Let $\mathcal{F}$ be a class of real-valued functions, $S = (\mathbf{x}_i, y_i)_{1\leqslant i\leqslant m}$ be a training set of size m generated i.i.d. with respect to a fixed unknown probability distribution $\mathcal{D}$. Let $\hat{\mathcal{F}}$, be an ϵ-cover of $\mathcal{F}$; that is for any $f \in \mathcal{F}$, there exists $\hat{f} \in \hat{\mathcal{F}}$ such that:

$$\frac{1}{m}\sum_{i=1}^{m}|f(\mathbf{x}_i) - \hat{f}(\mathbf{x}_i)| \leqslant \epsilon.$$

(a) Assuming $\sup_{f\in\mathcal{F}}\left(f^2(\mathbf{x}_i)\right)^{\frac{1}{2}} \leqslant r$, show that:

$$\hat{\mathfrak{R}}_S(\mathcal{F}) \leqslant \epsilon + r\frac{\sqrt{8\ln|\hat{\mathcal{F}}|}}{m}.$$

Hint: From the definition of the ϵ-cover, first note that $\hat{\mathcal{F}} \subseteq \mathcal{F}$ then use Lemma 2.1.

(b) Suppose that $C(\mathcal{F}, \epsilon, \|.\|_{1,S})$ is the size of minimal ϵ-cover of $\mathcal{F}$, deduce from the previous result that:

$$\hat{\mathfrak{R}}_S(\mathcal{F}) \leqslant \inf_{\epsilon>0}\left(\epsilon + r\frac{\sqrt{8\ln C(\mathcal{F}, \epsilon, \|.\|_{1,S})}}{m}\right).$$

7. Let $\mathcal{F}$ be the class of linear functions, with ℓ_2-norm of the weights bounded by B_2, $\mathcal{F} = \{\mathbf{x} \mapsto \langle \mathbf{x}, \mathbf{w}\rangle \mid \|\mathbf{w}\| \leqslant B\}$, and $S = (\mathbf{x}_i, y_i)_{1\leqslant i\leqslant m}$ be a training set of size m contained in a hypershpere of radius R, i.e. $\max_i \|\mathbf{x}_i\|_2 \leqslant R$.

(a) Show:

$$\hat{\mathfrak{R}}_S(\mathcal{F}) \leqslant \frac{2B_2}{m}\mathbb{E}_\sigma\left[\left\|\sum_{i=1}^{m}\sigma_i\mathbf{x}_i\right\|\right].$$

(b) Using Jensen's inequality show that:

$$\hat{\mathfrak{R}}_S(\mathcal{F}) \leqslant \frac{2B_2}{m}\left(\mathbb{E}_\sigma\left[\sum_{i,j=1}^{m}\sigma_i\sigma_j\langle\mathbf{x}_i, \mathbf{x}_j\rangle\right]\right)^{\frac{1}{2}}.$$

(c) By noticing that the random variables $\boldsymbol{\sigma} = \{\sigma_1, \ldots, \sigma_m\}$ can independently take values of -1 or $+1$ with equal probability (meaning that each variable either keeps the effect of the examples $\mathbf{x}_i$ and $\mathbf{x}_j$ unchanged or effectively swaps them). Since the points are generated independently, any swap results in an equally likely configuration. Therefore, averaging over all possible swaps will not change the overall expectation. Prove

$$\hat{\mathfrak{R}}_S(\mathcal{F}) \leqslant \frac{2B_2R}{\sqrt{m}}.$$

8. Let $\mathcal{F}$ be the class of linear functions, with ℓ_1-norm of the weights bounded by B_1, and $S = (\mathbf{x}_i, y_i)_{1 \leqslant i \leqslant m}$ be a training set of size m. Assume that $\forall i,\ \|\mathbf{x}_i\|_\infty \leqslant L$.

   (a) Show:

$$\hat{\mathfrak{R}}_S(\mathcal{F}) \leqslant \frac{2B_1}{m}\mathbb{E}_\sigma\left[\sup_{j\in\{1,\ldots,d\}} \sum_{i=1}^m \sigma_i x_{ij}\right],$$

   where d is the dimension of the input space, and x_{ij} is the jth component of $\mathbf{x}_i$.

   (b) Using Lemma 2.1 show that:

$$\begin{aligned}\hat{\mathfrak{R}}_S(\mathcal{F}) &\leqslant \frac{2B_1\sqrt{2\ln d}}{m}\sup_j\sqrt{\sum_{i=1}^m x_{ij}^2}\\ &\leqslant 2B_1L\sqrt{\frac{2\ln d}{m}}.\end{aligned}$$

9. Consider the class of linear functions

$$\mathcal{F} = \{x \mapsto \boldsymbol{w}^\top \mathbf{x} : \|\boldsymbol{w}\|_0 \leqslant s,\ \|\boldsymbol{w}\|_2 \leqslant B\},$$

   where $\|\boldsymbol{w}\|_0$ denotes the number of nonzero components of $\boldsymbol{w}$, bounded by sparsity level s, and $\|\boldsymbol{w}\|_2$ is the Euclidean norm bounded by B. Assume the input vectors satisfy $\|\mathbf{x}\|_2 \leqslant R$.

   (a) Provide an upper bound on the empirical Rademacher complexity $\hat{\mathfrak{R}}_S(\mathcal{F})$ in terms of s, B, R, and the sample size m.
   (b) Explain how sparsity affects the capacity of the function class and its generalization ability.
   (c) Consider the case where the dimension d is much larger than the sample size m. Show how the sparsity constraint leads to a bound that depends on s rather than d.

(d) Compare the Rademacher complexity bound for sparse linear predictors with that of non-sparse linear predictors (where $\|\boldsymbol{w}\|_0 \leqslant d$). What is the ratio between these bounds when $s \ll d$?
(e) Discuss how this result relates to compressed sensing and high-dimensional statistics. In particular, explain the connection to the restricted isometry property.

Hint: For part (a), consider decomposing the problem into bounding the Rademacher complexity over all possible support sets of size s, then use the bound for linear predictors with ℓ_2 constraints.

10. Consider the composite class of functions $\mathcal{H}$ defined as:

$$\mathcal{H} = \{f \circ g | f \in \mathcal{F}, g \in \mathcal{G}\},$$

where $\mathcal{F}$ is a class of functions with empirical Rademacher complexity bounded by:

$$\hat{\mathfrak{R}}_S(\mathcal{F}) \leqslant \frac{C}{\sqrt{m}},$$

and $\mathcal{G}$ is a class of 1-Lipschitz functions.

(a) Describe the implications of $\mathcal{F}$ having an empirical Rademacher complexity bounded by $\frac{C}{\sqrt{m}}$. How does this bound influence the generalization capability of functions in $\mathcal{F}$?
(b) Discuss the role of the 1-Lipschitz property in the functions of $\mathcal{G}$. How does this property affect the behavior of $g \in \mathcal{G}$ with respect to input changes?
(c) Show that

$$\hat{\mathfrak{R}}_S(\mathcal{H}) \leqslant \frac{C}{\sqrt{m}}.$$

11. Let $\mathcal{F}$ be a class of real-valued functions mapping from an input space $\mathcal{X}$ to $\mathbb{R}$. Consider a composite class of functions $\mathcal{G}$ defined as:

$$\mathcal{G} = \{g(x) = \sigma(f(x)) \mid f \in \mathcal{F}\},$$

where $\sigma(z) = \frac{1}{1+e^{-z}}$ is the sigmoid function.

(a) State the properties of the sigmoid function that are relevant to bounding the Rademacher complexity of $\mathcal{G}$. In particular, discuss its Lipschitz constant.

(b) Using Talagrand's lemma (Lemma 2.15), show that the empirical Rademacher complexity of $\mathcal{G}$ can be bounded in terms of the empirical Rademacher complexity of $\mathcal{F}$. Specifically, prove that:

$$\hat{\mathfrak{R}}_S(\mathcal{G}) \leqslant \frac{1}{4}\hat{\mathfrak{R}}_S(\mathcal{F}).$$

12. Let $\mathcal{F}$ be a class of functions mapping $\mathcal{X}$ to $\{1, 2, \ldots, K\}$, and let $\ell(f(x), y) = \mathbf{1}\{f(x) \neq y\}$ be the multiclass 0-1 loss. Consider the associated loss class $\mathcal{L} = \{\ell(f(\cdot), y) : f \in \mathcal{F}, y \in \{1, \ldots, K\}\}$.
    (a) Write the definition of the empirical Rademacher complexity $\hat{\mathfrak{R}}_S(\mathcal{L})$ for this multiclass case.
    (b) Suppose $\mathcal{F}$ is the class of all possible functions from $\mathcal{X}$ to $\{1, \ldots, K\}$, and $|S| = m$. What is the value of $\hat{\mathfrak{R}}_S(\mathcal{L})$?
    (c) Discuss how the complexity changes as K increases, and what this implies for multiclass learning.

Chapter 3
Descent Direction Optimization Algorithms

In this chapter, we are interested in the convergence properties of the descent direction algorithms. These algorithms are undoubtedly the most widely used for learning model parameters. For the most part, they are iterative methods that decrease the objective function to be minimized along a descent direction; defined as the direction whose scalar product with the gradient of the loss function is negative. The difficulty in using these algorithms lies in defining an appropriate step size, called the learning rate, for updating the weights along this direction, which would enable them to converge towards the minimum of the objective function. We will show the convergence of these algorithms towards the minimum of convex objective functions under certain conditions concerning this step size, known as Wolfe's conditions. We conclude this chapter by presenting other simple and efficient conventional optimization algorithms. We will begin this presentation by the definition of convex functions.

Definition 3.1 Let A be a convex subset of a real vector space,

- A function $\mathcal{L} : A \to \mathbb{R}$ is convex, if for any points $\boldsymbol{w}_1$ and $\boldsymbol{w}_2$ within A and for any positive real value $\theta \in [0, 1]$ the following inequality holds:

$$\mathcal{L}(\theta \boldsymbol{w}_1 + (1-\theta)\boldsymbol{w}_2) \leqslant \theta \mathcal{L}(\boldsymbol{w}_1) + (1-\theta)\mathcal{L}(\boldsymbol{w}_2). \tag{3.1}$$

- Furthermore, if $\mathcal{L}$ is differentiable on A, it is then convex if and only if:

$$\forall (\boldsymbol{w}_1, \boldsymbol{w}_2) \in A^2; \mathcal{L}(\boldsymbol{w}_2) \geqslant \mathcal{L}(\boldsymbol{w}_1) + \nabla \mathcal{L}(\boldsymbol{w}_1)^\top (\boldsymbol{w}_2 - \boldsymbol{w}_1), \tag{3.2}$$

where $\nabla \mathcal{L}(\boldsymbol{w}_1)$ is the gradient of $\mathcal{L}$ on $\boldsymbol{w}_1$ and the superscript $^\top$ is the sign of transpose.

(continued)

M.-R. Amini, *Advanced Supervised and Semi-supervised Learning*,
Cognitive Technologies, https://doi.org/10.1007/978-3-031-99928-4_3

Definition 3.1 (continued)

- Finally, if $\mathcal{L}$ is twice differentiable, it is convex on A if and only if its Hessian matrix, of second partial derivatives, is positive semidefinite on the interior of A.

Figure 3.1 illustrates the first two points of Definition (3.1) for a convex function $\mathcal{L} : \mathbb{R} \to \mathbb{R}$.

We will now study the geometric form of a convex loss function $\mathcal{L}$ around its minimum, assuming it is continuous and twice differentiable. To do this, we will consider a local quadratic approximation of this function in the vicinity of its minimum. This description will provide a better understanding of the optimization problem to be solved, which we will address in the next sections.

To simplify notation, let $\mathcal{L}(\boldsymbol{w})$ denote the convex surrogate on the empirical error of a function $h_{\boldsymbol{w}} : \mathcal{X} \to \mathcal{Y}$ with parameters $\boldsymbol{w}$ over a training set S (Chap. 2, Sect. 2.3.2).

Considering the Taylor expansion of order 2 of $\mathcal{L}(\boldsymbol{w})$ around its minimum at point $\boldsymbol{w}^*$, we have:

$$\mathcal{L}(\boldsymbol{w}) = \mathcal{L}(\boldsymbol{w}^*) + (\boldsymbol{w} - \boldsymbol{w}^*)^\top \underbrace{\nabla\mathcal{L}(\boldsymbol{w}^*)}_{=0} + \frac{1}{2}(\boldsymbol{w} - \boldsymbol{w}^*)^\top \mathbf{H}(\boldsymbol{w} - \boldsymbol{w}^*) + o(\|\boldsymbol{w} - \boldsymbol{w}^*\|^2), \tag{3.3}$$

where $\mathbf{H}$ denotes the Hessian matrix of the loss function $\mathcal{L}$, evaluated at point $\boldsymbol{w}^*$. The matrix $\mathbf{H}$ is symmetric and positive semidefinite. By Schwarz's theorem,

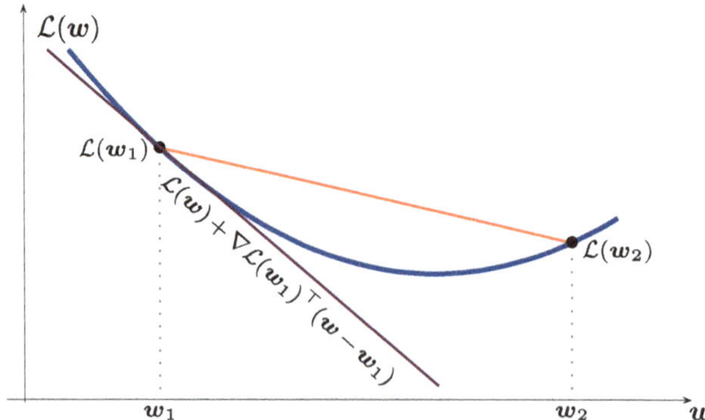

Fig. 3.1 For two points $\boldsymbol{w}_1$ and $\boldsymbol{w}_2$, the graph of the convex function $\mathcal{L}$ lies below the line segment connecting $(\boldsymbol{w}_1, \mathcal{L}(\boldsymbol{w}_1))$ and $(\boldsymbol{w}_2, \mathcal{L}(\boldsymbol{w}_2))$ (3.1) and above the tangent passing through $(\boldsymbol{w}_1, \mathcal{L}(\boldsymbol{w}_1))$ (3.2)

its eigenvectors $(\boldsymbol{v}_i)_{i=1}^d$ form an orthonormal basis. Let $(\lambda_i)_{i=1}^d$ be the set of eigenvalues of $\mathbf{H}$. We thus have:

$$\forall(i,j) \in \{1,\ldots,d\}^2, \mathbf{H}\boldsymbol{v}_i = \lambda_i \boldsymbol{v}_i, \text{ and } \boldsymbol{v}_i^\top \boldsymbol{v}_j = \begin{cases} +1 & \text{if } i = j, \\ 0 & \text{otherwise.} \end{cases} \tag{3.4}$$

Any vector $\boldsymbol{w} - \boldsymbol{w}^*$ can then be uniquely decomposed on this basis:

$$\boldsymbol{w} - \boldsymbol{w}^* = \sum_{i=1}^d q_i v_i. \tag{3.5}$$

Substituting this equality into (3.3) gives:

$$\mathcal{L}(\boldsymbol{w}) = \mathcal{L}(\boldsymbol{w}^*) + \frac{1}{2}\sum_{i=1}^d \lambda_i q_i^2. \tag{3.6}$$

Moreover, $\mathbf{H}$ is positive semidefinite and its eigenvalues are all positive, and by the definition of the global minimum for any vector $\boldsymbol{w}$ in the vicinity of $\boldsymbol{w}^*$ we have:

$$(\boldsymbol{w} - \boldsymbol{w}^*)^\top \mathbf{H}(\boldsymbol{w} - \boldsymbol{w}^*) = \sum_{i=1}^d \lambda_i q_i^2 = 2(\mathcal{L}(\boldsymbol{w}) - \mathcal{L}(\boldsymbol{w}^*)) \geqslant 0,$$

which implies that the contour lines (or isopleths) of $\mathcal{L}$ (formed by points in the space of weights on which $\mathcal{L}$ admits constant values) are ellipses (Fig. 3.2).

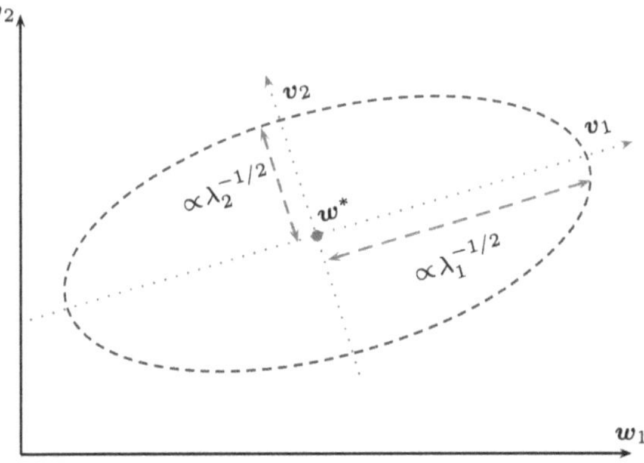

Fig. 3.2 Illustration of an elliptic contour line of a continuous and twice differentiable loss function, $\mathcal{L}(\boldsymbol{w})$ in the vicinity of its minimum $\boldsymbol{w}^*$. In the vicinity of this minimum, the axes of the ellipses are defined with respect to the eigenvectors of the Hessian matrix of $\mathcal{L}$ estimated at $\boldsymbol{w}^*$, and they are inversely proportional to the square roots of the associated eigenvalues

3.1 Gradient Algorithm

The gradient algorithm or *steepest descent* is undoubtedly the simplest optimization approach that could be used to find the parameters of a learning model [145].

3.1.1 Full Batch Mode

In the case of minimizing the global loss function, the $\boldsymbol{w}^{(0)}$ weights are initialized at random and the loss function is minimized by iteratively updating the weight vectors. This is done by moving one step in the opposite direction of the gradient of the loss function, which locally indicates the steepest direction of this function. Thus, at iteration t, the new weight values $\boldsymbol{w}^{(t+1)}$ are estimated from the weight values at the previous step $\boldsymbol{w}^{(t)}$ and the gradient of the loss function estimated at weight $\boldsymbol{w}^{(t)}$:

$$\forall t \in \mathbb{N}, \boldsymbol{w}^{(t+1)} = \boldsymbol{w}^{(t)} - \eta \nabla \mathcal{L}(\boldsymbol{w}^{(t)}), \tag{3.7}$$

where $\eta \in \mathbb{R}_+^*$ is a strictly positive real called the learning rate. Figure 3.3 illustrates geometrically how this update rule works for a fixed learning rate $\eta > 0$.

One of the difficulties in using this algorithm is the choice of the learning rate. If the step size is too large, the updating rule (3.7) may lead to oscillations around the minimum. If the step size is too small, convergence towards the minimum will be very slow and, in some cases, may quickly become impossible.

We will exhibit this in the following. Let $\mathcal{L}(\boldsymbol{w})$ be a continuous, twice differentiable loss function. Consider the second-order Taylor development of $\mathcal{L}(\boldsymbol{w})$ around its global minimum reached at $\boldsymbol{w}^*$ (3.3). Consider the derivative of the

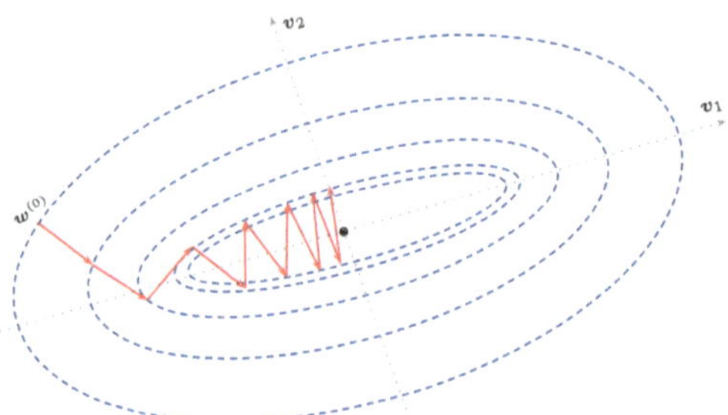

Fig. 3.3 Illustration of the minimization of a convex $\mathcal{L}(\boldsymbol{w})$ loss function with the gradient descent algorithm. Elliptical curves represent contour lines on which the error function admits constant values. The vectors $\boldsymbol{v}_1$ and $\boldsymbol{v}_2$ represent the eigenvectors of the Hessian (the minimum sought is at the center of the axes). Note the oscillating motion of the weights found around the minimum, and note that the opposite of the gradient of the estimated loss function at these points does not generally point to the same minimum

approximation of $\mathcal{L}(\boldsymbol{w})$ at a point $\boldsymbol{w}$ in the vicinity of $\boldsymbol{w}^*$:

$$\nabla\mathcal{L}(\boldsymbol{w}) = \mathbf{H}(\boldsymbol{w} - \boldsymbol{w}^*). \tag{3.8}$$

Taking the decomposition of the vector $\boldsymbol{w} - \boldsymbol{w}^*$ on the orthonormal basis $(\boldsymbol{v}_i)_{i=1}^d$ consisting of the eigenvectors of the $\mathbf{H}$ estimated at point $\boldsymbol{w}^*$ (3.5) and the relation (3.4), we get:

$$\nabla\mathcal{L}(\boldsymbol{w}) = \sum_{i=1}^{d} q_i\lambda_i\boldsymbol{v}_i. \tag{3.9}$$

For $t \in \mathbb{N}^*$, let $\boldsymbol{w}^{(t-1)}$ and $\boldsymbol{w}^{(t)}$ be the weight vectors obtained after $t-1$ and t applications of the rule (3.7). According to the relation (3.5), we have:

$$\boldsymbol{w}^{(t)} - \boldsymbol{w}^{(t-1)} = \sum_{i=1}^{d}\left(q_i^{(t)} - q_i^{(t-1)}\right)\boldsymbol{v}_i = -\eta\nabla\mathcal{L}(\boldsymbol{w}^{(t-1)}) = -\eta\sum_{i=1}^{d} q_i^{(t-1)}\lambda_i\boldsymbol{v}_i, \tag{3.10}$$

where $(q_i^{(t-1)})_{i=1}^d$ and $(q_i^{(t)})_{i=1}^d$ denote the respective coordinates of the vectors $\boldsymbol{w}^{(t-1)} - \boldsymbol{w}^*$ and $\boldsymbol{w}^{(t)} - \boldsymbol{w}^*$ in the orthonormal basis $(\boldsymbol{v}_i)_{i=1}^d$. According to the property of orthonormality of eigenvectors, multiplying (3.10) on the right by $\boldsymbol{v}_i$, $\forall i$, we get:

$$\forall i, q_i^{(t)} - q_i^{(t-1)} = -\eta q_i^{(t-1)}\lambda_i. \tag{3.11}$$

Hence:

$$\forall i, q_i^{(t)} = (1 - \eta\lambda_i)q_i^{(t-1)}, \tag{3.12}$$

thus, after t updates to the weight vector, we have:

$$\forall i, q_i^{(t)} = (1 - \eta\lambda_i)^t q_i^{(0)}. \tag{3.13}$$

In the case where $\forall i \in \{1, \ldots, d\}, |1 - \eta\lambda_i| < 1$, the coordinates $q_i^{(t)}$ all converge to 0, which amounts to the convergence of the weight vector $\boldsymbol{w}^{(t)}$ to the minimum $\boldsymbol{w}^*$; since, again according to the orthonormality of eigenvectors, we have:

$$\forall i, \boldsymbol{v}_i^\top(\boldsymbol{w}^{(t)} - \boldsymbol{w}^*) = q_i^{(t)}. \tag{3.14}$$

The global condition on the acceptable values of the fixed learning rate η for convergence of the weight vector to $\boldsymbol{w}^*$ according to the betting rule (3.7) is then:

$$\eta < \frac{2}{\lambda_{max}}, \tag{3.15}$$

where λ_{max} is the largest eigenvalue of $\mathbf{H}$.

3.1.2 Online Mode

When dealing with data collections containing millions or even billions of examples, batch algorithms will not be able to scale. This realization has motivated much work in the fields of learning and optimization to design online-type algorithms, which process one example at a time and can therefore be significantly more efficient in time and space, and also more practical than the full *batch* mode [19]. Consider the decomposition of the loss function $\mathcal{L}$ into a sum of convex partial loss functions estimated on a training set S:

$$\mathcal{L}(\boldsymbol{w}) = \frac{1}{m}\sum_{i=1}^{m} \ell(h_{\boldsymbol{w}}(\boldsymbol{w}_i), y_i). \tag{3.16}$$

The stochastic gradient algorithm (Algorithm 5) is an iterative process that randomly selects a training example $(\boldsymbol{w}_t, y_t)$ at each iteration t, and updates the weight vector according to the opposite of the gradient of the partial loss function at that point.

Algorithm 5 is called a first-order optimization method, since it uses only gradient information (first order derivates of the loss) to update the parameters. The convergence rates of first-order convex optimization methods are established by the complexity theory of optimization problems [123], which states, in particular, that when a problem is solved by such a method, and the dimension of the problem is large, the speed of convergence can be no better than a sub-linear speed proportional to $t^{-1/2}$, where t is the number of iterations. This theoretical result implies, in particular, that first-order online optimization methods will never be able to solve large-scale problems with high accuracy in a reasonable time. To compensate for this, these computational techniques guarantee a convergence rate proportional to $t^{-1/2}$, which depends little or not at all on the problem's dimension [122]. So, when we are aiming for a moderately accurate solution, these methods will be able to solve very large-scale problems, which is not the case with *batch* methods.

Input :

- A partial loss function $(\boldsymbol{w}, y) \mapsto \ell(h_{\boldsymbol{w}}(\boldsymbol{w}), y)$;
- Maximum number of iterations T;
- Precision $\epsilon > 0$;
- Learning rate $\eta > 0$;

Initialisation:

- Random initialization of the weights $\boldsymbol{w}^{(0)}$;
- $t \leftarrow 0$;

repeat

- Choose randomly an example $(\boldsymbol{w}_t, y_t) \in S$;
- $\boldsymbol{w}^{(t+1)} \leftarrow \boldsymbol{w}^{(t)} - \eta \nabla \ell(h_{\boldsymbol{w}^{(t)}}(\boldsymbol{w}_t), y_t)$;
- $t \leftarrow t + 1$;

until $\left(||\nabla \ell(h_{\boldsymbol{w}^{(t)}}(\boldsymbol{w}_t), y_t|| \leqslant \epsilon\right) \vee (t > T)$;
Output : Final updated weights $\boldsymbol{w}^{(t)}$.

Algorithme 5: Stochastic gradient descent

3.2 Quasi-Newton Method

To avoid the oscillatory movements of the weight vectors around the minimizer of the loss function, induced by the gradient algorithm, one solution would be to iteratively update these vectors by choosing the descent direction that points to the loss function's minimum (called Newton's direction) with an optimal learning rate (see Fig. 3.4). In Sect. 3.3, we present a simple and effective procedure for automatically finding the optimal rate.

In the present section, we present a procedure for approximating Newton's direction, called the quasi-Newton method [18, 48, 64].

3.2.1 *Newton Direction*

The starting point is the quadratic approximation around the minimum of the loss function (3.3), which we introduced earlier in this section. From this approximation, we know that the gradient of any weight vector $\boldsymbol{w}$ in the neighborhood of the loss function minimizer, $\boldsymbol{w}^*$, can be estimated by:

$$\nabla \mathcal{L}(\boldsymbol{w}) = \mathbf{H}(\boldsymbol{w} - \boldsymbol{w}^*). \tag{3.17}$$

Thus, we have $\boldsymbol{w}^* = \boldsymbol{w} - \mathbf{H}^{-1}\nabla \mathcal{L}(\boldsymbol{w})$, or $-\mathbf{H}^{-1}\nabla \mathcal{L}(\boldsymbol{w})$ which is the Newton's direction.

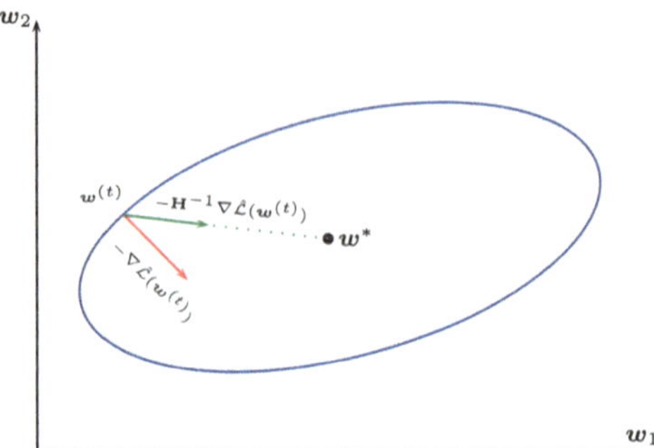

Fig. 3.4 Illustration of the quasi Newton method on an elliptic contour line. The red arrow indicates the standard gradient descent update from the current position $\boldsymbol{w}^{(t)}$ using the direction of the opposite of the gradient, $\mathbf{p}_t = -\nabla\mathcal{L}(\boldsymbol{w}^{(t)})$. The green arrow depicts the Newton's direction $\mathbf{p}_t = -\mathbf{H}^{-1}\nabla\mathcal{L}(\boldsymbol{w}^{(t)})$ towards the minimizer $\boldsymbol{w}^*$

Since the quadratic approximation of the loss function is not exact, it is necessary to iteratively evaluate the Newton direction each time the weight vector is updated. However, for high-dimensional problems, this approach quickly becomes impractical, given the complexity of evaluating the Hessian and its inversion.

3.2.2 Broyden-Fletcher-Goldfarb-Shanno Formula

Alternative approaches have been proposed to approximate the inverse of the Hessian in the estimation of the descent direction. The best-known of these is the quasi-Newton method, which is a special case of the variable metric [44, 55] methods. This approach generates a sequence of matrices $\mathbf{B}_t$ whose limit at infinity reaches the inverse of the Hessian:

$$\lim_{t\to+\infty} \mathbf{B}_t = \mathbf{H}^{-1}. \tag{3.18}$$

At the first iteration, we consider a symmetric positive definite matrix, usually the identity matrix $\mathbf{B}_0 = \mathbf{Id}_d$, and we construct the following approximations $(\mathbf{B}_t)_{t\in\mathbb{N}^*}$ so that they also remain symmetric positive definite.[1] The matrices $(\mathbf{B}_t)_{t\in\mathbb{N}^*}$ are then generated according to the condition of Eq. (3.17), which leads to the quasi-Newton equation or secant method:

$$\boldsymbol{w}^{(t)} - \boldsymbol{w}^{(t-1)} = \mathbf{H}^{-1}\left(\nabla\mathcal{L}(\boldsymbol{w}^{(t)}) - \nabla\mathcal{L}(\boldsymbol{w}^{(t-1)})\right). \tag{3.19}$$

[1] This constraint is justified by the fact that at each iteration $t \in \mathbb{N}$, we want the direction $\mathbf{p}_t \propto \boldsymbol{w}^{(t+1)} - \boldsymbol{w}^{(t)}$ to be a descent direction (along which the cost function decreases).

Since $\mathbf{B}_t$ is an approximation of $\mathbf{H}^{-1}$, we want $\mathbf{B}_t$ to also satisfy:

$$\boldsymbol{w}^{(t)} - \boldsymbol{w}^{(t-1)} = \mathbf{B}_t \left(\nabla\mathcal{L}(\boldsymbol{w}^{(t)}) - \nabla\mathcal{L}(\boldsymbol{w}^{(t-1)})\right). \tag{3.20}$$

The preceding system of equations, known as the quasi-Newton condition, has n equations with n^2 unknowns, and admits a very large number of solutions. Among existing update formulas, the best known is undoubtedly that of *Broyden-Fletcher-Goldfarb-Shanno* (BFGS) which, at each iteration t, searches for the new matrix $\mathbf{B}_{t+1}$ as the solution to the problem [129, p. 136]:

$$\min_{\mathbf{B}} \frac{1}{2} \|\mathbf{B} - \mathbf{B}_t\|^2,$$

$$\text{s.t. } \mathbf{B}\mathbf{g}_t = \mathbf{v}_t \text{ and } \mathbf{B}^\top = \mathbf{B}.$$

with:

$$\mathbf{v}_t = \boldsymbol{w}^{(t+1)} - \boldsymbol{w}^{(t)}, \tag{3.21}$$

$$\mathbf{g}_t = \nabla\mathcal{L}(\boldsymbol{w}^{(t+1)}) - \nabla\mathcal{L}(\boldsymbol{w}^{(t)}). \tag{3.22}$$

The matrix norm used in this case is a norm of the form $\|X\|_{\mathbf{A}} = \|\mathbf{A}^{\frac{1}{2}} X \mathbf{A}^{\frac{1}{2}}\|$ where $\mathbf{A}$ is a symmetric invertible matrix verifying $\mathbf{g}_t = \mathbf{A}\mathbf{v}_t$. The BFGS formula is one of the solutions to the previous optimization problem which leads to the following update [135, p. 57]:

$$\forall t \in \mathbb{N}^*, \mathbf{B}_{t+1} = \mathbf{B}_t + \frac{\mathbf{v}_t \mathbf{v}_t^\top}{\mathbf{v}_t^\top \mathbf{g}_t} - \frac{(\mathbf{B}_t \mathbf{g}_t)\mathbf{g}_t^\top \mathbf{B}_t}{\mathbf{g}_t^\top \mathbf{B}_t \mathbf{g}_t} + (\mathbf{g}_t^\top \mathbf{B}_t \mathbf{g}_t)\mathbf{u}_{t+1}\mathbf{u}_{t+1}^\top, \tag{3.23}$$

where:

$$\mathbf{u}_t = \frac{\mathbf{v}_t}{\mathbf{v}_t^\top \mathbf{g}_t} - \frac{\mathbf{B}_t \mathbf{g}_t}{\mathbf{g}_t^\top \mathbf{B}_t \mathbf{g}_t}. \tag{3.24}$$

We can easily prove that the definition (3.23) guarantees that all $\mathbf{B}_t, t \in \mathbb{N}$ satisfy the quasi-Newton condition (3.20). Indeed, for a given iteration $t \in \mathbb{N}$, let us multiply (3.23) on the right by $\mathbf{g}_t = \nabla\mathcal{L}(\boldsymbol{w}^{(t+1)}) - \nabla\mathcal{L}(\boldsymbol{w}^{(t)})$:

$$\mathbf{B}_{t+1}\mathbf{g}_t = \cancel{\mathbf{B}_t \mathbf{g}_t} + \frac{\mathbf{v}_t \cancel{\mathbf{v}_t^\top \mathbf{g}_t}}{\cancel{\mathbf{v}_t^\top \mathbf{g}_t}} - \frac{\cancel{\mathbf{B}_t \mathbf{g}_t} \cancel{\mathbf{g}_t^\top \mathbf{B}_t \mathbf{g}_t}}{\cancel{\mathbf{g}_t^\top \mathbf{B}_t \mathbf{g}_t}} + \underbrace{(\mathbf{g}_t^\top \mathbf{B}_t \mathbf{g}_t)\mathbf{u}_{t+1}\mathbf{u}_{t+1}^\top \mathbf{g}_t}_{\mathbf{A}_t}.$$

Let us expand the last term of the previous equality using the definition of $\mathbf{u}_t$ (3.24)

$$\mathbf{A}_t = (\mathbf{g}_t^\top \mathbf{B}_t \mathbf{g}_t) \frac{\mathbf{v}_t \cancel{\mathbf{v}_t^\top \mathbf{g}_t}}{(\mathbf{v}_t^\top \mathbf{g}_t)^{\cancel{2}}} - \frac{\mathbf{v}_t (\mathbf{g}_t^\top \mathbf{B}_t \mathbf{g}_t)}{\mathbf{v}_t^\top \mathbf{g}_t} - \frac{\mathbf{B}_t \mathbf{g}_t \cancel{\mathbf{v}_t^\top \mathbf{g}_t}}{\cancel{\mathbf{v}_t^\top \mathbf{g}_t}} + \frac{\mathbf{B}_t \mathbf{g}_t \cancel{\mathbf{g}_t^\top \mathbf{B}_t \mathbf{g}_t}}{\cancel{\mathbf{g}_t^\top \mathbf{B}_t \mathbf{g}_t}} = 0,$$

which gives:

$$\mathbf{B}_{t+1}(\nabla\mathcal{L}(\boldsymbol{w}^{(t+1)}) - \nabla\mathcal{L}(\boldsymbol{w}^{(t)})) = \mathbf{B}_{t+1}\mathbf{g}_t = \mathbf{v}_t = \boldsymbol{w}^{(t+1)} - \boldsymbol{w}^{(t)}.$$

The last point to verify is that the quasi-Newton direction defined as $\mathbf{p}_t = -\mathbf{B}_t\nabla\mathcal{L}(\boldsymbol{w}^{(t)})$ is indeed a descent direction. To do this, simply show that the matrices $(\mathbf{B})_{t\in\mathbb{N}^*}$ are positive definite. The proof here is by recurrence, demonstrated in the following proposition.

Proposition 3.1 *Suppose that at a given iteration,* $\mathbf{B}_t$ *is positive definite and* $\mathbf{v}_t^\top\mathbf{g}_t > 0$. *Then* $\mathbf{B}_{t+1}$ *is positive definite.*

Proof By expanding (3.23) we get:

$$\begin{aligned}
\mathbf{B}_{t+1} &= \mathbf{B}_t + \frac{\mathbf{v}_t\mathbf{v}_t^\top}{\mathbf{v}_t^\top\mathbf{g}_t} + (\mathbf{g}_t^\top\mathbf{B}_t\mathbf{g}_t)\frac{\mathbf{v}_t\mathbf{v}_t^\top}{(\mathbf{v}_t^\top\mathbf{g}_t)^2} - \frac{\mathbf{v}_t\mathbf{g}_t^\top\mathbf{B}_t}{\mathbf{v}_t^\top\mathbf{g}_t} - \frac{\mathbf{B}_t\mathbf{g}_t\mathbf{v}_t^\top}{\mathbf{v}_t^\top\mathbf{g}_t}\\
&= \mathbf{B}_t - \mathbf{B}_t\frac{\mathbf{g}_t\mathbf{v}_t^\top}{\mathbf{v}_t^\top\mathbf{g}_t} - \frac{\mathbf{v}_t\mathbf{g}_t^\top\mathbf{B}_t}{\mathbf{v}_t^\top\mathbf{g}_t} + \frac{\mathbf{v}_t(\mathbf{g}_t^\top\mathbf{B}_t\mathbf{g}_t)\mathbf{v}_t^\top}{(\mathbf{v}_t^\top\mathbf{g}_t)^2} + \frac{\mathbf{v}_t\mathbf{v}_t^\top}{\mathbf{v}_t^\top\mathbf{g}_t}\\
&= \left(\mathbf{B}_t - \frac{\mathbf{v}_t\mathbf{g}_t^\top\mathbf{B}_t}{\mathbf{v}_t^\top\mathbf{g}_t}\right)\left(\mathbf{Id} - \frac{\mathbf{g}_t\mathbf{v}_t^\top}{\mathbf{v}_t^\top\mathbf{g}_t}\right) + \frac{\mathbf{v}_t\mathbf{v}_t^\top}{\mathbf{v}_t^\top\mathbf{g}_t}\\
&= \left(\mathbf{Id} - \frac{\mathbf{v}_t\mathbf{g}_t^\top}{\mathbf{v}_t^\top\mathbf{g}_t}\right)\mathbf{B}_t\left(\mathbf{Id} - \frac{\mathbf{g}_t\mathbf{v}_t^\top}{\mathbf{v}_t^\top\mathbf{g}_t}\right) + \frac{\mathbf{v}_t\mathbf{v}_t^\top}{\mathbf{v}_t^\top\mathbf{g}_t},
\end{aligned}$$

where $\mathbf{Id}$ is the identity matrix of dimension $d \times d$. Suppose now that the matrix $\mathbf{B}_t$ is positive definite. We then have for any vector $\boldsymbol{w} \in \mathbb{R}^d$:

$$\begin{aligned}
\boldsymbol{w}^\top\mathbf{B}_{t+1}\boldsymbol{w} &= \boldsymbol{w}^\top\left(\mathbf{Id} - \frac{\mathbf{v}_t\mathbf{g}_t^\top}{\mathbf{v}_t^\top\mathbf{g}_t}\right)\mathbf{B}_t\left(\mathbf{Id} - \frac{\mathbf{g}_t\mathbf{v}_t^\top}{\mathbf{v}_t^\top\mathbf{g}_t}\right)\boldsymbol{w} + \frac{\boldsymbol{w}^\top\mathbf{v}_t\mathbf{v}_t^\top\boldsymbol{w}^\top}{\mathbf{v}_t^\top\mathbf{g}_t}\\
&= \boldsymbol{w}'^\top\mathbf{B}_t\boldsymbol{w}' + \frac{(\boldsymbol{w}^\top\mathbf{v}_t)^2}{\mathbf{v}_t^\top\mathbf{g}_t} \geqslant 0,
\end{aligned}$$

where $\boldsymbol{w}' = \boldsymbol{w}^\top\left(\mathbf{Id} - \frac{\mathbf{v}_t\mathbf{g}_t^\top}{\mathbf{v}_t^\top\mathbf{g}_t}\right)$. Since $\mathbf{B}_t$ is positive definite and $\mathbf{v}_t^\top\mathbf{g}_t > 0$. This shows that $\mathbf{B}_{t+1}$ is positive semidefinite.

Let us verify under which condition $\boldsymbol{w}^\top\mathbf{B}_{t+1}\boldsymbol{w} = 0$. We have:

$$\boldsymbol{w}^\top\mathbf{B}_{t+1}\boldsymbol{w} = 0 \Leftrightarrow \boldsymbol{w}'^\top\mathbf{B}_t\boldsymbol{w}' + \frac{(\boldsymbol{w}^\top\mathbf{v}_t)^2}{\mathbf{v}_t^\top\mathbf{g}_t} = 0,$$

which is equivalent to $w'^{\top}\mathbf{B}_t w' = 0$ and $w^{\top}\mathbf{v}_t = 0$. As $\mathbf{B}_t$ is positive definite, the conditions above are equivalent to $w^{\top}w' = w^{\top}w - \underbrace{w^{\top}\mathbf{v}_t}_{=0} \frac{w\mathbf{g}_t^{\top}}{\mathbf{v}_t^{\top}\mathbf{g}_t} = 0$, hence $w = 0$ and which proves that $\mathbf{B}_{t+1}$ is positive definite.

□

Input :

- A continuous, twice differentiable, convex loss function to be minimized $w \mapsto \mathcal{L}(w)$;
- A given precision $\epsilon > 0$;

Initialization:

- Randomly initialize the weights $w^{(0)}$;
- $\mathbf{B}_0 \leftarrow \mathbf{Id}_d$; $\mathbf{p}_0 \leftarrow -\nabla\mathcal{L}(w^{(0)})$;
- $t \leftarrow 0$;

repeat

- Find the optimal learning rate η_t, along the descent direction $\mathbf{p}_t$; // For example, using the line search method (3.3)
- Update $w^{(t+1)} \leftarrow w^{(t)} + \eta_t \mathbf{p}_t$;
- Estimate $\mathbf{B}_{t+1}$; // BFGS formula (3.23)
- Define the new descent direction $\mathbf{p}_{t+1} = -\mathbf{B}_{t+1}\nabla\mathcal{L}(w^{(t+1)})$;
- $t \leftarrow t + 1$;

until $|\mathcal{L}(w^{(t)}) - \mathcal{L}(w^{(t-1)})| \leqslant \epsilon\mathcal{L}(w^{(t-1)})$;
Output : $w^{(t)}$.

Algorithme 6: Quasi-Newton

Algorithm 6 gives an iterative method for jointly estimating the matrices $(\mathbf{B}_t)_{t\in\mathbb{N}^*}$ and the weight vectors $(w^{(t)})_{t\in\mathbb{N}^*}$ using a procedure for calculating the optimal learning rate applied at each iteration, which we present in the next section. This is a descent algorithm conditionally that the initial matrix $\mathbf{B}_0$ is symmetric positive definite and that at each iteration the condition:

$$(\nabla\mathcal{L}(w^{(t+1)}) - \nabla\mathcal{L}(w^{(t)}))^{\top}(w^{(t+1)} - w^{(t)}) > 0,$$

is satisfied. We can easily show that this will be the case if the η_t step is chosen at each iteration, respecting Wolfe's conditions [185], which we will explain in the next section. We also note that at the first iteration, the update rule applied by Algorithm 6 is the same as that of the gradient algorithm. Nevertheless, the major drawback of this algorithm, frequently cited in the literature, is the updating and storage of the $\mathbf{B}$ matrix of size $d \times d$, which can become disabling for high-dimensional problems.

3.3 Line Search

As discussed in the previous section, the primary challenge with the gradient algorithm lies in selecting an appropriate learning rate.Starting from a current weight vector $\boldsymbol{w}^{(t)}$, a natural solution to this problem is to go along a descent direction $\mathbf{p}_t$.

Definition 3.2 (Descent Direction) Given an objective function $\mathcal{L}$, a vector $\mathbf{p}_t$ is called a descent direction from $\boldsymbol{w}^{(t)}$ if it satisfies:

$$\mathbf{p}_t^\top \nabla \mathcal{L}(\boldsymbol{w}^{(t)}) < 0.$$

Using such a descent direction, the idea is to employ a maximum admissible value of the learning rate and iteratively decrease the rate until reaching an acceptable solution of the weights. This idea led to the development of a simple optimization method called line search, which went on to become the forerunner of many other successful optimization techniques.

3.3.1 *Wolfe's Conditions*

To find the sequence $(\boldsymbol{w}^{(t)})_{t\in\mathbb{N}}$ following the update rule (3.7), the decrease condition of the loss function:

$$\forall t \in \mathbb{N}, \mathcal{L}(\boldsymbol{w}^{(t+1)}) < \mathcal{L}(\boldsymbol{w}^{(t)}), \tag{3.25}$$

although necessary, does not guarantee convergence of this sequence to the minimizer of $\mathcal{L}$. Indeed, there are two situations where the condition (3.25) can be satisfied without reaching the minimizer of $\mathcal{L}$. We will illustrate these two cases by considering a toy example in dimension $d = 1$: $\mathcal{L}(\boldsymbol{w}) = \boldsymbol{w}^2$ with $\boldsymbol{w}^{(0)} = 2$.

1. The first case occurs when the decrease of $\mathcal{L}$ is too small compared to the jump lengths in the weight space. For example, let us take as descent directions the sequence $(\mathbf{p}_t = (-1)^{t+1})_{t\in\mathbb{N}^*}$ and as learning rate the sequence $(\eta_t = (2 + \frac{3}{2^{t+1}}))_{t\in\mathbb{N}^*}$. The sequence of jumps in weight space will then be:

$$\forall t \in \mathbb{N}^*, \boldsymbol{w}^{(t)} = (-1)^t(1 + 2^{-t}).$$

Figure 3.5 illustrates this situation.

We can see from this example that each $\mathbf{p}_t$ is a direction of descent. At each iteration $t \in \mathbb{N}^*$, the condition (3.25) is verified, but $\lim_{t\to+\infty} \boldsymbol{w}^{(t)} = \pm 1$ is not the minimizer of $\mathcal{L}$.

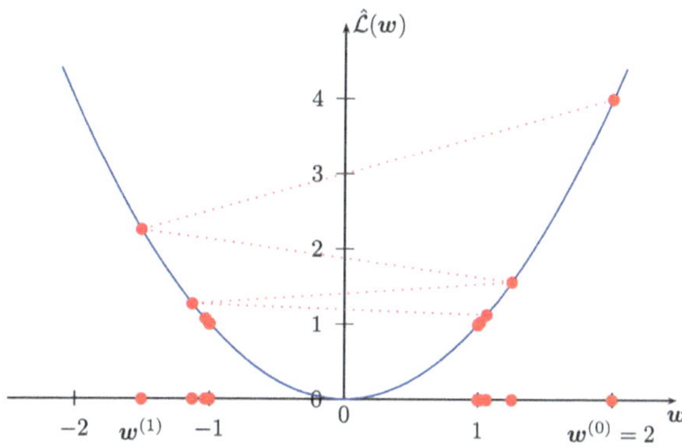

Fig. 3.5 Illustration of the non-convergent case of a decreasing sequence of weights $(\boldsymbol{w}^{(t)})_{t\in\mathbb{N}}$ towards the minimizer of an objective function $\mathcal{L}$, when the decrease is too small compared to the jump lengths

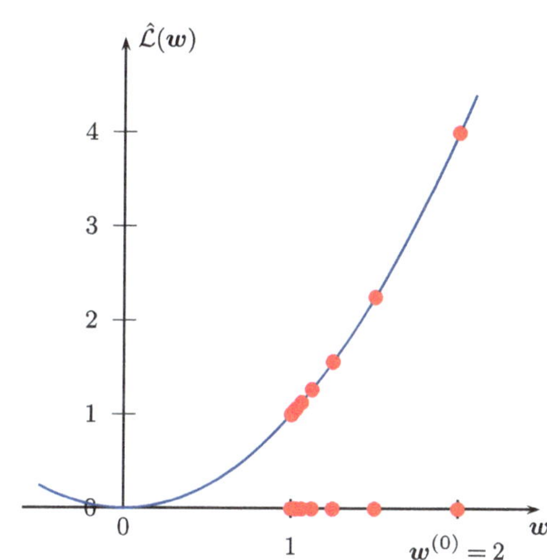

Fig. 3.6 Illustration of the non-convergent case of a decreasing sequence of weights $(\boldsymbol{w}^{(t)})_{t\in\mathbb{N}}$ towards the minimizer of an objective function $\mathcal{L}$, when the jumps are too small compared to the initial decrease rate

To avoid this problem, we add the constraint that the average decay rate from $\mathcal{L}(\boldsymbol{w}^{(t)})$ to $\mathcal{L}(\boldsymbol{w}^{(t+1)})$ is at least a fraction of the decay rate in that direction. In other words, we require that, for a given $\alpha \in (0, 1)$, the learning rate at a given iteration $\eta_t > 0$ satisfies the following condition (known as the Armijo's condition):

$$\forall t \in \mathbb{N}^*, \mathcal{L}(\boldsymbol{w}^{(t)} + \eta_t \mathbf{p}_t) \leqslant \mathcal{L}(\boldsymbol{w}^{(t)}) + \alpha \eta_t \mathbf{p}_t^\top \nabla \mathcal{L}(\boldsymbol{w}^{(t)}). \tag{3.26}$$

2. The second problem arises when the jumps from the weight vector to the minimizer are too small compared with the initial rate of decay of $\mathcal{L}$. To illustrate this case, let us take the following example, where the sequence of descent directions is: $(\mathbf{p}_t = -1)_{t\in\mathbb{N}^*}$ and the sequence of learning rates is $(\eta_t = (2^{-t+1}))_{t\in\mathbb{N}^*}$. The sequence of jumps in the weight space will then be:

$$\forall t \in \mathbb{N}^*, \boldsymbol{w}^{(t)} = (1 + 2^{-t}).$$

In this example, we see that at each iteration $t \in \mathbb{N}^*$ the condition (3.25) is verified, but $\lim_{t\to+\infty} \boldsymbol{w}^{(t)} = 1$ is not the minimizer of $\mathcal{L}$. Figure 3.6 shows this case.

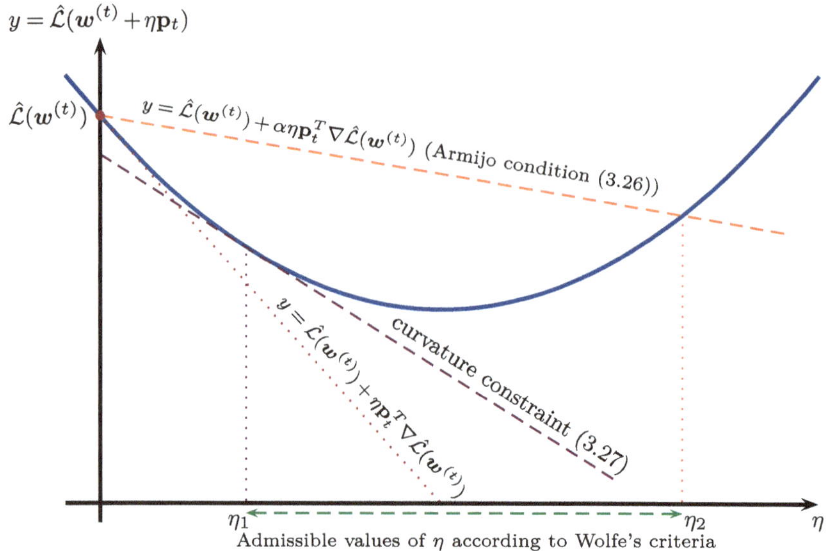

Fig. 3.7 The admissible values of the learning rate according to the Wolfe's conditions (3.26) and (3.27) on a toy example. The slope of the tangent to the curve $\eta \mapsto \mathcal{L}(\boldsymbol{w}^{(t)} + \eta \mathbf{p}_t)$ at the point $\eta = 0$ is $\mathbf{p}_t^\top \nabla \mathcal{L}(\boldsymbol{w}^{(t)}) < 0$, for a fixed value of $\alpha > 0$, the constraint (3.26) is verified when $\eta \in (0, \eta_2]$, and for a given value of $\alpha < \beta < 1$, the curvature constraint (3.27) is verified for $\eta \geqslant \eta_1$

The constraint used in this case is a curvature constraint, and it stipulates that the rate of decrease of the objective function $\mathcal{L}$ at the point $\boldsymbol{w}^{(t+1)}$ along the direction $\mathbf{p}_t$ must be at least equal to a fraction $\beta \in (\alpha, 1)$ of the rate of decrease in the same direction at the point $\boldsymbol{w}^{(t+1)}$. Mathematically, this can be expressed as follows.

$$\forall t \in \mathbb{N}^*, \mathbf{p}_t^\top \nabla \mathcal{L}(\boldsymbol{w}^{(t)} + \eta_t \mathbf{p}_t) \geqslant \beta \mathbf{p}_t^\top \nabla \mathcal{L}(\boldsymbol{w}^{(t)}). \tag{3.27}$$

Thus, β is a parameter that controls the acceptable reduction in the rate of decrease. This condition helps to maintain a sufficient level of progress towards minimizing the objective function, preventing the optimization process from stalling or converging too slowly. By enforcing this curvature constraint, the line search method ensures that each update step contributes meaningfully to the overall optimization goal.

The two conditions (3.26) and (3.27) are widely recognized as Wolfe's conditions [185] and are illustrated in Fig. 3.7. It can be demonstrated that these conditions hold in a more general context, as outlined in the following lemma.

Lemma 3.2 (Existence of Learning Rate Satisfying Wolfe's Conditions) *Let $\mathbf{p}_t$ be a descent direction from $\mathcal{L}$ at the point $\boldsymbol{w}^{(t)}$. Assume that the function $\psi_t : \eta \mapsto \mathcal{L}(\boldsymbol{w}^{(t)} + \eta \mathbf{p}_t)$ is differentiable and lower bounded. Then there exists a learning rate η_t satisfying Wolfe's conditions (3.26) and (3.27).*

Proof Let E be the set of steps verifying the Armijo's condition (3.26):

$$E = \{a \in \mathbb{R}_+ \mid \forall \eta \in]0, a], \mathcal{L}(\boldsymbol{w}^{(t)} + \eta \mathbf{p}_t) \leqslant \mathcal{L}(\boldsymbol{w}^{(t)}) + \alpha \eta \mathbf{p}_t^\top \nabla \mathcal{L}(\boldsymbol{w}^{(t)})\}.$$

Since $\mathbf{p}_t$ is a descent direction of $\mathcal{L}$ at the point $\boldsymbol{w}^{(t)}$, i.e. $\mathbf{p}_t^\top \nabla \mathcal{L}(\boldsymbol{w}^{(t)}) < 0$, then for any $\alpha < 1$ there exists $\bar{a} > 0$ such that:

$$\forall \eta \in]0, \bar{a}], \mathcal{L}(\boldsymbol{w}^{(t)} + \eta \mathbf{p}_t) < \mathcal{L}(\boldsymbol{w}^{(t)}) + \alpha \eta \mathbf{p}_t^\top \nabla \mathcal{L}(\boldsymbol{w}^{(t)}).$$

We therefore have $E \neq \emptyset$. Moreover, since the function ψ_t is lower bounded, the largest step in E, $\hat{\eta}_t = \sup E$, exists. By continuity of ψ_t, we then have:

$$\mathcal{L}(\boldsymbol{w}^{(t)} + \hat{\eta}_t \mathbf{p}_t) < \mathcal{L}(\boldsymbol{w}^{(t)}) + \alpha \hat{\eta}_t \mathbf{p}_t^\top \nabla \mathcal{L}(\boldsymbol{w}^{(t)}). \tag{3.28}$$

Let $(\eta_n)_{n \in \mathbb{N}}$ be a sequence of steps converging to $\hat{\eta}_t$ by upper values, i.e. $\forall n \in \mathbb{N}, \eta_n > \hat{\eta}_t$ and $\lim\limits_{n \to +\infty} \eta_n = \hat{\eta}_t$.

Since none of the terms of the sequence $(\eta_n)_{n \in \mathbb{N}}$ is an element of E, we have:

$$\forall n \in \mathbb{N}, \mathcal{L}(\boldsymbol{w}^{(t)} + \eta_n \mathbf{p}_t) > \mathcal{L}(\boldsymbol{w}^{(t)}) + \alpha \eta_n \mathbf{p}_t^\top \nabla \mathcal{L}(\boldsymbol{w}^{(t)}). \tag{3.29}$$

Moreover, going to the limit when n tends to $+\infty$, and as $\hat{\eta}_t \in E$, we get:

$$\mathcal{L}(\boldsymbol{w}^{(t)} + \hat{\eta}_t \mathbf{p}_t) \leqslant \mathcal{L}(\boldsymbol{w}^{(t)}) + \alpha \hat{\eta}_t \mathbf{p}_t^\top \nabla \mathcal{L}(\boldsymbol{w}^{(t)}). \tag{3.30}$$

From the two previous inequalities, we get:

$$\mathcal{L}(\boldsymbol{w}^{(t)} + \hat{\eta}_t \mathbf{p}_t) = \mathcal{L}(\boldsymbol{w}^{(t)}) + \alpha \hat{\eta}_t \mathbf{p}_t^\top \nabla \mathcal{L}(\boldsymbol{w}^{(t)}).$$

Subtracting $\mathcal{L}(\boldsymbol{w}^{(t)} + \hat{\eta}_t \mathbf{p}_t)$ from the inequality (3.29), using the previous equality, and dividing both sides by $(\eta_n - \hat{\eta}_t) > 0, \forall n \in \mathbb{N}$ gives:

$$\frac{\mathcal{L}(\boldsymbol{w}^{(t)} + \eta_n \mathbf{p}_t) - \mathcal{L}(\boldsymbol{w}^{(t)} + \hat{\eta}_t \mathbf{p}_t)}{\eta_n - \hat{\eta}_t} > \alpha \mathbf{p}_t^\top \nabla \mathcal{L}(\boldsymbol{w}^{(t)}). \tag{3.31}$$

Then, by taking the limit as n approaches to $+\infty$, we ultimately obtain:

$$\mathbf{p}_t^\top \nabla \mathcal{L}(\boldsymbol{w}^{(t)} + \hat{\eta}_t \mathbf{p}_t) \geqslant \alpha \mathbf{p}_t^\top \nabla \mathcal{L}(\boldsymbol{w}^{(t)}) \geqslant \beta \mathbf{p}_t^\top \nabla \mathcal{L}(\boldsymbol{w}^{(t)}),$$

where $\beta \in (\alpha, 1)$ and $\mathbf{p}_t^\top \nabla \mathcal{L}(\boldsymbol{w}^{(t)}) < 0$. The $\hat{\eta}_t$ step thus satisfies both Wolfe's conditions [185]. □

Consider now an objective function $\mathcal{L}$ having a L-Lipschitz continuous gradient defined as follows.

Definition 3.3 (L-Lipschitz Continuous Gradient) The objective function $\mathcal{L}$ has a L-Lipschitz continuous gradient, with $L > 0$ if and only if:

$$\forall(\boldsymbol{w}, \boldsymbol{w}') \in \mathbb{R}^d \times \mathbb{R}^d;\ ||\nabla \mathcal{L}(\boldsymbol{w}) - \nabla \mathcal{L}(\boldsymbol{w}')|| \leqslant L||\boldsymbol{w} - \boldsymbol{w}'||. \tag{3.32}$$

It can be shown that any algorithm generating the sequence $(\boldsymbol{w}^{(t)})_{t\in\mathbb{N}}$ respecting Wolfe's conditions, (3.26) and (3.27), is convergent [24, 129, 185]. This result is known as Zoutendijk's theorem [198] and is stated as follows:

Theorem 3.3 (Zoutendijk's Theorem [198]) *Let $\mathcal{L} : \mathbb{R}^d \rightarrow \mathbb{R}$ be a differentiable function having a lower bounded L-Lipschitz continuous gradient. Let $\mathfrak{A}$ be an algorithm generating the sequence $(\boldsymbol{w}^{(t)})_{t\in\mathbb{N}}$ defined by:*

$$\forall t \in \mathbb{N},\ \boldsymbol{w}^{(t+1)} = \boldsymbol{w}^{(t)} + \eta_t \mathbf{p}_t, \tag{3.33}$$

where $\mathbf{p}_t$ is a descent direction of $\mathcal{L}$ and η_t a learning rate verifying both Wolfe's conditions (3.26) and (3.27). Considering the angle θ_t between the descent direction $\mathbf{p}_t$ and the gradient direction:

$$\cos(\theta_t) = \frac{-\mathbf{p}_t^\top \nabla \mathcal{L}(\boldsymbol{w}^{(t)})}{||\mathcal{L}(\boldsymbol{w}^{(t)})|| \times ||\mathbf{p}_t||},$$

the series defined by:

$$\sum_t \cos^2(\theta_t)||\nabla \mathcal{L}(\boldsymbol{w}^{(t)})||^2$$

is convergent.

Proof With the curvature constraint (3.27), we have:

$$\forall t,\ \mathbf{p}_t^\top \nabla \mathcal{L}(\boldsymbol{w}^{(t+1)}) \geqslant \beta \left(\mathbf{p}_t^\top \nabla \mathcal{L}(\boldsymbol{w}^{(t)})\right),$$

hence, by subtracting $\mathbf{p}_t^\top \nabla \mathcal{L}(\boldsymbol{w}^{(t)})$ from the two terms of the inequality:

$$\forall t, \mathbf{p}_t^\top (\nabla \mathcal{L}(\boldsymbol{w}^{(t+1)}) - \nabla \mathcal{L}(\boldsymbol{w}^{(t)})) \geqslant (\beta - 1) \left(\mathbf{p}_t^\top \nabla \mathcal{L}(\boldsymbol{w}^{(t)})\right).$$

Furthermore, using the lipschitzian gradient property of $\mathcal{L}$ and the update rule (3.33), we get:

$$\begin{aligned}
\mathbf{p}_t^\top (\nabla \mathcal{L}(\boldsymbol{w}^{(t+1)}) - \nabla \mathcal{L}(\boldsymbol{w}^{(t)})) &\leqslant ||\nabla \mathcal{L}(\boldsymbol{w}^{(t+1)}) - \nabla \mathcal{L}(\boldsymbol{w}^{(t)})|| \times ||\mathbf{p}_t|| && (3.34)\\
&\leqslant L||\boldsymbol{w}^{(t+1)} - \boldsymbol{w}^{(t)}|| \times ||\mathbf{p}_t|| && (3.35)\\
&\leqslant L\eta_t||\mathbf{p}_t||^2. && (3.36)
\end{aligned}$$

Combining the two previous inequalities, we get:

$$\forall t, 0 \leqslant (\beta - 1)(\mathbf{p}_t^\top \nabla \mathcal{L}(\boldsymbol{w}^{(t)})) \leqslant L\eta_t||\mathbf{p}_t||^2.$$

For:

$$\eta_t \geqslant \frac{\beta - 1}{L} \frac{\mathbf{p}_t^\top \nabla \mathcal{L}(\boldsymbol{w}^{(t)})}{||\mathbf{p}_t||^2} > 0, \tag{3.37}$$

we have from the Armijo's condition (3.26) and (3.37):

$$\begin{aligned}
\mathcal{L}(\boldsymbol{w}^{(t)}) - \mathcal{L}(\boldsymbol{w}^{(t+1)}) &\geqslant -\alpha\eta_t \mathbf{p}_t^\top \nabla \mathcal{L}(\boldsymbol{w}^{(t)}) && (3.38)\\
&\geqslant \alpha \frac{1-\beta}{L} \frac{(\mathbf{p}_t^\top \nabla \mathcal{L}(\boldsymbol{w}^{(t)}))^2}{||\mathbf{p}_t||^2} && (3.39)\\
&\geqslant \alpha \frac{1-\beta}{L} \cos^2(\theta_t)||\nabla \mathcal{L}(\boldsymbol{w}^{(t)})||^2 \geqslant 0. && (3.40)
\end{aligned}$$

Finally, since the function $\mathcal{L}$ is lower bounded, the series with general term $\mathcal{L}(\boldsymbol{w}^{(t)}) - \mathcal{L}(\boldsymbol{w}^{(t+1)}) > 0$ is convergent since its partial sum of rank t, $\left(\mathcal{L}(\boldsymbol{w}^{(0)}) - \mathcal{L}(\boldsymbol{w}^{(t+1)})\right)$ is increased. According to the Comparison test theorem for two series with positive terms, the series $\sum_t \cos^2(\theta_t)||\nabla \mathcal{L}(\boldsymbol{w}^{(t)})||^2$ is also convergent. □

The previous theorem implies that if the descent direction and gradient are not orthogonal from a certain fixed iteration, i.e.:

$$\exists \kappa > 0, \forall t \geqslant T, \cos^2(\theta_t) \geqslant \kappa,$$

the series $\sum_t ||\nabla \mathcal{L}(\boldsymbol{w}^{(t)})||^2$ converges and the sequence $(\nabla \mathcal{L}(\boldsymbol{w}^{(t)}))_t$ tends to 0, when t tends to infinity. This result thus ensures the global convergence of the descent algorithm respecting both Wolfe's conditions [185].

3.3.2 Line Search Algorithm Based on Backtracking Strategy

In this section, we present a classical implementation of the line search technique based on a backtracking strategy that avoids taking η_t steps that are too small, and where verification of the curvature condition (3.27) is no longer necessary (Algorithm 7).

For a value of $\alpha \in (0, 1)$, a current weight vector $\boldsymbol{w}^{(t)}$ and a descent direction $\mathbf{p}_t$ (Definition 3.2), the algorithm evaluates the cost function at the point $\boldsymbol{w}^{(t)} + \mathbf{p}_t$ ($\eta_t = 1$). If the condition (3.26) is verified, the algorithm stops the search. Otherwise, the step η_t is reduced (backtracking phase) by interpolating the function $g : \eta \mapsto \mathcal{L}(\boldsymbol{w}^{(t)} + \eta \mathbf{p}_t)$ with a polynomial of degree 2 and choosing a new step value that minimizes this polynomial [46]. The choice of interpolation with a parabola is justified by the fact that after the first iteration of the algorithm we know the values of $g(0) = \mathcal{L}(\boldsymbol{w}^{(t)})$, $g'(0) = \mathbf{p}_t^\top \nabla \mathcal{L}(\boldsymbol{w}^{(t)})$ and $g(1) = \mathcal{L}(\boldsymbol{w}^{(t)} + \mathbf{p}_t)$, and that the minimizer of this polynomial can be calculated analytically. Indeed, the expression of the interpolation polynomial knowing these values is:

$$\eta \mapsto [g(1) - g(0) - g'(0)]\eta^2 + g'(0)\eta + g(0), \tag{3.41}$$

and its minimum is reached at:

$$\eta_m = \frac{-g'(0)}{2(g(1) - g(0) - g'(0))}. \tag{3.42}$$

Since $g(1) > g(0) + \alpha g'(0) > g(0) + g'(0)$ and $g'(0) < 0$, we have $\eta_m \in (0, \frac{1}{2}]$. To prevent the chosen step size from being too small, a lower bound b_{inf} is imposed so that, when $\eta_m \leqslant b_{inf}$, we continue the search from $\eta_m = b_{inf}$. If, with this new step value, the condition (3.26) is verified, the algorithm stops the search; otherwise, this time we interpolate the function g with a polynomial of degree three, since we know a new value of g at the point $\eta = \eta_m$.

In subsequent iterations, the new η values are estimated on a cubic interpolation that fits the function g, using $g(0)$, $g'(0)$, $g(\eta_{p_1})$ and $g(\eta_{p_2})$ where η_{p_1} and η_{p_2} are the step values estimated in the two previous iterations. The expression of the polynomial of degree three in this case is:

$$\eta \mapsto a\eta^3 + b\eta^2 + g'(0)\eta + g(0),$$

where:

$$\begin{pmatrix} a \\ b \end{pmatrix} = \frac{1}{\eta_{p_1} - \eta_{p_2}} \times \begin{bmatrix} \frac{1}{\eta_{p_1}^2} & \frac{-1}{\eta_{p_2}^2} \\ \frac{-\eta_{p_2}}{\eta_{p_1}^2} & \frac{\eta_{p_1}}{\eta_{p_2}^2} \end{bmatrix} \begin{pmatrix} g(\eta_{p_1}) - g(0) - g'(0)\eta_{p_1} \\ g(\eta_{p_2}) - g(0) - g'(0)\eta_{p_2} \end{pmatrix}. \tag{3.43}$$

Input :

- A loss function $\boldsymbol{w} \mapsto \mathcal{L}(\boldsymbol{w})$;
- Current weights $\boldsymbol{w}^{(t)}$;
- Descent direction $\mathbf{p}_t$;
- The gradient of the loss function at the point $\boldsymbol{w}^{(t)}$, $\nabla\mathcal{L}(\boldsymbol{w}^{(t)})$;
- $\alpha \in (0, \frac{1}{4})$;
- $0 < \beta < \gamma < 1$;

Initialization:

- $\eta_t \leftarrow 1$;

while $\mathcal{L}(\boldsymbol{w}^{(t)} + \eta_t\mathbf{p}_t) > \mathcal{L}(\boldsymbol{w}^{(t)}) + \alpha\eta_t\mathbf{p}_t^\top\nabla\mathcal{L}(\boldsymbol{w}^{(t)})$ **do**
  $\eta_t \leftarrow \delta\eta_t$; // With $\delta \in [\beta, \gamma]$ newly selected at each iteration

Output : $\boldsymbol{w}^{(t+1)} \leftarrow \boldsymbol{w}^{(t)} + \eta_t\mathbf{p}_t$.

Algorithme 7: Backtraking linesearch

And its minimum is reached at the point:

$$\frac{-b + \sqrt{b^2 - 3ag'(0)}}{3a}. \tag{3.44}$$

We can show that if $\alpha < \frac{1}{4}$, $b^2 - 3ag'(0)$ is always positive.

3.4 Conjugate Gradient Method

Iterative adaptation of the learning rate with the line search method does not fully prevent the oscillatory movements of the weight vector around the minimizer that we mentioned in Sect. 3.1. To illustrate this problem, let us assume that the descent direction is always the opposite of the gradient calculated for each new weight vector [9]. At iteration t and with the current weight vector $\boldsymbol{w}^{(t)}$, we have the descent direction $\mathbf{p}_t = -\nabla\mathcal{L}(\boldsymbol{w}^{(t)})$ and the learning rate η_t chosen with the line search algorithm is the one for which the partial derivative of the function $\eta \mapsto \mathcal{L}(\boldsymbol{w}^{(t)} + \eta\mathbf{p}_t)$ would be zero:

$$\frac{\partial}{\partial\eta}\mathcal{L}\left(\underbrace{\boldsymbol{w}^{(t)} + \eta\mathbf{p}_t}_{\boldsymbol{w}^{(t+1)}}\right) = \mathbf{p}_t^\top\nabla\mathcal{L}(\boldsymbol{w}^{(t+1)}) = 0. \tag{3.45}$$

We can see from this relationship that the gradient from the cost function to the new weight vector $\boldsymbol{w}^{(t+1)}$ is orthogonal to the previous descent direction $\mathbf{p}_t = -$

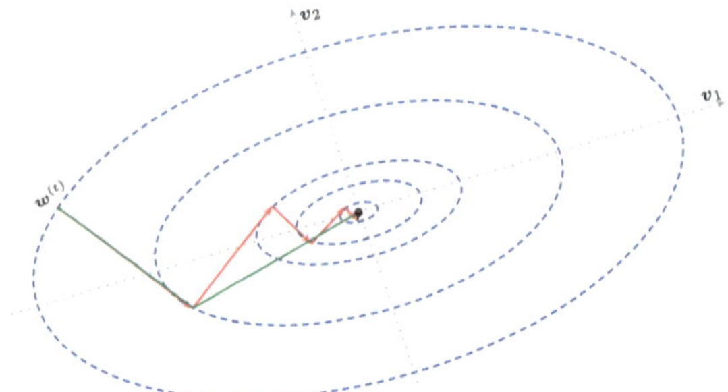

Fig. 3.8 Illustration of how the line search algorithm works using the opposite of the gradient (red) and the conjugate gradient method (green)

$\nabla\mathcal{L}(\boldsymbol{w}^{(t)})$. As we approach the minimizer $\boldsymbol{w}^*$, the weight vectors found start to oscillate, without making any significant progress towards $\boldsymbol{w}^*$ (Fig. 3.8).

3.4.1 Conjugate Directions

A simple solution to this problem is to choose the new descent direction $\mathbf{p}_{t+1}$ at the point $\boldsymbol{w}^{(t+1)}$ in such a way that the gradient at the point $\boldsymbol{w}^{(t+1)} + \eta\mathbf{p}_{t+1}$ remains orthogonal to the old descent direction $\mathbf{p}_t$, i.e.:

$$\mathbf{p}_t^\top \nabla\mathcal{L}(\boldsymbol{w}^{(t+1)} + \eta\mathbf{p}_{t+1}) = 0. \tag{3.46}$$

Consider the Taylor expansion of order 1 of $\eta \mapsto \nabla\mathcal{L}(\boldsymbol{w}^{(t+1)} + \eta\mathbf{p}_{t+1})$:

$$\nabla\mathcal{L}(\boldsymbol{w}^{(t+1)} + \eta\mathbf{p}_{t+1}) = \nabla\mathcal{L}(\boldsymbol{w}^{(t+1)}) + \eta\mathbf{H}^{(t+1)}\mathbf{p}_{t+1},$$

where $\mathbf{H}^{(t+1)}$ is the estimated Hessian matrix at point $\boldsymbol{w}^{(t+1)}$. Multiplying the previous Taylor expansion on the left by the direction $\mathbf{p}_t^\top$, we obtain from (3.45) and (3.46)

$$\mathbf{p}_t^\top \mathbf{H}^{(t+1)}\mathbf{p}_{t+1} = 0. \tag{3.47}$$

Definition 3.4 (Conjugate Directions) The descent directions $\mathbf{p}_t$ and $\mathbf{p}_{t+1}$ that satisfy Eq. (3.47) are called conjugate directions.

At the neighborhood of the loss function minimizer where the quadratic approximation holds. Suppose $\{\mathbf{p}_t, t \in [\![0, d-1]\!]\}$ is a sequence of d conjugate directions in pairs:

$$\forall (t,t') \in [\![0, d-1]\!]^2, t \neq t', \mathbf{p}_t^\top \mathbf{H}\mathbf{p}_{t'} = 0, \tag{3.48}$$

defined from the first weight vector $\boldsymbol{w}^{(0)}$ where we can apply the quadratic approximation (3.3). Since $\mathbf{H}$ is symmetric positive definite, we can show that the directions $\mathbf{p}_t$ are linearly independent and form a basis of $\mathbb{R}^d$ [77].

We can then decompose the vector $\boldsymbol{w}^* - \boldsymbol{w}^{(0)}$ according to this basis:

$$\boldsymbol{w}^* - \boldsymbol{w}^{(0)} = \sum_{t=0}^{d-1} \eta_t \mathbf{p}_t. \tag{3.49}$$

The values of the coefficients $(\eta_t)_{t=0}^{d-1}$ can also be deduced from this relationship, by multiplying (3.49) on the left by $\mathbf{p}_t^\top \mathbf{H}$, $\forall t \in [\![0, d-1]\!]$, which using the equality in (3.48) gives the definition of the adaptive learning rate:

$$\forall t, \eta_t = \frac{\mathbf{p}_t^\top \mathbf{H}(\boldsymbol{w}^* - \boldsymbol{w}^{(0)})}{\mathbf{p}_t^\top \mathbf{H}\mathbf{p}_t}. \tag{3.50}$$

Assuming:

$$\boldsymbol{w}^{(t)} = \boldsymbol{w}^{(0)} + \sum_{i=0}^{t-1} \eta_i \mathbf{p}_i, \tag{3.51}$$

we have the following update rule:

$$\forall t \in [\![0, d-1]\!], \boldsymbol{w}^{(t+1)} = \boldsymbol{w}^{(t)} + \eta_t \mathbf{p}_t. \tag{3.52}$$

Now multiply (3.51) on the left by $\mathbf{p}_t^\top \mathbf{H}$. According to the property of mutual conjugation, we obtain $(\mathbf{p}_t)_{t=0}^{d-1}$ (3.48):

$$\mathbf{p}_t^\top \mathbf{H}\boldsymbol{w}^{(t)} = \mathbf{p}_t^\top \mathbf{H}\boldsymbol{w}^{(0)}.$$

From (3.50) and (3.17), we have:

$$\forall t, \eta_t = -\frac{\mathbf{p}_t^\top \nabla \mathcal{L}(\boldsymbol{w}^{(t)})}{\mathbf{p}_t^\top \mathbf{H}\mathbf{p}_t}. \tag{3.53}$$

The following lemma proves the orthogonality of the current gradient vector with all previous descent directions:

Lemma 3.4 (Orthogonality of Gradient to Previous Descent Directions) *With the quadratic approximation and the definition of the optimal learning rate (3.53), we can show that the current gradient vector is orthogonal to all previous descent directions.*

Proof Indeed, according to this approximation, (3.17) gives the estimate of the gradient at a point $\boldsymbol{w}$ in the vicinity of $\boldsymbol{w}^*$, which for a weight vector $\boldsymbol{w}^{(t)}$ and the vector $\boldsymbol{w}^{(t+1)}$, obtained by applying the update rule (3.52) gives:

$$\forall t, \nabla\mathcal{L}(\boldsymbol{w}^{(t+1)}) - \nabla\mathcal{L}(\boldsymbol{w}^{(t)}) = \mathbf{H}(\underbrace{\boldsymbol{w}^{(t+1)} - \boldsymbol{w}^{(t)}}_{\eta_t \mathbf{p}_t}). \tag{3.54}$$

Multiplying on the left by $\mathbf{p}_t$ and according to the definition of the η_t given in (3.53), we get:

$$\forall t, \mathbf{p}_t^\top(\nabla\mathcal{L}(\boldsymbol{w}^{(t+1)}) - \nabla\mathcal{L}(\boldsymbol{w}^{(t)})) = -\mathbf{p}_t^\top\nabla\mathcal{L}(\boldsymbol{w}^{(t)}),$$

which gives:

$$\forall t, \mathbf{p}_t^\top\nabla\mathcal{L}(\boldsymbol{w}^{(t+1)}) = 0.$$

For a given index $t \in [\![0, d-1]\!]$, we have from the previous relation and the mutual conjugation of the directions of descent:

$$\forall t', \forall t, t < t', \mathbf{p}_t^\top\nabla\mathcal{L}(\boldsymbol{w}^{(t')}) = 0. \tag{3.55}$$

So, if the descent directions are mutually conjugate, after d updates following the rule (3.52) and with the learning rate defined in (3.53), we arrive at a point where all the coordinates of the gradient vector at this point will be zero on the basis formed by the d directions; and this point can only be the minimum of the loss function. □

3.4.2 Conjugate Gradient Algorithm

The necessary condition for arriving at the previous result (3.55) is to have descent directions $(\mathbf{p}_t)_{t=0}^{d-1}$ mutually conjugate. To achieve this, let us consider the sequence of directions defined by:

$$\begin{cases} \mathbf{p}_0 = -\nabla\mathcal{L}(\boldsymbol{w}^{(0)}), \\ \mathbf{p}_{t+1} = -\nabla\mathcal{L}(\boldsymbol{w}^{(t+1)}) + \beta_t \mathbf{p}_t \quad \text{if } t \geqslant 0. \end{cases} \tag{3.56}$$

where the first descent direction is equal to the opposite of the gradient vector $\mathbf{p}_0 \;=\; -\mathcal{L}(\boldsymbol{w}^{(0)})$ and subsequent descent directions are defined as a linear combination of the current gradient vector and the previous descent direction. Assuming:

$$\forall t, \beta_t = \frac{\mathbf{p}_t^\top \mathbf{H} \nabla \mathcal{L}(\boldsymbol{w}^{(t+1)})}{\mathbf{p}_t^\top \mathbf{H} \mathbf{p}_t}. \tag{3.57}$$

It is straightforward to check that iterative application of the rule (3.56) with the coefficients $(\beta_t)_{t=0}^{d-1}$ defined in (3.57) ensures that the directions of descent will be mutually conjugate. Furthermore, we have the interesting result that the gradients calculated at all weights found according to the rule (3.52) are mutually orthogonal. Indeed, according to the definition given in (3.56), the direction of descent at iteration t can be calculated as a linear combination of all the gradients calculated up to that iteration, i.e.:

$$\forall t \in [\![1, d-1]\!], \mathbf{p}_t = -\nabla \mathcal{L}(\boldsymbol{w}^{(t)}) + \sum_{i=0}^{t-1} \alpha_i \nabla \mathcal{L}(\boldsymbol{w}^{(i)}). \tag{3.58}$$

Using the orthogonality of a gradient vector with all previous directions of descent that have led to the weight where this gradient is estimated, and multiplying (3.58) by $\nabla^\top \mathcal{L}(\boldsymbol{w}^{(t')}), t' > t$, we get:

$$\forall t, \forall t' \in [\![1, d-1]\!], t' > t, \nabla^\top \mathcal{L}(\boldsymbol{w}^{(t')}) \nabla \mathcal{L}(\boldsymbol{w}^{(t)}) = \sum_{i=0}^{t-1} \alpha_i \nabla^\top \mathcal{L}(\boldsymbol{w}^{(t')}) \nabla \mathcal{L}(\boldsymbol{w}^{(i)}). \tag{3.59}$$

Since by definition $\mathbf{p}_0 = -\nabla \mathcal{L}(\boldsymbol{w}^{(0)})$, we always have according to (3.55) $\forall t' \in [\![1, d-1]\!], \nabla^\top \mathcal{L}(\boldsymbol{w}^{(t')}) \nabla \mathcal{L}(\boldsymbol{w}^{(0)}) = 0$. By induction and from Eq. (3.58) we finally have:

$$\forall t, \forall t' \in [\![1, d-1]\!], t' > t, \nabla^\top \mathcal{L}(\boldsymbol{w}^{(t')}) \nabla \mathcal{L}(\boldsymbol{w}^{(t)}) = 0. \tag{3.60}$$

The results obtained above lead to simple expressions for the coefficients $(\beta_t)_{t=0}^{d-1}$, which do not require estimation of the Hessian. Indeed, according to (3.54), we have [77]:

$$\forall t, \beta_t = \frac{\nabla^\top \mathcal{L}(\boldsymbol{w}^{(t+1)}) \mathbf{H} \mathbf{p}_t}{\mathbf{p}_t^\top \mathbf{H} \mathbf{p}_t} = \frac{\nabla^\top \mathcal{L}(\boldsymbol{w}^{(t+1)}) (\nabla \mathcal{L}(\boldsymbol{w}^{(t+1)}) - \nabla \mathcal{L}(\boldsymbol{w}^{(t)}))}{\mathbf{p}_t^\top (\nabla \mathcal{L}(\boldsymbol{w}^{(t+1)}) - \nabla \mathcal{L}(\boldsymbol{w}^{(t)}))}. \tag{3.61}$$

Input :

- A loss function $\boldsymbol{w} \mapsto \mathcal{L}(\boldsymbol{w})$;
- A precision $\epsilon > 0$;

Initialization:

- Random initialization of weights $\boldsymbol{w}^{(0)}$;
- $\mathbf{p}_0 \leftarrow -\nabla\mathcal{L}(\boldsymbol{w}^{(0)})$
- $t \leftarrow 0$;

repeat

- Run Algorithm 7 to find the optimal learning rate η_t;
- Update the weights $\boldsymbol{w}^{(t+1)} = \boldsymbol{w}^{(t)} + \eta_t\mathbf{p}_t$; // (3.52)
- Set $\beta_t = \dfrac{||\nabla\mathcal{L}(\boldsymbol{w}^{(t+1)})||^2}{||\nabla\mathcal{L}(\boldsymbol{w}^{(t)})||^2}$; // Fletcher-Reeves Formula (3.63)
- Find the new descent direction $\mathbf{p}_{t+1} = -\nabla\mathcal{L}(\boldsymbol{w}^{(t+1)}) + \beta_t\mathbf{p}_t$; // (3.56)
- $t \leftarrow t+1$;

until $|\mathcal{L}(\boldsymbol{w}^{(t)}) - \mathcal{L}(\boldsymbol{w}^{(t-1)})| \leqslant \epsilon\mathcal{L}(\boldsymbol{w}^{(t-1)})$;
Output : $\boldsymbol{w}^{(t)}$.

Algorithme 8: Conjugate Gradient

Using (3.55) and (3.56), we obtain a new expression for the coefficients $(\beta_t)_{t=0}^{d-1}$, known as Ribiere-Polak formula [136]:

$$\forall t, \beta_t = \frac{\nabla^\top\mathcal{L}(\boldsymbol{w}^{(t+1)})(\nabla\mathcal{L}(\boldsymbol{w}^{(t+1)}) - \nabla\mathcal{L}(\boldsymbol{w}^{(t)}))}{\nabla^\top\mathcal{L}(\boldsymbol{w}^{(t)})\nabla\mathcal{L}(\boldsymbol{w}^{(t)})}. \tag{3.62}$$

Finally, using mutual orthogonality between gradient vectors (3.60), we arrive at the practical and commonly used expression of Fletcher-Reeves formula [56]:

$$\forall t, \beta_t = \frac{\nabla^\top\mathcal{L}(\boldsymbol{w}^{(t+1)})\nabla\mathcal{L}(\boldsymbol{w}^{(t+1)})}{\nabla^\top\mathcal{L}(\boldsymbol{w}^{(t)})\nabla\mathcal{L}(\boldsymbol{w}^{(t)})}. \tag{3.63}$$

From these expressions, we can then define a simple algorithm for minimizing a continuous doubly derivable loss function (Algorithm 8).

Starting with a randomly chosen initial weight vector $\boldsymbol{w}^{(0)}$, we set the first descent direction to the opposite of the gradient vector calculated at this point $\mathbf{p}_0 = -\nabla\mathcal{L}(\boldsymbol{w}^{(0)})$, then repeat the update of the weight vector and the calculation of a new descent direction using an existing formula for estimating the β coefficient (3.62) or (3.63) until the relative deviation between two values of the cost function reaches a desired precision $\epsilon > 0$.

The conjugate gradient method is used not only for its computational efficiency but also for its robust convergence properties [130]. Unlike traditional gradient descent, which may require numerous iterations to converge—particularly for ill-conditioned problems—the conjugate gradient method typically achieves convergence in significantly fewer iterations [83]. This efficiency is attributed to the method's use of conjugate directions, which ensure that each new search direction is orthogonal to the previous ones, thereby enabling a more comprehensive exploration of the solution space.

As it has been seen, the method's efficiency stems from its exploitation of conjugate directions, allowing it to converge in at most d iterations for d-dimensional quadratic problems under exact arithmetic. However, the conjugate gradient method can be adapted to various types of optimization problems, including nonlinear optimization, through the incorporation of techniques such as line search and restarts. These adaptations help maintain the algorithm's efficiency and convergence properties even when dealing with non-quadratic loss functions. This versatility makes the conjugate gradient method ideal for high-dimensional machine learning tasks, where the optimization landscape can be complex and challenging.

For non-quadratic loss functions, restarts are particularly beneficial as they prevent divergence by reinitializing the search direction to the negative gradient after a set number of iterations [69]. This periodic reset helps the algorithm avoid getting trapped in suboptimal regions of the solution space.

It is also important to note that the choice of the β coefficient, whether using the Ribiere-Polak (3.62) or the Fletcher-Reeves (3.63) formula, significantly affects the convergence properties and numerical stability of the algorithm. The selection of an appropriate β coefficient can enhance the algorithm's performance and robustness, making it more effective for a wide range of optimization problems.

3.5 Nesterov's Acceleration Method

Nesterov's acceleration method [124, 126] is a powerful technique to speed up the convergence of gradient-based optimization algorithms for smooth convex functions. It improves upon the classical gradient descent by incorporating a momentum term that anticipates future gradients, achieving an optimal convergence rate among first-order methods.

For a convex function $\mathcal{L} : \mathbb{R}^d \to \mathbb{R}$ with an L-Lipschitz continuous gradient (Definition 3.32), Nesterov's method introduces an auxiliary sequence $\boldsymbol{u}^{(t)}$ to the gradient descent update rule (3.7), updating the iterates as follows:

$$\boldsymbol{u}^{(t)} = \boldsymbol{w}^{(t)} + \beta_t \left(\boldsymbol{w}^{(t)} - \boldsymbol{w}^{(t-1)} \right), \tag{3.64}$$

$$\boldsymbol{w}^{(t+1)} = \boldsymbol{u}^{(t)} - \frac{1}{L} \nabla \mathcal{L}(\boldsymbol{u}^{(t)}), \tag{3.65}$$

where β_t is a momentum parameter defined by:

$$\beta_{t+1} = \frac{n_t - 1}{n_{t+1}}, \quad \text{with} \quad n_0 = 0, \quad n_{t+1} = \frac{1 + \sqrt{1 + 4n_t^2}}{2}. \tag{3.66}$$

Note that, $n_1 = 1$ and $n_t \geqslant \frac{1+\sqrt{4n_{t-1}^2}}{2} \geqslant \frac{t+1}{2}$ for all $t \geqslant 1$.

Furthermore, the vector $\boldsymbol{u}^{(t)}$ serves as a lookahead point that incorporates momentum from previous updates, enabling the algorithm to accelerate towards the optimum. This approach reduces the oscillations typical in gradient descent, leading to faster convergence.

The momentum term $\beta_t \left(\boldsymbol{w}^{(t)} - \boldsymbol{w}^{(t-1)}\right)$ smooths the update trajectory by leveraging the direction of past steps, which helps dampen the oscillations that often occur in standard gradient descent-especially in regions with high curvature. By considering both the current and previous updates, Nesterov's method reduces abrupt changes in direction and stabilizes the optimization path. This leads to faster and more stable convergence, making the method particularly effective for ill-conditioned or highly non-linear loss landscapes [187].

Figure 3.9 illustrates the update of the weights using the lookahead point compared to the gradient descent update rule.

Based in this, the following theorem states that the error in the objective function value, measured as $\mathcal{L}(\boldsymbol{w}^{(t)}) - \mathcal{L}(\boldsymbol{w}^*)$, decreases at a rate of $O(1/t^2)$, instead of a rate of $O(1/t)$ for the standard gradient descent method.

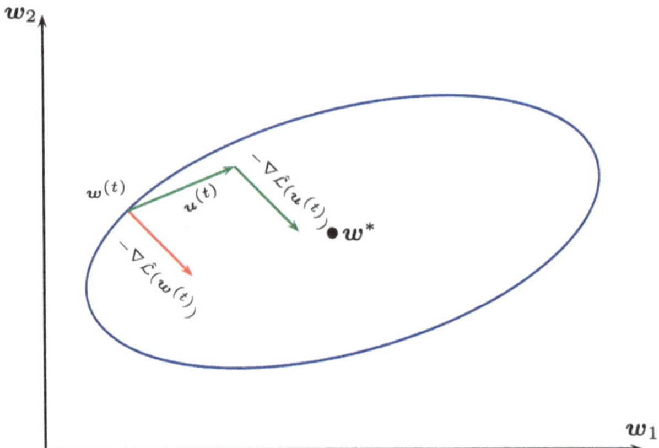

Fig. 3.9 Illustration of gradient descent and the Nesterov's accelerated gradient method on an elliptic contour line. The red arrow indicates the standard gradient descent update from the current position $\boldsymbol{w}^{(t)}$ moving in the direction opposite to the gradient. The green arrows depict the Nesterov accelerated gradient update, which includes a lookahead step to compute the gradient at an intermediate position $\boldsymbol{u}^{(t)} = \boldsymbol{w}^{(t)} + \beta_t \left(\boldsymbol{w}^{(t)} - \boldsymbol{w}^{(t-1)}\right)$. The update is then performed in the direction opposite to the gradient at this lookahead point

The proof of this result is based on the smoothness property of L-Lipschitz continuous gradient functions stated below.

Lemma 3.5 (Lipschitz Smoothness Condition) *For any function $\mathcal{L}: \mathbb{R}^d \to \mathbb{R}$ with an L-Lipschitz continuous gradient, the following smoothness property holds for any points $\boldsymbol{w}$ and $\boldsymbol{v}$ in its domain:*

$$\mathcal{L}(\boldsymbol{w}) \leqslant \mathcal{L}(\boldsymbol{v}) + \langle \nabla \mathcal{L}(\boldsymbol{v}), \boldsymbol{w} - \boldsymbol{v} \rangle + \frac{L}{2} \|\boldsymbol{w} - \boldsymbol{v}\|^2. \tag{3.67}$$

Proof For any points $\boldsymbol{w}$ and $\boldsymbol{v}$ in the domain of definition of $\mathcal{L}$, consider the following real-valued function defined in [0, 1]:

$$\phi(\gamma) = \mathcal{L}(\boldsymbol{v} + \gamma(\boldsymbol{w} - \boldsymbol{v})), \gamma \in [0, 1].$$

By the chain rule we have:

$$\phi'(\gamma) = \langle \nabla \mathcal{L}(\boldsymbol{v} + \gamma(\boldsymbol{w} - \boldsymbol{v})), \boldsymbol{w} - \boldsymbol{v} \rangle.$$

Applying the fundamental theorem of calculus for the scalar function Φ:

$$\Phi(1) - \Phi(0) = \int_0^1 \Phi'(\gamma) d\gamma,$$

we obtain:

$$\mathcal{L}(\boldsymbol{w}) = \mathcal{L}(\boldsymbol{v}) + \int_0^1 \langle \nabla \mathcal{L}(\boldsymbol{v} + \gamma(\boldsymbol{w} - \boldsymbol{v})), \boldsymbol{w} - \boldsymbol{v} \rangle d\gamma.$$

By subtracting $\langle \nabla \mathcal{L}(\boldsymbol{v}), \boldsymbol{w} - \boldsymbol{v} \rangle$ from both sides of the previous equality, we obtain:

$$\mathcal{L}(\boldsymbol{w}) - \mathcal{L}(\boldsymbol{v}) - \langle \nabla \mathcal{L}(\boldsymbol{v}), \boldsymbol{w} - \boldsymbol{v} \rangle = \int_0^1 \left(\langle \nabla \mathcal{L}(\boldsymbol{v} + \gamma(\boldsymbol{w} - \boldsymbol{v})), \boldsymbol{w} - \boldsymbol{v} \rangle - \langle \nabla \mathcal{L}(\boldsymbol{v}), \boldsymbol{w} - \boldsymbol{v} \rangle \right) d\gamma.$$

By leveraging the bilinearity of the dot product, applying the Cauchy-Schwarz inequality, and utilizing the L-Lipschitz condition of the gradient (Definition (3.3)), we derive:

$$\begin{aligned} \langle \nabla \mathcal{L}(\boldsymbol{v} + \gamma(\boldsymbol{w} - \boldsymbol{v})) - \nabla \mathcal{L}(\boldsymbol{v}), \boldsymbol{w} - \boldsymbol{v} \rangle &\leqslant \|\nabla \mathcal{L}(\boldsymbol{v} + \gamma(\boldsymbol{w} - \boldsymbol{v})) - \nabla \mathcal{L}(\boldsymbol{v})\| \|\boldsymbol{w} - \boldsymbol{v}\| \\ &\leqslant L\gamma \|\boldsymbol{w} - \boldsymbol{v}\|^2. \end{aligned}$$

Hence,

$$\mathcal{L}(\boldsymbol{w}) - \mathcal{L}(\boldsymbol{v}) - \langle \nabla \mathcal{L}(\boldsymbol{v}), \boldsymbol{w} - \boldsymbol{v} \rangle \leqslant \int_0^1 L\gamma \|\boldsymbol{w} - \boldsymbol{v}\|^2 d\gamma.$$

□

Theorem 3.6 (Convergence Rate of Nesterov's Method [125]) *For a convex function $\mathcal{L}$ with L-Lipschitz continuous gradient, the sequence $\{\mathbf{w}^{(t)}\}$ generated by Nesterov's acceleration satisfies:*

$$\mathcal{L}(\boldsymbol{w}^{(t)}) - \mathcal{L}(\boldsymbol{w}^*) \leqslant \frac{2L\|\boldsymbol{w}^{(0)} - \boldsymbol{w}^*\|^2}{(t+1)^2},$$

where $\mathbf{w}^*$ is a minimizer of $\mathcal{L}$.

Proof By applying Lemma (3.5) with $\boldsymbol{w} = \boldsymbol{w}^{(t+1)}$ and $\boldsymbol{v} = \boldsymbol{u}^{(t)}$, and using the update rule (3.65), we obtain:

$$\mathcal{L}(\boldsymbol{w}^{(t+1)}) \leqslant \mathcal{L}(\boldsymbol{u}^{(t)}) + \langle \nabla \mathcal{L}(\boldsymbol{u}^{(t)}), \underbrace{\boldsymbol{w}^{(t+1)} - \boldsymbol{u}^{(t)}}_{=-\frac{1}{L}\nabla\mathcal{L}(\boldsymbol{u}^{(t)})} \rangle + \frac{L}{2} \| \underbrace{\boldsymbol{w}^{(t+1)} - \boldsymbol{u}^{(t)}}_{=-\frac{1}{L}\nabla\mathcal{L}(\boldsymbol{u}^{(t)})} \|^2.$$

Hence,

$$\mathcal{L}(\boldsymbol{w}^{(t+1)}) \leqslant \mathcal{L}(\boldsymbol{u}^{(t)}) - \frac{1}{2L}\|\nabla \mathcal{L}(\boldsymbol{u}^{(t)})\|^2.$$

Using the previous inequality, the convexity of $\mathcal{L}$ (i.e. 3.2), and the update rule (3.65), we get:

$$\begin{aligned}
\mathcal{L}(\boldsymbol{w}^{(t+1)}) - \mathcal{L}(\boldsymbol{w}^{(t)}) &= \mathcal{L}(\boldsymbol{w}^{(t+1)}) - \mathcal{L}(\boldsymbol{u}^{(t)}) + \mathcal{L}(\boldsymbol{u}^{(t)}) - \mathcal{L}(\boldsymbol{w}^{(t)}) \\
&\leqslant -\frac{1}{2L}\|\nabla \mathcal{L}(\boldsymbol{u}^{(t)})\|^2 + \langle \nabla \mathcal{L}(\boldsymbol{u}^{(t)}), \boldsymbol{u}^{(t)} - \boldsymbol{w}^{(t)} \rangle \\
&= -\frac{L}{2}\|\boldsymbol{u}^{(t)} - \boldsymbol{w}^{(t+1)}\|^2 + L\langle \boldsymbol{u}^{(t)} - \boldsymbol{w}^{(t+1)}, \boldsymbol{u}^{(t)} - \boldsymbol{w}^{(t)} \rangle.
\end{aligned} \tag{3.68}$$

Similarly at the minimizer $\boldsymbol{w}^*$, we have:

$$\begin{aligned}\mathcal{L}(\boldsymbol{w}^{(t+1)}) - \mathcal{L}(\boldsymbol{w}^*) &= \mathcal{L}(\boldsymbol{w}^{(t+1)}) - \mathcal{L}(\boldsymbol{u}^{(t)}) + \mathcal{L}(\boldsymbol{u}^{(t)}) - \mathcal{L}(\boldsymbol{w}^*) \\ &\leqslant -\frac{1}{2L}\|\nabla\mathcal{L}(\boldsymbol{u}^{(t)})\|^2 + \langle\nabla\mathcal{L}(\boldsymbol{u}^{(t)}), \boldsymbol{u}^{(t)} - \boldsymbol{w}^*\rangle \\ &= -\frac{L}{2}\|\boldsymbol{u}^{(t)} - \boldsymbol{w}^{(t+1)}\|^2 + L\langle\boldsymbol{u}^{(t)} - \boldsymbol{w}^{(t+1)}, \boldsymbol{u}^{(t)} - \boldsymbol{w}^*\rangle.\end{aligned} \tag{3.69}$$

By defining the optimality gap $\Delta_t = \mathcal{L}(\boldsymbol{w}^{(t)}) - \mathcal{L}(\boldsymbol{w}^*)$, we have $\mathcal{L}(\boldsymbol{w}^{(t+1)}) - \mathcal{L}(\boldsymbol{w}^{(t)}) = \Delta_{t+1} - \Delta_t$. Utilizing this definition, we multiply the first inequality (3.68) by $n_t(n_t - 1)$ and the second (3.69) by n_t. By adding these results, we obtain:

$$\begin{aligned}n_t(n_t - 1)\,(\Delta_{t+1} - \Delta_t) + n_t\Delta_{t+1} \leqslant -\frac{L}{2}n_t^2\|\boldsymbol{u}^{(t)} - \boldsymbol{w}^{(t+1)}\|^2 + \\ L\langle\boldsymbol{u}^{(t)} - \boldsymbol{w}^{(t+1)}, n_t(n_t - 1)(\boldsymbol{u}^{(t)} - \boldsymbol{w}^{(t)}) + n_t(\boldsymbol{u}^{(t)} - \boldsymbol{w}^*)\rangle.\end{aligned}$$

By rearranging the terms, we get:

$$n_t^2\Delta_{t+1} - (n_t^2 - n_t)\Delta_t \leqslant A_t, \tag{3.70}$$

where

$$\begin{aligned}A_t = \frac{L}{2}\Big[2\langle n_t(\boldsymbol{u}^{(t)} - \boldsymbol{w}^{(t+1)}), n_t\boldsymbol{u}^{(t)} - (n_t - 1)\boldsymbol{w}^{(t)} - \boldsymbol{w}^*\rangle \\ -\|n_t(\boldsymbol{u}^{(t)} - \boldsymbol{w}^{(t+1)})\|^2\Big].\end{aligned}$$

Using the identity $2\langle a, b\rangle - \|a\|^2 = \|b\|^2 - \|b - a\|^2$, we obtain:

$$A_t = \frac{L}{2}\left(\|n_t\boldsymbol{u}^{(t)} - (n_t - 1)\boldsymbol{w}^{(t)} - \boldsymbol{w}^*\|^2 - \|n_t\boldsymbol{w}^{(t+1)} - (n_t - 1)\boldsymbol{w}^{(t)} - \boldsymbol{w}^*\|^2\right). \tag{3.71}$$

Furthermore, by the definition of $\boldsymbol{u}^{(t)}$ and β_{t+1} (3.66):

$$\boldsymbol{u}^{(t+1)} = \boldsymbol{w}^{(t+1)} + \beta_{t+1}(\boldsymbol{w}^{(t+1)} - \boldsymbol{w}^{(t)}) = \boldsymbol{w}^{(t+1)} + \frac{n_t - 1}{n_{t+1}}(\boldsymbol{w}^{(t+1)} - \boldsymbol{w}^{(t)}),$$

we have:

$$n_{t+1}\boldsymbol{u}^{(t+1)} - (n_{t+1} - 1)\boldsymbol{w}^{(t+1)} = n_t\boldsymbol{w}^{(t+1)} - (n_t - 1)\boldsymbol{w}^{(t)}. \tag{3.72}$$

Substituting (3.72) into (3.71), we obtain:

$$A_t = \frac{L}{2}\Big(\|n_t\boldsymbol{u}^{(t)} - (n_t - 1)\boldsymbol{w}^{(t)} - \boldsymbol{w}^*\|^2 - \|n_{t+1}\boldsymbol{u}^{(t+1)}$$
$$-(n_{t+1} - 1)\boldsymbol{w}^{(t+1)} - \boldsymbol{w}^*\|^2\Big).$$

Given the definition of n_t (3.66), where we have $n_t^2 - n_t = n_{t-1}^2$, we sum the inequalities (3.70) over t. Noting that both sides telescope, and using the initial conditions $n_0 = 0, n_1 = 1$, we obtain:

$$\begin{aligned} n_t^2 \Delta_{t+1} &\leqslant \frac{L}{2}\|\boldsymbol{w}^{(0)} - \boldsymbol{w}^*\|^2 - \frac{L}{2}\|n_{t+1}\boldsymbol{u}^{(t+1)} - (n_{t+1} - 1)\boldsymbol{w}^{(t+1)} - \boldsymbol{w}^*\|^2 \\ &\leqslant \frac{L}{2}\|\boldsymbol{w}^{(0)} - \boldsymbol{w}^*\|^2. \end{aligned}$$

Finally, since $n_t \geqslant \frac{t+1}{2}$ for all $t \geqslant 1$, it follows that:

$$\mathcal{L}(\boldsymbol{w}^{(t+1)}) - \mathcal{L}(\boldsymbol{w}^*) = \Delta_{t+1} \leqslant \frac{2L\|\boldsymbol{w}^{(0)} - \boldsymbol{w}^*\|^2}{(t+1)^2}.$$

□

Nesterov's acceleration is often implemented in training deep neural networks (Chap. 4) as the Nesterov accelerated gradient (NAG) optimizer, which improves convergence speed and generalization compared to vanilla gradient descent [199].

However, careful consideration must be given to the selection of the learning rate η and the momentum parameter β_t. These parameters play a crucial role in ensuring the stability of the training process and avoiding potential divergence. Improperly chosen values can lead to suboptimal performance or even failure to converge [101, 106, 107]. To mitigate these risks, adaptive variants of the NAG optimizer can be employed. These variants automatically adjust the learning rate and momentum parameters based on the observed gradients, thereby enhancing the robustness and reliability of the optimization process.

3.6 Subgradients

For convex but possibly nondifferentiable functions, the classical gradient does not always exist. The concept of subgradient generalizes the gradient to nonsmooth convex functions.

Let $\mathcal{L} : \mathbb{R}^d \to \mathbb{R}$ be a convex function. A vector $\mathbf{g} \in \mathbb{R}^d$ is a subgradient of $\mathcal{L}$ at $\boldsymbol{w} \in \mathbb{R}^d$ if it satisfies the following inequality:

$$\mathcal{L}(\mathbf{z}) \geqslant \mathcal{L}(\boldsymbol{w}) + \langle \mathbf{g}, \mathbf{z} - \boldsymbol{w} \rangle, \quad \forall \mathbf{z} \in \mathbb{R}^d. \tag{3.73}$$

Considering the definition of an epigraph

Definition 3.5 (Epigraph) The epigraph of a function $\mathcal{L} : \mathbb{R}^d \to \mathbb{R}$ is defined as the set of points lying on or above its graph, i.e.,

$$\text{epi}(\mathcal{L}) = \{(\boldsymbol{w}, \mu) \in \mathbb{R}^{d+1} \mid \mu \geqslant \mathcal{L}(\boldsymbol{w})\}.$$

Inequality (3.73) implies that the subgradient provides a supporting hyperplane to the epigraph of the function at the point $\boldsymbol{w}$, ensuring that the function lies above this hyperplane.

The set of all subgradients at $\boldsymbol{w}$ is called the subdifferential and is denoted by $\partial\mathcal{L}(\boldsymbol{w})$. The subdifferential captures all possible directions of ascent or descent at a given point, providing a comprehensive view of the local behavior of the function.

The main properties of the subgradient are the following:

- If $\mathcal{L}$ is differentiable at $\boldsymbol{w}$, then the subdifferential $\partial\mathcal{L}(\boldsymbol{w})$ is a singleton set containing only the gradient, i.e., $\partial\mathcal{L}(\boldsymbol{w}) = \{\nabla\mathcal{L}(\boldsymbol{w})\}$.
- The subdifferential $\partial\mathcal{L}(\boldsymbol{w})$ is a nonempty, closed, and convex set for all $\boldsymbol{w}$ in the interior of the domain of $\mathcal{L}$. This property ensures that there is always a well-defined direction for optimization algorithms to follow.
- The subgradient inequality implies that any subgradient defines a global linear underestimator of $\mathcal{L}$. This underestimator provides a lower bound on the function value, which is crucial for proving convergence and analyzing the performance of optimization algorithms.

Subgradients enable the optimization of nonsmooth convex functions via iterative algorithms. The basic subgradient method updates the iterates as follows:

$$\boldsymbol{w}^{(k+1)} = \boldsymbol{w}^{(k)} - \eta_k \mathbf{g}^{(k)}, \tag{3.74}$$

where $\mathbf{g}^{(k)} \in \partial\mathcal{L}(\boldsymbol{w}^{(k)})$ is any subgradient of $\mathcal{L}$ at $\boldsymbol{w}^{(k)}$, and $\eta_k > 0$ is a learning rate. The choice of the learning rate is critical for the convergence of the algorithm, with common strategies including constant, diminishing, and adaptive learning rates.

Although convergence is generally slower than gradient methods for smooth functions, subgradient methods are fundamental for nonsmooth optimization problems. They are widely used in various applications, including machine learning, signal processing, and operations research, where nonsmooth objective functions are common. The robustness and simplicity of subgradient methods make them a valuable tool in the optimization toolkit.

3.7 Proximal Operators

In convex optimization, the proximal operator plays a central role in handling nonsmooth functions efficiently. Given a proper, closed, convex function $\mathcal{L} : \mathbb{R}^d \to \mathbb{R} \cup \{+\infty\}$, the proximal operator of $\mathcal{L}$ with parameter $\eta > 0$ is defined as:

$$\text{prox}_{\eta\mathcal{L}}(\mathbf{v}) = \underset{\boldsymbol{w}\in\mathbb{R}^d}{\text{argmin}} \left\{ \mathcal{L}(\boldsymbol{w}) + \frac{1}{2\eta}\|\boldsymbol{w} - \mathbf{v}\|^2 \right\}. \tag{3.75}$$

Intuitively, $\text{prox}_{\eta\mathcal{L}}(\mathbf{v})$ finds a point $\boldsymbol{w}$ that balances minimizing $\mathcal{L}$ and staying close to $\mathbf{v}$.

The proximal operator possesses several important properties. First, it is well-defined: since $\mathcal{L}$ is proper, closed, and convex, the optimization problem that defines $\text{prox}_{\eta\mathcal{L}}(\mathbf{v})$ admits a unique solution. Moreover, the proximal operator is firmly non-expansive, which means that for any vectors $\mathbf{v}, \mathbf{u} \in \mathbb{R}^d$, the following inequality holds:

$$\|\text{prox}_{\eta\mathcal{L}}(\mathbf{v}) - \text{prox}_{\eta\mathcal{L}}(\mathbf{u})\|^2 \leqslant \langle \text{prox}_{\eta\mathcal{L}}(\mathbf{v}) - \text{prox}_{\eta\mathcal{L}}(\mathbf{u}), \mathbf{v} - \mathbf{u} \rangle. \tag{3.76}$$

This property ensures that the proximal operator does not amplify differences between inputs, which is important for the stability of iterative algorithms.

Furthermore, when $\mathcal{L}$ is differentiable with an L-Lipschitz continuous gradient, the proximal operator approximates a gradient descent step for sufficiently small η:

$$\text{prox}_{\eta\mathcal{L}}(\mathbf{v}) \approx \mathbf{v} - \eta\nabla\mathcal{L}(\mathbf{v}), \tag{3.77}$$

recovering the gradient descent update.

Finally, the proximal operator satisfies the Moreau decomposition relating $\mathcal{L}$ and its convex conjugate $\mathcal{L}^*$:

$$\mathbf{v} = \text{prox}_{\eta\mathcal{L}}(\mathbf{v}) + \eta\, \text{prox}_{\mathcal{L}^*/\eta}\left(\frac{\mathbf{v}}{\eta}\right). \tag{3.78}$$

This identity is useful for deriving proximal operators of complicated functions via their conjugates.

To Sum Up

1. *The Gradient algorithm updates weights along a descent direction, where the dot product with the gradient of the loss function is negative.*
2. *In full batch mode, the Gradient algorithm uses the entire dataset to ensure accuracy, while in online mode, it processes single data points for computational efficiency.*
3. *The first Wolfe condition, also known as the Armijo's condition or sufficient decrease condition, ensures that the objective function decreases sufficiently at each iteration.*
4. *The second Wolfe condition, known as the curvature condition, ensures that the step size is not too small, preventing excessively conservative steps.*
5. *Both Wolfe conditions ensure that the chosen step size along a descent direction results in a significant reduction of the objective function while maintaining the positivity of the gradient's projection onto the descent direction.*
6. *If the loss function is convex and has an L-Lipschitz continuous gradient, then selecting a learning rate according to Wolfe's conditions ensures that the gradient descent algorithm converges to the minimizer of the loss function.*
7. *The Quasi-Newton method enhances gradient descent by approximating the Hessian matrix, using the Newton direction and the BFGS formula for improved convergence.*
8. *The line search optimizes step size along the descent direction, employing Wolfe's conditions for sufficient decrease and a backtracking strategy for robust step size.*
9. *The conjugate gradient method is effective for large-scale problems, utilizing conjugate directions, ensuring each update captures unique information for efficient convergence.*
10. *The Nesterov accelerated gradient (NAG) method enhances traditional gradient descent by incorporating a momentum term that anticipates future gradients. This method introduces an auxiliary sequence that acts as a lookahead point, effectively smoothing the optimization trajectory and reducing oscillations.*
11. *By leveraging past gradients to inform current updates, NAG achieves faster convergence and improved stability, particularly in ill-conditioned or highly non-linear optimization landscapes.*
12. *The proximal gradient method is particularly useful for optimization problems where the objective function can be decomposed into a smooth and a non-smooth part.*

3.8 Exercises

1. Show that the absolute value function $\mathcal{L}(w) = |w|$ is convex using the definition of convexity.
2. Show that $\mathcal{L}(\boldsymbol{w}) = \frac{1}{2}\boldsymbol{w}^\top \mathbf{Q}\boldsymbol{w} + \mathbf{c}^\top \boldsymbol{w} + d$ is convex if $\mathbf{Q}$ is positive semidefinite.
3. Prove that $\mathcal{L}(\boldsymbol{w}) = \log\left(\sum_{i=1}^{d} e^{w_i}\right)$ is convex on $\mathbb{R}^d$.
4. Prove that ℓ_2-regularized logistic loss:

$$\mathcal{L}(\boldsymbol{w}) = \sum_{i=1}^{m} \ln\left(1 + e^{-y_i \boldsymbol{w}^\top \mathbf{x}_i}\right) + \frac{\lambda}{2}\|\boldsymbol{w}\|^2,$$

   is convex in $\boldsymbol{w}$.
5. For $\mathcal{L}(\boldsymbol{w}) = w_1^2 + w_1 w_2 + w_2^2$, starting at $\boldsymbol{w}^{(0)} = (1, 2)^\top$ with descent direction $\mathbf{p} = (-1, -1)^\top$:

   (a) Compute the learning rate η using backtracking line search (Algorithm 7, start with $\eta_{\max} = 10$, reducing factor $\delta = 0.5$).
   (b) Verify Wolfe's conditions for $\alpha = 10^{-4}$ and $\beta = 0.9$.
   (c) Confirm that the third iteration gives $\eta = 2.5$ and $\boldsymbol{w}^{(3)} = (-1.5, -0.5)^\top$.

6. Show that if $\mathcal{L} : \mathbb{R}^d \to \mathbb{R}$ is convex, then $g(\boldsymbol{w}) = \mathcal{L}(\mathbf{A}^\top \boldsymbol{w} + \mathbf{b})$ is convex for any matrix $\mathbf{A}$ and vector $\mathbf{b}$.
   Hint: Use the definition of convexity or gradient inequality.
7. Let $\mathcal{L} : \mathbb{R}^d \to \mathbb{R}$ be strictly convex and differentiable. Show that the error of the first order approximation of $\mathcal{L}$ (also called the Bregman divergence):

$$D_{\mathcal{L}}(\boldsymbol{w}, \boldsymbol{w}') = \mathcal{L}(\boldsymbol{w}) - \mathcal{L}(\boldsymbol{w}') - \langle \nabla \mathcal{L}(\boldsymbol{w}'), \boldsymbol{w} - \boldsymbol{w}' \rangle,$$

   is convex in $\boldsymbol{w}$.
8. Consider the projection onto the unit simplex:

$$\Pi_\Delta(\mathbf{w}) = \underset{\mathbf{w}' \in \Delta}{\operatorname{argmin}} \|\mathbf{w}' - \mathbf{w}\|^2, \quad \text{where} \quad \Delta = \left\{ \mathbf{w}' \geqslant 0 \mid \sum_{i=1}^{d} w_i' = 1 \right\}.$$

   Derive the projection operator and show that this is a convex optimization problem.
9. Consider a convex optimization problem where the variable $\boldsymbol{w}$ belongs to a convex set. If the Hessian of the objective function, $\mathcal{L}$ is positive semidefinite, ensuring that the function is convex. Show that any two points $\boldsymbol{w}_0$ and $\boldsymbol{w}_1$ that minimize the objective function can be connected by a path $\boldsymbol{w}(\tau) = (1-\tau)\boldsymbol{w}_0 + \tau \boldsymbol{w}_1$ for $\tau \in [0, 1]$, such that every point on this path is also a minimizer of the objective function.
   Hint: Consider the convexity of the objective function, which implies that the set of minimizers is convex. Given two minimizers $\mathbf{w}_0$ and $\mathbf{w}_1$, construct the line

segment $w(\tau) = (1-\tau)w_0 + \tau w_1$ *for* $\tau \in [0, 1]$. *Use the definition of convexity to show that the objective function evaluated at any point on this segment is less than or equal to the maximum of its values at the endpoints.*

10. Let, $\mathcal{L} : \mathbb{R}^d \to \mathbb{R}$ be a convex function with L-Lipschitz continuous gradient (Definition 3.3) and minimizer $w^\star$. Suppose further that $\mathcal{L}$ is μ-strongly convex; that is for all w and w', it satisfies:

$$\mathcal{L}(w) \geqslant \mathcal{L}(w') + \langle \nabla \mathcal{L}(w'), w - w' \rangle + \frac{\mu}{2}\|w - w'\|^2.$$

(a) Show that after t updates of the gradient descent algorithm, we have:

$$\|w^{(t)} - w^\star\|^2 \leqslant \left(1 - \frac{\mu}{L}\right)^t \|w^{(0)} - w^\star\|^2.$$

(b) Interpret how the condition number $\kappa = \frac{L}{\mu}$ affects convergence speed.

Hint: Show that each iteration reduces the distance to the minimizer by a factor of $\left(1 - \frac{\mu}{L}\right)$, *then conclude by recursively applying this contraction.*

11. Find the convex conjugate of the log-sum-exp function:

$$\mathcal{L}(\mathbf{w}) = \log\left(\sum_{i=1}^{d} e^{w_i}\right).$$

Hint: Use the definition of the convex conjugate $f^*(w') = \sup_w\{w'^\top w - \mathcal{L}(w)\}$.

12. For the ℓ_1 regularization term $\mathcal{L}(\mathbf{w}) = \lambda\|\mathbf{w}\|_1$, show that its proximal operator is given by the soft-thresholding function:

$$\text{prox}_{\eta\mathcal{L}}(\mathbf{w}) = \text{sign}(\mathbf{w}) \odot \max\left(|\mathbf{w}| - \eta\lambda, 0\right),$$

where $\odot$ denotes element-wise multiplication.

13. Let, $\mathcal{L} : \mathbb{R}^d \to \mathbb{R}$ be a convex function with L-Lipschitz continuous gradient (Definition 3.3) and minimizer $w^\star$.

(a) Show that for any $w^{(0)}$:

$$\min_w \mathcal{L}(w) \leqslant \mathcal{L}(w^{(0)}) - \frac{1}{2L}\|\nabla \mathcal{L}(w^{(0)})\|^2.$$

(b) Prove the inequality:

$$D_{\mathcal{L}}(w, w^{(0)}) \geqslant \frac{1}{2L}\|\nabla \mathcal{L}(w) - \nabla \mathcal{L}(w^{(0)})\|^2,$$

where $D_{\mathcal{L}}$ is the Bregman divergence of $\mathcal{L}$.

(c) Prove that for all $\boldsymbol{w}, \boldsymbol{w}'$ we have:

$$\langle \nabla \mathcal{L}(\boldsymbol{w}) - \nabla \mathcal{L}(\boldsymbol{w}'), \boldsymbol{w} - \boldsymbol{w}' \rangle \geqslant \frac{1}{L} \| \nabla \mathcal{L}(\boldsymbol{w}) - \nabla \mathcal{L}(\boldsymbol{w}') \|^2.$$

(d) From the last inequality deduce the upper-bound:

$$D_{\mathcal{L}}(\boldsymbol{w}, \boldsymbol{w}^{(0)}) \leqslant \frac{L}{2} \| \boldsymbol{w} - \boldsymbol{w}^{(0)} \|^2.$$

Hint: For the first part, consider the quadratic upper bound derived from the L-Lipschitz continuous gradient property and minimize it to find a point that provides a tighter bound on the minimum value of the function. For the second part, use the definition of the Bregman divergence and relate it to the gradient differences, utilizing the Lipschitz continuity.

14. For the stochastic gradient descent (SGD) algorithm, show that under convexity assumptions and bounded gradient variance $\mathbb{E}[\|\nabla \mathcal{L}(\mathbf{w})\|^2] \leqslant G^2$, the expected suboptimality after t iterations satisfies:

$$\mathbb{E}[\mathcal{L}(\bar{\mathbf{w}}_t)] - \mathcal{L}(\mathbf{w}^\star) \leqslant \frac{\|\mathbf{w}^{(0)} - \mathbf{w}^\star\|^2 + G^2\sqrt{t}}{2\sqrt{t}},$$

where $\bar{\mathbf{w}}_t = \frac{1}{t}\sum_{k=1}^{t} \mathbf{w}^{(k)}$.

Hint: Use a step size $\eta_k = \frac{1}{\sqrt{k}}$ and bound the expected cumulative error.

15. Consider the Lasso problem:

$$\min_{\boldsymbol{w}} \frac{1}{2} \| \langle \mathbf{x}, \boldsymbol{w} \rangle - \boldsymbol{w}' \|^2 + \lambda \| \boldsymbol{w} \|_1.$$

Derive the closed-form update rule for coordinate descent on the j-th coordinate w_j.

16. Show that a function $\mathcal{L}$ is convex if and only if its epigraph $\mathrm{epi}(\mathcal{L}) = \{(\boldsymbol{w}, t) \mid \mathcal{L}(\boldsymbol{w}) \leqslant t\}$ is a convex set.
17. For the ℓ_1-norm $\mathcal{L}(\boldsymbol{w}) = \|\boldsymbol{w}\|_1$, characterize the subdifferential $\partial \mathcal{L}(\boldsymbol{w})$ at a point $\boldsymbol{w}$ where some coordinates satisfy $w_i = 0$.
18. Prove that for any convex function f and its proximal operator prox_f, the following decomposition holds:

$$\boldsymbol{w} = \mathrm{prox}_{\eta f}(\boldsymbol{w}) + \eta \mathrm{prox}_{f^*/\eta}(\boldsymbol{w}/\eta),$$

where f^* is the convex conjugate of f.

19. Let, $\mathcal{L} : \mathbb{R}^d \to \mathbb{R}$ be a convex function with L-Lipschitz continuous gradient (Definition 3.3).

(a) Prove that for all w, $w^{(0)}$:

$$\mathcal{L}(w) \leqslant \mathcal{L}(w^{(0)}) + \langle \nabla \mathcal{L}(w^{(0)}), w - w^{(0)} \rangle + \frac{L}{2} \| w - w^{(0)} \|^2.$$

(b) Use this inequality to derive the convergence rate for gradient descent with learning rate $\eta \leqslant \frac{1}{L}$:

$$\mathcal{L}(w^{(t)}) - \mathcal{L}(w^{\star}) \leqslant \frac{2L \| w^{(0)} - w^{\star} \|^2}{t+4},$$

where $w^{\star}$ is the minimizer of $\mathcal{L}$.

Hint: Consider the Taylor expansion of $\mathcal{L}$ around $\mathbf{w}^{(0)}$ and bound the remainder term using the L-Lipschitz continuous gradient condition. For the convergence rate, apply the derived inequality to the gradient descent update rule, ensuring that the learning rate η is chosen appropriately to satisfy $\eta \leqslant \frac{1}{L}$. Use the convexity of $\mathcal{L}$ to relate the gradient to the difference between the current iterate and the minimizer $\mathbf{w}^{\star}$. Sum the resulting inequalities over iterations and simplify.

20. For the hinge loss function $\mathcal{L}(\mathbf{w}) = \max(0, 1 - y\mathbf{w}^{\top} w)$:

(a) Compute a subgradient at a point $\mathbf{w}_0$ where $1 - y\mathbf{w}_0^{\top} w = 0$.
(b) Show that the hinge loss is convex but not differentiable everywhere.

21. In many machine learning applications, such as portfolio selection, resource allocation, or risk-aware prediction, one often needs to balance expected performance against uncertainty or variability.

Consider a decision vector $w \in \mathbb{R}^d$ representing weights assigned to d different features, or assets. The goal is to find w that minimizes a quadratic risk measure while ensuring a minimum expected performance level.

Let $\boldsymbol{\Sigma} \in \mathbb{R}^{d \times d}$ be a positive semidefinite covariance matrix quantifying the uncertainty or variability associated with the components of w. For example, in portfolio optimization, $\boldsymbol{\Sigma}$ measures the variance and covariance of asset returns; in ensemble learning, it might represent model prediction variances and correlations.

Let $\boldsymbol{\mu} \in \mathbb{R}^d$ be a vector of expected returns, or utility values corresponding to each component of w. We require the weighted expected performance to be at least a threshold $r_{\min}$, i.e.,

$$\boldsymbol{\mu}^{\top} w \geqslant r_{\min}.$$

Additionally, the weights must sum to one to represent a proper allocation or convex combination:

$$\mathbf{1}^{\top} w = 1.$$

The optimization problem is then formulated as:

$$\min_{\boldsymbol{w} \in \mathbb{R}^d} \quad \boldsymbol{w}^\top \boldsymbol{\Sigma} \boldsymbol{w} \quad \text{subject to} \quad \mathbf{1}^\top \boldsymbol{w} = 1, \quad \boldsymbol{\mu}^\top \boldsymbol{w} \geqslant r_{\min}.$$

(a) Explain why this problem is a convex optimization problem.
(b) Show that the objective function $\boldsymbol{w} \mapsto \boldsymbol{w}^\top \boldsymbol{\Sigma} \boldsymbol{w}$ is convex when $\boldsymbol{\Sigma}$ is positive semidefinite.
(c) Argue that the constraints define a convex feasible set.
(d) Discuss how this formulation can be interpreted as a trade-off between minimizing uncertainty (risk) and achieving a desired expected performance in machine learning models or resource allocation.

Hints: Recall that quadratic forms with positive semidefinite matrices are convex functions. Linear equality and inequality constraints define affine and convex sets. Convex objective plus convex feasible set implies the problem is convex and can be efficiently solved using convex optimization techniques.

22. Let $\mathcal{L}(\boldsymbol{w}) = \frac{1}{2}\boldsymbol{w}^\top \mathbf{A}\boldsymbol{w}$ be a quadratic function, where $\mathbf{A} \in \mathbb{R}^{d\times d}$ is symmetric positive definite with eigenvalues in $[m, L]$, $0 < m \leqslant L$. Consider the Nesterov accelerated gradient updates (3.64) and (3.65).

(a) Show that for the quadratic function above, the gradient is given by $\nabla\mathcal{L}(\boldsymbol{w}) = \mathbf{A}\boldsymbol{w}$, and the unique minimizer is $\boldsymbol{w}^* = \mathbf{0}$.
(b) Write the Nesterov accelerated gradient (NAG) update explicitly for this quadratic case, and show that the iteration can be expressed as a linear recurrence involving $\mathbf{A}$, $\boldsymbol{w}^{(t)}$, and $\boldsymbol{w}^{(t-1)}$.
(c) Let $\kappa = L/m$ denote the condition number of $\mathbf{A}$. By considering the eigendecomposition of $\mathbf{A}$, analyze the convergence of each coordinate (eigendirection) separately. Show that the convergence rate of NAG is governed by the largest eigenvalue and the momentum parameter.
(d) Recall that standard gradient descent with step size $1/L$ satisfies

$$\|\boldsymbol{w}^{(t)}\| \leqslant \left(1 - \frac{m}{L}\right)^t \|\boldsymbol{w}^{(0)}\|,$$

for all t. Show that, with optimal choice of β_t, NAG achieves the accelerated rate

$$\|\boldsymbol{w}^{(t)}\| \leqslant 2\left(\frac{\sqrt{\kappa}-1}{\sqrt{\kappa}+1}\right)^t \|\boldsymbol{w}^{(0)}\|,$$

and explain why this is a significant improvement when κ is large [124].
(e) Discuss how the momentum term $\beta_t(\boldsymbol{w}^{(t)} - \boldsymbol{w}^{(t-1)})$ helps accelerate convergence, particularly in ill-conditioned problems (i.e., when κ is large), and why this makes NAG preferable to standard gradient descent in such settings.

Chapter 4
Deep Learning

In this chapter, we will outline the main algorithms that have their origins in work on the numerical modeling of neurons in the nervous system for learning an association rule between examples and their desired outputs. After a brief history of the development of artificial neural networks, we present in Sect. 4.1.2, the Perceptron, the first neuro-inspired model based on the work of neuroscientists and psychologists. Variants of this model are presented in the Sects. 4.1.3 and 4.2. In Sect. 4.3, we present multi-layer Perceptrons, which are universal approximators capable of performing complex learning tasks. We will also describe the backpropagation error algorithm, which underpins deep network learning. In Sect. 4.4, we will present two popular networks for image classification and learning in sequences. We will conclude this chapter with considerations that have been found over the years to optimize the learning of these models.

The development of artificial neural networks has been marked by three main phases. The first dates back to the work of McCulloch and Pitts [113], who attempted to determine the mathematical properties of living neurons discovered some fifty years earlier by Santiago Ramón y Cajal, a Spanish neuroscientist and pathologist. McCullogh and Pitts began to study a model, called a formal neuron, which for an input vector produced a binary output based on a linear combination of the vector's characteristics.

The important result of their work was that these neurons were capable of calculating certain logical functions, which brought them close to a universal Turing machine [171]. Towards the end of the 1940s, other researchers tried to explain the mechanisms of learning, memory and conditioning using groups of cells. For example, in the law bearing his name, Hebb explained the processes involved in learning as a function of experience [75]. He proposed a model in which cells learn to modify the intensity of the connections linking them according to their simultaneous activity. At the same time, Minsky designed SNARC, the first real model of a neural network capable of learning by correcting the intensity of its synapses [117]. On this basis, Rosenblatt, a psychologist, created a model called *Perceptron* (Sect. 4.1.2)

M.-R. Amini, *Advanced Supervised and Semi-supervised Learning*,
Cognitive Technologies, https://doi.org/10.1007/978-3-031-99928-4_4

which aimed to make an artificial neural network learn perceptual categorizations [143]. The result of this work aroused great enthusiasm among researchers from different scientific fields, notably Novikoff, a mathematician who demonstrated that, if the training examples are linearly separable in the vector space in which they are represented, then the Perceptron model finds any separating hyperplane in a finite number of weight updates (Sect. 4.1.2.1) [131].

In the same years, electrical engineers Widrow and Hoff [183] proposed a learning rule close to that of the Perceptron and built a model called *Adaptive Linear Neuron* (Adaline) to learn noise attenuation in telephone lines (Sect. 4.1.3). The conclusion of all this work was *Learning Machines* by Nilsson [128], which laid the mathematical foundations of machine learning for pattern recognition. Throughout this period, neural network research proceeded without any real questioning of the methods employed or the results obtained. Towards the end of the 1960s, the machine learning community began to synthesize this research. Minsky and Pappert [118] have shown the limits of these models. For example, a Perceptron is not capable of learning the logical function XOR, because, unlike the AND function, it is not linearly separable, so an artificial neuron cannot be a Turing machine. This book marked the first winter of work on artificial neural networks.

In the years that followed, several authors explored the subject. Anderson and Rosenfeld [7] have collected a good number of articles corresponding to this period. Then, at the end of the eighties, the field took off again. Hopfield [84] on the one hand, Rumelhart and Hinton [145] on the other, showed that certain learning rules (for example, the backpropagation technique) (Sect. 4.3.2) enabled more powerful neural networks called *multi-layer Perceptrons* (MLP, Sect. 4.3) to learn complex functions that the Perceptron could not.

The important result of these studies was that, under certain conditions, the MLP is a universal approximator that can approximate any real function to a given accuracy. During these years, neural networks had become an attractive field bringing together psychologists, physicists, mathematicians, biologists and computer scientists, with the creation of several international conferences and journals; among the most popular were the conference *Neural Information Processing Systems*[1] (NeurIPS) or the journal *Neural Computation*.[2] Various models derived from MLP have been applied to different prediction problems. These include the convolutional neural networks proposed by LeCun [102], which were designed to recognize visual patterns directly from image pixels. This model was successfully applied to the recognition of digits on checks, and is the basis of many of the models subsequently proposed for pattern recognition in images. New types of neural networks, called recurrent networks, with architectures involving loops on their neurons in different layers were also proposed for prediction tasks on sequences [80].

At that time, these black-box models were becoming increasingly complex, requiring a lot of training data and powerful machines to learn them. The introduction of SVMs (Chap. 5), derived from the statistical learning theory of Vapnik and

---

[1] https://neurips.cc/.

[2] https://www.mitpressjournals.org/loi/neco.

Chervonenkis [176], marked a turning point in the development of neural networks, since, unlike the latter, the solution found by SVMs is unique and their operation is explained according to the principle of structural risk minimization (Chap. 1, Sect. 1.3.1). The Caltech 101 competition in 2004 marked the second winter of neural networks. The task was to categorize an image into 101 classes from a training database of 3,000 images, but this was not enough to learn the convolutional networks that were among the competing systems.

The renaissance of these models came a few years later with the invention of GPUs, and the *ImageNet Large-Scale Visual Recognition Challenge*[3] competition (ILSVRC), based on the new ImageNet collection[4] created in 2009. At that time, the ImageNet database contained almost 15 million high-resolution images classified according to 22,000 categories. In 2012, ILSVRC's winning system was a convolutional neural network called AlexNet [97], which beat the runner-up by over 10.8 % on the error measure employed. This system was trained for 6 days using 2 Titan GPUs (whereas, on 16-core Xeon CPUs, more than 40 days would have been needed). With the *BigData* phenomenon, ever larger data collections, and unceasing advances in computing processors, the use of neural networks has become widespread in many application fields, such as automatic natural language processing or information retrieval. With increasingly complex architectures, the term deep networks or *deep learning* has become increasingly popular.

Deep learning's success in achieving human-like performance in various tasks, along with its widespread adoption and the advancements it has brought to the field, has led to it being commonly referred to as artificial intelligence. In 2018, Yoshua Bengio, Geoffrey Hinton and Yann LeCun, were awarded the Turing Award, often referred to as the "Nobel Prize of Computing", for their pioneering work in deep learning. Also, John Hopfield and Geoffrey Hinton were awarded the 2024 Nobel Prize in Physics for their groundbreaking work in developing machine learning technology using artificial neural networks. In Chemistry, the prize was awarded to David Baker for his pioneering work in computational protein design, and to Demis Hassabis and John Jumper for developing an AI algorithm that solved the long-standing challenge of predicting protein structures. This recognition highlights the profound impact of their research on the advancement of artificial intelligence.

4.1 Models Based on the Formal Neuron

4.1.1 Formal Representation

The first mathematical model of the biological neuron was proposed by McCulloch and Pitts [113]. This model is based on a simple formulation of a neuron with binary or real inputs and a binary output (Fig. 4.1).

---

[3] http://www.image-net.org/challenges/LSVRC/.

[4] http://www.image-net.org/.

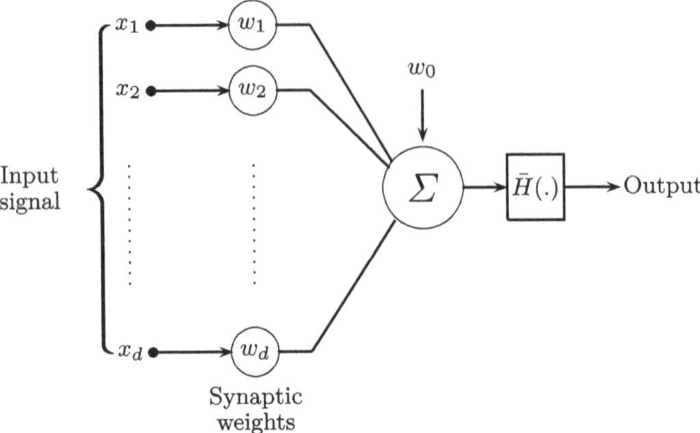

Fig. 4.1 Illustration of a formal neuron proposed by McCulloch and Pitts [113]

The function of this model is to perform a weighted sum of the d inputs using its synaptic coefficients, $\bar{\boldsymbol{w}} = (w_1, \ldots, w_d)^\top$, and a threshold, w_0, then transform this weighed sum using an activation function.

Definition 4.1 (Activation Function) A transfer function, in the context of neural networks, is a mathematical function, $\bar{H} : \mathbb{R} \to \mathbb{R}$ that maps the weighted sum of inputs of a neuron to an output value.

The output of the formal neuron is given by:

$$\bar{H}\left(\langle \bar{\boldsymbol{w}}, \mathbf{x}\rangle + w_0\right) = \bar{H}\left(\sum_{j=1}^{d} w_j x_j + w_0\right),$$

where the activation function, $\bar{H}(.)$, is the Heaviside function defined as:

$$\bar{H} : x \mapsto \begin{cases} 1 & \text{if } x \geqslant 0 \\ 0 & \text{otherwise.} \end{cases}$$

Hence, for a given input $\mathbf{x}$, if the dot product $\langle \bar{\boldsymbol{w}}, \mathbf{x}\rangle$ is greater than the threshold $-\,w_0$, the output of the formal neuron is equal 1; otherwise, it is equal to 0. Other activation functions more commonly used in the literature are presented in Sect. 4.5.1.

4.1.2 Perceptron

Rosenblatt adapted the formal neuron to the case of neurons in the visual system, and called it a Perceptron in reference to the perceptive system [143]. Starting with a light-bulb array, called a Retina, where shapes could be displayed by switching certain bulbs on and off, Rosenblatt had linked light sensors randomly arranged on this array to the association areas, $(\Phi_i)_{1 \leqslant i \leqslant d}$, whose proportion of bulbs they measured as switched on. These association areas formed the input to the formal neuron, with the aim of finding synaptic weights and the threshold that would distinguish a particular shape drawn on the Retina (e.g. a 7) from other shapes. This model is illustrated in Fig. 4.2.

In this model, the activation function, $\bar{H}$, is linear and the input signals obtained after the association areas thus correspond to the characteristics of the shape displayed on the retina. The threshold, w_0 , is one of the parameters of the model to be learned and is associated with a fixed input commonly called the bias $x_0 = 1$. In this case, the output estimated by the model with parameters $\boldsymbol{w} = (\bar{\boldsymbol{w}}, w_0)$, for a given input $\mathbf{x}$ is:

$$
\begin{aligned}
h_{\boldsymbol{w}} : \mathbb{R}^d &\rightarrow \mathbb{R} \\
\mathbf{x} &\mapsto \langle \bar{\boldsymbol{w}}, \mathbf{x} \rangle + w_0.
\end{aligned}
$$

The Perceptron algorithm finds the model parameters, by minimizing the distance of misclassified examples to the decision frontier.

In the case where the classification function is linear, the distance of any point $\mathbf{x}$ from the separating hyperplane is proportional to the output of the linear model for that point. This is shown in Fig. 4.3.

Indeed, let $\mathbf{x}$ be a point in vector space, and if we consider $\mathbf{x}_p$ the orthogonal project of $\mathbf{x}$ onto the separating hyperplane, we have the following equality

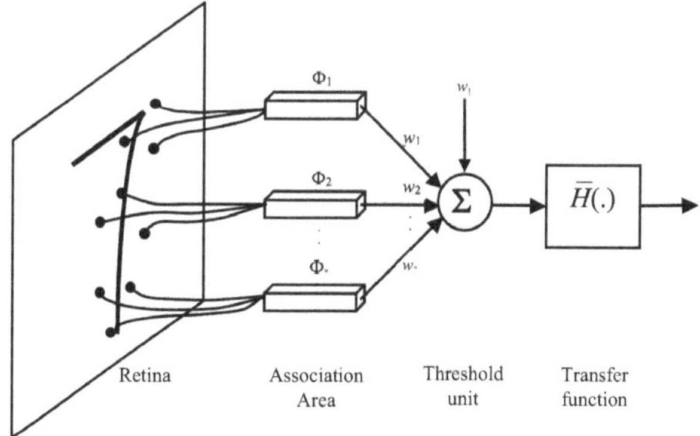

Fig. 4.2 Illustration of the Perceptron model proposed by Rosenblatt [143]

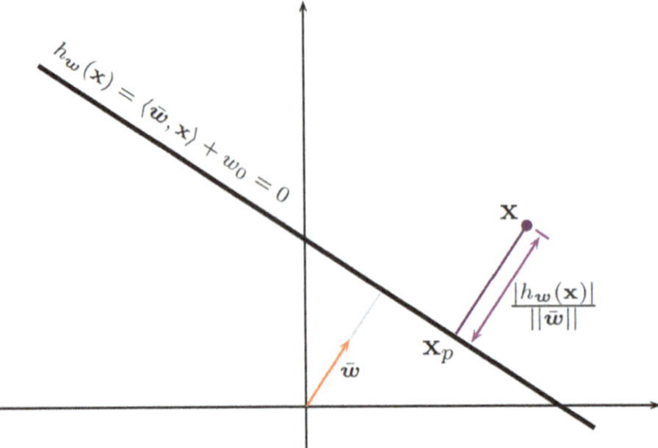

Fig. 4.3 Illustration of the distance of an example **x**, in a plane, to the hyperplane (line in this case) of equation $h_{\boldsymbol{w}}(\mathbf{x}) = \langle \bar{\boldsymbol{w}}, \mathbf{x}\rangle + w_0 = 0$

according to the relation $\mathbf{x} = \mathbf{x}_p + \underbrace{(\mathbf{x} - \mathbf{x}_p)}_{\mathbf{u}}$:

$$h_{\boldsymbol{w}}(\mathbf{x}) = \langle \bar{\boldsymbol{w}}, \mathbf{x}_p + \mathbf{u}\rangle + w_0.$$

According to the bilinearity of the scalar product, this is written:

$$h_{\boldsymbol{w}}(\mathbf{x}) = \underbrace{\langle \bar{\boldsymbol{w}}, \mathbf{x}_p\rangle + w_0}_{h_{\boldsymbol{w}}(\mathbf{x}_p)} + \langle \bar{\boldsymbol{w}}, \mathbf{u}\rangle.$$

Since $\mathbf{x}_p$ belongs to the separating hyperplane (i.e. $h_{\boldsymbol{w}}(\mathbf{x}_p) = 0$) and the vectors $\mathbf{u}$ and $\boldsymbol{w}$ are perpendicular to the hyperplane, we have:

$$|h_{\boldsymbol{w}}(\mathbf{x})| = ||\bar{\boldsymbol{w}}|| \times ||\mathbf{u}||,$$

where $||\mathbf{u}|| = \frac{|h_{\boldsymbol{w}}(\mathbf{x})|}{||\bar{\boldsymbol{w}}||}$ is the distance of $\mathbf{x}$ to the hyperplane.

Definition 4.2 (Margin) The margin of a linear classifier $h_{\boldsymbol{w}}$ on a training set $S = (\mathbf{x}_i, y_i)_{i=1}^m$, ρ, is defined as the smallest distance of the examples in this set with respect to the separating hyperplane defined by $h_{\boldsymbol{w}}$:

$$\rho = \min_{i\in\{1,\dots,m\}} \frac{|h_{\boldsymbol{w}}(\mathbf{x}_i)|}{||\bar{\boldsymbol{w}}||}. \tag{4.1}$$

Thus, since the distance of an example from the separator hyperplane is proportional to the prediction of the linear model for that example and, for a misclassified example, the product of this prediction and the example's class label is negative, the Perceptron learning algorithm attempts to minimize the following objective function:

$$\mathcal{L}(\boldsymbol{w}) = -\sum_{i' \in \mathcal{I}} y_{i'}(\langle \bar{\boldsymbol{w}}, \mathbf{x}_{i'} \rangle + w_0), \tag{4.2}$$

where $\mathcal{I}$ is the index of training examples misclassified by the current model with parameters $\boldsymbol{w} = (\bar{\boldsymbol{w}}, w_0)$. Assuming that the set $\mathcal{I}$ is fixed, the gradient of $\mathcal{L}$ is equal to:

$$\frac{\partial \mathcal{L}(\boldsymbol{w})}{\partial w_0} = -\sum_{i' \in \mathcal{I}} y_{i'}, \tag{4.3}$$

$$\nabla \mathcal{L}(\bar{\boldsymbol{w}}) = -\sum_{i' \in \mathcal{I}} y_{i'} \mathbf{x}_{i'}. \tag{4.4}$$

Weights are updated using the gradient algorithm (Chap. 3, Sect. 3.1). In practice, this learning is carried out using a *online* algorithm (Algorithm 9) where, at each iteration, an example is randomly chosen and the vector of weights $\boldsymbol{w}$ is updated if the classifier gives a wrong class prediction for this example, i.e.

$$\forall (\mathbf{x}, y), \text{ if } y(\langle \bar{\boldsymbol{w}}, \mathbf{x} \rangle + w_0) \leqslant 0 \text{ then } \begin{pmatrix} w_0 \\ \bar{\boldsymbol{w}} \end{pmatrix} \leftarrow \begin{pmatrix} w_0 \\ \bar{\boldsymbol{w}} \end{pmatrix} + \eta \begin{pmatrix} y \\ y\mathbf{x} \end{pmatrix}, \tag{4.5}$$

Input :

- Training set $S = (\mathbf{x}_i, y_i)_{1 \leqslant i \leqslant m}$;
  Learning rate $\eta > 0$; Maximum number of iterations T; Number of updates k;

Initialization:

- $\boldsymbol{w}^{(0)} = (\bar{\boldsymbol{w}}^{(0)}, w_0^{(0)}) = (\mathbf{0}, 0)$; $t \leftarrow 0$; $k \leftarrow 0$;

while $t \leqslant T$ **do**
  Random choice $(\mathbf{x}, y) \in S$;
  if $y \times \left(\langle \bar{\boldsymbol{w}}^{(k)}, \mathbf{x} \rangle + w_0^{(k)}\right) \leqslant 0$ **then**
    $w_0^{(k+1)} \leftarrow w_0^{(k)} + \eta \times y$;
    $\bar{\boldsymbol{w}}^{(k+1)} \leftarrow \bar{\boldsymbol{w}}^{(k)} + \eta \times y \times \mathbf{x}$;
    $k \leftarrow k + 1$;
  $t \leftarrow t + 1$;

Output : Model parameters $\boldsymbol{w}^{(k)}$.

Algorithme 9: Perceptron

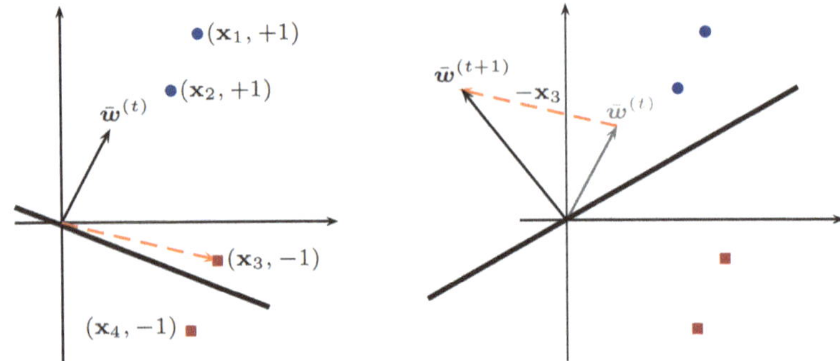

Fig. 4.4 Illustration of the Perceptron algorithm's update rule for $w_0 = 0$. The current synaptic weights $\bar{\boldsymbol{w}}^{(t)}$, which correspond to the normal vector of the separating hyperplane, point to the positive half-space where examples of class $y = +1$ are expected to lie. The observation $\mathbf{x}_3$ of class $y_3 = -1$ is misclassified (i.e., $y_3 h_{\bar{\boldsymbol{w}}^{(t)}}(\mathbf{x}_3) < 0$). The update of the normal vector, $\bar{\boldsymbol{w}}^{(t+1)} \leftarrow \bar{\boldsymbol{w}}^{(t)} + y_3\mathbf{x}_3$, adjusts the separating hyperplane so that $\mathbf{x}_3$ will be in the negative half-space

where $\eta > 0$ is the learning step. Figure 4.4 illustrates geometrically the algorithm's update rule on a toy example.

4.1.2.1 Perceptron Convergence Theorem

This algorithm, although simple, performs well in certain scenarios. This foundational algorithm serves as the precursor of many other learning algorithms, notably the multi-layer Perceptron presented in Sect. 4.3. The Perceptron's significance is underscored by a key theoretical result: if a classification problem is linearly separable, the Perceptron algorithm is guaranteed to identify a separating hyperplane within a finite number of updates. This means that for datasets where a linear boundary can perfectly distinguish between different classes, the Perceptron will successfully converge to a solution. There are various proofs for this result; in the following, we present the one proposed by Novikoff [131].

Theorem 4.1 (Perceptron Convergence for Linearly Separable Data [131]) *Consider a two-class classification problem and let $(\mathbf{x}_i, y_i)_{i=1}^m$ be a sequence of m training examples contained in a hypersphere of radius R; $\forall i, ||\mathbf{x}_i|| \leqslant R$. Assume that the decision boundaries pass through the origin (i.e. $w_0 = 0$ and $\boldsymbol{w} = \bar{\boldsymbol{w}}$) and that there exists a hyperplane of normal vector $\boldsymbol{w}^*$ separating the examples of the two classes (the problem is linearly separable). Let $\rho = \min_{i\in\{1,\ldots,m\}}\left(y_i\left\langle\frac{\boldsymbol{w}^*}{||\boldsymbol{w}^*||}, \mathbf{x}_i\right\rangle\right)$ the positive margin of this classifier. Let us set the initial values of the weight vector to 0 (i.e. $\boldsymbol{w}^{(0)} = 0$)*

(continued)

Theorem 4.1 (continued)
and fix the learning step $\eta = 1$. The number k of updates to the weight vector to reach $\mathbf{w}^*$ from $\mathbf{w}^{(0)}$ is then bounded by:

$$k \leqslant \left\lceil \left(\frac{R}{\rho}\right)^2 \right\rceil,$$

where $\lceil z \rceil$ designates the smallest integer that is greater than or equal to k.

Proof Since the hyperplane with normal vector $\boldsymbol{w}^*$ perfectly separates the examples of the two classes, i.e. $\forall i \in \{1, \ldots, m\}$, $y_i \langle \boldsymbol{w}^*, \mathbf{x}_i \rangle > 0$, it follows that $\rho > 0$. Furthermore, after k updates with the Perceptron learning rule (4.5), and with $w_0 = 0$ and $\eta = 1$, we have:

$$||\boldsymbol{w}^{(k)}||^2 = ||\boldsymbol{w}^{(k-1)} + y^{k)} \times \mathbf{x}^{(k)}||^2,$$

where $(\mathbf{x}^{(k)}, y^{(k)})$ is the kth misclassified example of the sequence after $k - 1$ weight vector updates (i.e. $y^{(k)} \times \langle \boldsymbol{w}^{(k-1)}, \mathbf{x}^{(k)} \rangle \leqslant 0$). With the condition $\forall i \in \{1, \ldots, m\}$, $||\mathbf{x}_i|| \leqslant R$ we get:

$$\begin{aligned} ||\boldsymbol{w}^{(k)}||^2 &= ||\boldsymbol{w}^{k-1)}||^2 + 2y^{(k)} \times \langle \boldsymbol{w}^{(k-1)}, \mathbf{x}^{(k)} \rangle + ||\mathbf{x}^{(k)}||^2 \\ &\leqslant ||\boldsymbol{w}^{(k-1)}||^2 + R^2. \end{aligned}$$

With the same reasoning, we also have $||\boldsymbol{w}^{(k-1)}||^2 \leqslant ||\boldsymbol{w}^{(k-2)}||^2 + R^2$, ... and $||\boldsymbol{w}^{(1)}||^2 \leqslant R^2$, hence:

$$||\boldsymbol{w}^{(k)}||^2 \leqslant k \times R^2. \tag{4.6}$$

With the Perceptron learning rule, the additivity of scalar product and previous initializations, we also have:

$$\left\langle \frac{\boldsymbol{w}^*}{||\boldsymbol{w}^*||}, \boldsymbol{w}^{(k)} \right\rangle = \left\langle \frac{\boldsymbol{w}^*}{||\boldsymbol{w}^*||}, \boldsymbol{w}^{(k-1)} \right\rangle + \left\langle \frac{\boldsymbol{w}^*}{||\boldsymbol{w}^*||}, y^{(k)} \times \mathbf{x}^{(k)} \right\rangle,$$

or, according to the definition of ρ:

$$\left\langle \frac{\boldsymbol{w}^*}{||\boldsymbol{w}^*||}, \boldsymbol{w}^{(k)} \right\rangle \geqslant \left\langle \frac{\boldsymbol{w}^*}{||\boldsymbol{w}^*||}, \boldsymbol{w}^{(k-1)} \right\rangle + \rho.$$

And since $\left\langle \frac{\boldsymbol{w}^*}{||\boldsymbol{w}^*||}, \boldsymbol{w}^{(k-1)} \right\rangle \geqslant \left\langle \frac{\boldsymbol{w}^*}{||\boldsymbol{w}^*||}, \boldsymbol{w}^{(k-2)} \right\rangle + \rho$, ... and this up to $\left\langle \frac{\boldsymbol{w}^*}{||\boldsymbol{w}^*||}, \boldsymbol{w}^{(1)} \right\rangle \geqslant \rho$ we can therefore deduce:

$$\underbrace{\left\| \frac{\boldsymbol{w}^*}{||\boldsymbol{w}^*||} \right\|}_{=1} \times \left\| \boldsymbol{w}^{(k)} \right\| \geqslant \left\langle \frac{\boldsymbol{w}^*}{||\boldsymbol{w}^*||}, \boldsymbol{w}^{(k)} \right\rangle \geqslant k \times \rho. \tag{4.7}$$

From the inequalities (4.6) and (4.7), we then have:

$$k^2 \times \rho^2 \leqslant ||\boldsymbol{w}^{(k)}||^2 \leqslant k \times R^2.$$

Hence the number of updates is bounded by:

$$k \leqslant \left\lceil \left(\frac{R}{\rho} \right)^2 \right\rceil.$$

□

We note that, in the case where the problem is not linearly separable, there is no guarantee that the algorithm will converge, but that, in the opposite case, the maximum number of iterations does not depend on the dimension of the problem. In addition, the margin plays a central role in machine learning; indeed, if we consider two linearly separable training sets enclosed in two hyperspheres of the same radius, the number of iterations to find the separating hyperplane will be lower for the set with the larger margin. This idea has given rise to a simple extension of Algorithm 9 where the classifier weights are modified for poorly classified examples and also for those whose margins relative to the current separating hyperplane are below a fixed threshold value. We will see in Chap. 5, that margin maximization is also at the heart of support vector machines (SVMs). We note that, in practice, the existence of a training set is not necessary and that, for on-line learning of the Perceptron, examples can be processed sequentially or on-the-fly.

4.1.2.2 Margin Based Perceptron and Link with the ERM Principle

For a fixed threshold value $\rho > 0$ and a training set S, with this variant the aim is to find the parameters $\boldsymbol{w}^*$ of a linear model $h_{\boldsymbol{w}^*} : \mathbf{x} \mapsto \left\langle \mathbf{x}, \bar{\boldsymbol{w}}^* \right\rangle + w_0^*$, for which the distances of the misclassified examples of the two classes from the two decision frontiers of equations $h_{\boldsymbol{w}^*}(\mathbf{x}) - y\rho = 0, y \in \{-1, +1\}$ would be minimum [59]. We shall see in Chap. 5 that this objective, illustrated in Fig. 4.5, is the basis for the development of support vector machines.

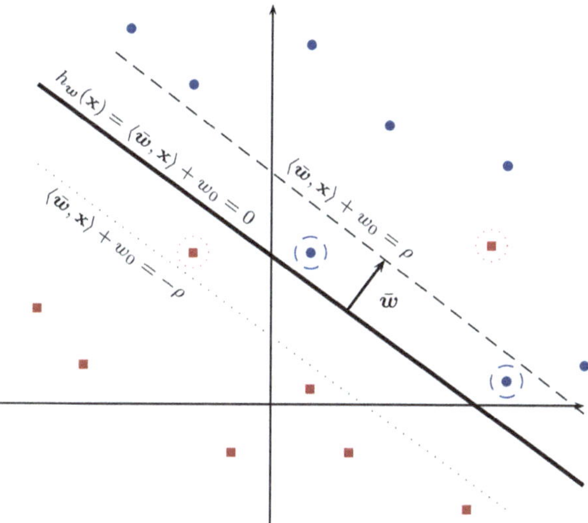

Fig. 4.5 Illustration of the objective of the margin Perceptron. The normal vector of the decision boundary points to the half-space containing positive examples (solid circles). Examples involved in modifying model weights are circled

The sum of the average distances of the misclassified examples in the training set by these two boundaries is:

$$\mathcal{L}'(\boldsymbol{w}) = \sum_{i' \in \mathcal{I}'} -y_{i'}(\langle \bar{\boldsymbol{w}}, \mathbf{x}_{i'} \rangle + w_0 - y_{i'}\rho) = \sum_{i' \in \mathcal{I}'} \rho - y_{i'}(\langle \bar{\boldsymbol{w}}, \mathbf{x}_{i'} \rangle + w_0), \tag{4.8}$$

where $\mathcal{I}'$ is the set of indices of examples misclassified by these two decision frontiers. In the case where $\rho = 1$, this minimization is equivalent to minimizing the empirical Hinge loss on the training set..

$$\mathcal{L}(\boldsymbol{w}) = \frac{1}{m} \sum_{i=1}^{m} \lfloor 1 - y_i(\langle \bar{\boldsymbol{w}}, \mathbf{x}_i \rangle + w_0) \rfloor_+, \tag{4.9}$$

where $\lfloor z \rfloor_+ = z$ if $z > 0$ and 0 otherwise. The Hinge loss provides an upper bound for the 0/1 loss (see Fig. 2.1, Chap. 2). In on-line mode, the update rule of weights is:

$$\forall (\mathbf{x}, y), \text{ if } y(\langle \bar{\boldsymbol{w}}, \mathbf{x} \rangle + w_0) \leqslant \rho \text{ then } \begin{pmatrix} w_0 \\ \bar{\boldsymbol{w}} \end{pmatrix} \leftarrow \begin{pmatrix} w_0 \\ \bar{\boldsymbol{w}} \end{pmatrix} + \eta \begin{pmatrix} y \\ y\mathbf{x} \end{pmatrix}. \tag{4.10}$$

4.1.3 Adaline

In parallel with Rosenblatt's work, Widrow and Hoff proposed the ADAptive LInear NEuron (Adaline) model [183], whose formal representation is analogous to that of the Perceptron.

4.1.4 Link with the ERM Principle

This algorithm finds the parameters $\boldsymbol{w} = (\bar{\boldsymbol{w}}, w_0)$ of a linear model $h_{\boldsymbol{w}}$ by minimizing the mean square error between the class labels of the examples of a training set S and the class predictions of the model for these examples:

$$\mathcal{L}(\boldsymbol{w}) = \frac{1}{m}\sum_{i=1}^{m}(y_i - h_{\boldsymbol{w}}(\mathbf{x}_i))^2. \tag{4.11}$$

We note that the quadratic loss $\ell^{\text{cvx}}(h_{\boldsymbol{w}}(\mathbf{x}), y) = (y - h_{\boldsymbol{w}}(\mathbf{x}))^2$ is a continuous convex bound of the classification error $\ell(h_{\boldsymbol{w}}, \mathbf{x}, y) = \mathbb{1}_{y \times h_{\boldsymbol{w}}(\mathbf{x}) \leqslant 0}$. Indeed, by developing ℓ^{cvx} and taking into account the fact that $y \in \{-1, +1\}$, we get:

$$\begin{aligned}\ell^{\text{cvx}}(h_{\boldsymbol{w}}(\mathbf{x}), y) &= (y \times h_{\boldsymbol{w}}(\mathbf{x}))^2 - 2(y \times h_{\boldsymbol{w}}(\mathbf{x})) + 1 \\ &= (1 - y h_{\boldsymbol{w}}(\mathbf{x}))^2 \geqslant \mathbb{1}_{y \times h_{\boldsymbol{w}}(\mathbf{x}) \leqslant 0}.\end{aligned}$$

The objective function (4.11) is thus a convex bound on the empirical classification error, and its minimization could be achieved with an unconstrained convex optimization technique presented in Chap. 3.

We also note that minimizing the quadratic loss (4.11) reduces the previous classification problem to linear regression with desired outputs in $\{-1, +1\}$, in the least squares sense. Indeed, posing $\boldsymbol{y} = (y_i)_{i=1}^m$ the vector of classes and $\mathbf{X} = (\tilde{\mathbf{x}}_1, \tilde{\mathbf{x}}_2, \ldots, \tilde{\mathbf{x}}_m)$ the matrix of training data, where $\tilde{\mathbf{x}} = (1, \mathbf{x})^\top$ is the vector composed of the 1 value bias and the $\mathbf{x}$ vector, the mean of the squared deviations between predicted and observed values with respect to the $\boldsymbol{w}$ parameters (4.11) is written:

$$\mathcal{L}(\boldsymbol{w}) = \frac{1}{m}(\boldsymbol{y} - \mathbf{X}^\top \boldsymbol{w})^\top (\boldsymbol{y} - \mathbf{X}^\top \boldsymbol{w}),$$

and cancelling its gradient gives the exact least-squares solution of the weights:

$$\boldsymbol{w}^* = (\mathbf{X}\mathbf{X}^\top)^{-1}\mathbf{X}\boldsymbol{y}.$$

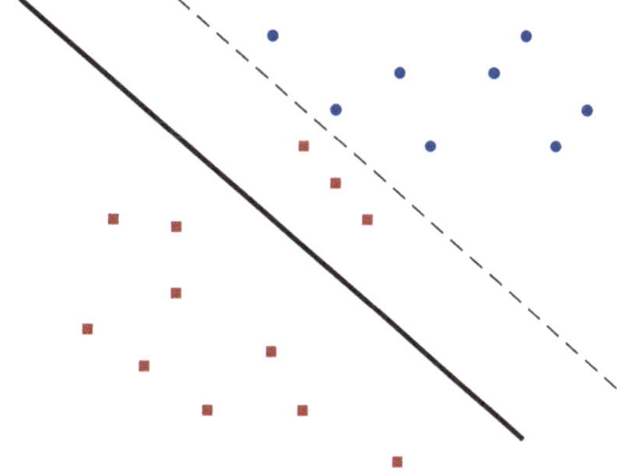

Fig. 4.6 Illustration of the solutions found by the Perceptron (dotted line) and Adaline (solid line) algorithms for a linearly separable classification problem

Nevertheless, solving this equation exactly poses numerical problems when the dimension of the input space d becomes too large, which is generally the case with real applications.

4.1.5 Differences with Perceptron

As in the case of the Perceptron, Widrow and Hoff proposed an algorithm based on the stochastic gradient technique to learn the model parameters [183]. Thus, for a learning rate $\eta > 0$ and a randomly drawn example $(\mathbf{x}, y)$, the update rule can be written:

$$\forall(\mathbf{x}, y), \begin{pmatrix} w_0 \\ \bar{\boldsymbol{w}} \end{pmatrix} \leftarrow \begin{pmatrix} w_0 \\ \bar{\boldsymbol{w}} \end{pmatrix} + \eta(y - h_{\boldsymbol{w}}(\mathbf{x})) \begin{pmatrix} 1 \\ \mathbf{x} \end{pmatrix}. \tag{4.12}$$

The major difference with the Perceptron learning rule (4.5) is that weights are updated on all training examples, misclassified or not. Figure 4.6 illustrates the difference in behavior between the Perceptron and the Adaline on a toy example. By minimizing the number of errors, the Perceptron finds the hyperplane that best separates the classes, at the risk of misclassifying noisy data. Adaline makes errors, but the solution found is generally more robust.

4.2 Logistic Regression

Logistic regression was introduced by statisticians in the late 1960s [169] and was later popularized by Anderson [6]. It is an attempt to overcome the restrictive assumptions often associated with probabilistic methods for classification. Before presenting the specifics of logistic regression, it is beneficial to understand the broader context of generative models for classification.

4.2.1 Generative Models for Classification

Generative models provide a framework for understanding how data is generated, which can be useful for classification tasks. These models for multi-class classification $k \in \{1, \ldots, K\}$ assume that each example $\mathbf{x}$ is generated by a mixture model with parameters Θ. The probability of an example $\mathbf{x}$ given the parameters Θ is expressed as:

$$P(\mathbf{x} \mid \Theta) = \sum_{k=1}^{K} P(y = k) P(\mathbf{x} \mid y = k, \Theta). \tag{4.13}$$

The goal is to find the parameters Θ that best explain the observations. This is achieved by maximizing the complete log-likelihood of the data $S = \{(\mathbf{x}_i, y_i); i \in \{1, \ldots, m\}\}$:

$$L(S, \Theta) = \ln \prod_{i=1}^{m} P(\mathbf{x}_i, y_i \mid \Theta). \tag{4.14}$$

Classical density functions used in these models are Gaussian density functions:

$$P(\mathbf{x} \mid y = k, \Theta) = \frac{1}{(2\pi)^{\frac{d}{2}} |\Sigma_k|^{\frac{1}{2}}} e^{-\frac{1}{2}(\mathbf{x}-\mu_k)^\top \Sigma_k^{-1} (\mathbf{x}-\mu_k)},$$

where $\Theta = \{(\mu_k, \Sigma_k); k \in \{1, \ldots, K\}\}$. Once the parameters Θ are estimated, the generative model can be used for classification by applying Bayes' rule:

$$\forall \mathbf{x}, y^* = \operatorname*{argmax}_k P(y = k \mid \mathbf{x}) \propto \operatorname*{argmax}_k P(y = k) \times P(\mathbf{x} \mid y = k, \Theta).$$

However, in most real-life applications, the distributional assumption over data does not hold, which can be a significant problem.

4.2.2 Presentation of the Model

This limitation motivates the use of discriminative models like logistic regression, which do not rely on such assumptions except that the log-odds of the probability of the classes are linear in the input features:

$$\ln \frac{P(y = 1 \mid \mathbf{x})}{P(y = 0 \mid \mathbf{x})} = \langle \bar{\boldsymbol{w}}, \mathbf{x} \rangle + w_0 = h_{\boldsymbol{w}}(\mathbf{x}). \tag{4.15}$$

The logistic regression model is used to model the posterior probability of classes via a sigmoid function defined as follows.

Definition 4.3 (Sigmoid Function) The sigmoid function σ is defined as:

$$\sigma : \mathbb{R} \to]0, 1[, \quad z \mapsto \frac{1}{1+e^{-z}}. \tag{4.16}$$

The derivative of the sigmoid function can be written in terms of the function itself.

$$\sigma'(z) = \frac{\partial \sigma}{\partial z} = \sigma(z)(1-\sigma(z)). \tag{4.17}$$

From (4.15) and Definition (4.3), the posterior probability can be expressed as:

$$P(y \mid \mathbf{x}) = (\sigma(h_{\boldsymbol{w}}(\mathbf{x})))^{y}(1-\sigma(h_{\boldsymbol{w}}(\mathbf{x})))^{1-y}.$$

The model parameters $\boldsymbol{w}$ are found by maximizing the complete log-likelihood, which, assuming that m training examples are generated independently, is written as:

$$L(\boldsymbol{w}, \Theta) = \ln \prod_{i=1}^{m} P(\mathbf{x}_i, y_i \mid \Theta, \boldsymbol{w}) = \ln \prod_{i=1}^{m} P(y_i \mid \mathbf{x}_i, \boldsymbol{w}) + \ln \prod_{i=1}^{m} P(\mathbf{x}_i \mid \Theta).$$

In the context of logistic regression, when maximizing the complete log-likelihood, the parameters Θ of the mixture model are not relevant. This is because our focus is not on modeling the data itself, but rather on estimating the posterior probabilities for classification. Therefore, the only important parameters are those of the logistic model, denoted as $\boldsymbol{w}$, which are involved in this estimation process.

$$L(\boldsymbol{w}, \Theta) = \sum_{i=1}^{m} \ln \left[(\sigma(h_{\boldsymbol{w}}(\mathbf{x}_i)))^{y_i}(1-\sigma(h_{\boldsymbol{w}}(\mathbf{x}_i)))^{1-y_i} \right]. \tag{4.18}$$

Logistic regression is particularly useful for binary classification problems, where the goal is to predict one of two possible outcomes. It can be extended to multi-class classification using techniques like one-vs-rest or softmax regression.

4.2.3 Formal Model

The formal model associated with the logistic model is identical to that representing the Perceptron or Adaline with the major difference being the transfer function, which in this case is the sigmoid function $F : z \mapsto \frac{1}{1+e^{-z}}$. The logistic model's predictions for examples far from the decision frontier are bounded between 0 and 1, whereas they are not for the Adaline , which is therefore more sensitive to misclassified examples (noise) far from the decision boundary. Indeed, from (4.12), we can see that the impact of an example $\mathbf{x}$ in updating weights is proportional to the model's output value for that example, and that this output value grows linearly as a function of the example's distance from the hyperplane. This difference is illustrated in Fig. 4.7.

4.2.4 Link with the ERM Principle

The maximization of the log-likelihood (4.18) is equivalent to the minimization of the empirical logistic loss in the case where $\forall i,\, y_i \in \{-1, +1\}$:

$$\mathcal{L}(\boldsymbol{w}) = \frac{1}{m}\sum_{i=1}^{m} \ln(1 + e^{-y_i h_{\boldsymbol{w}}(\mathbf{x}_i)}).$$

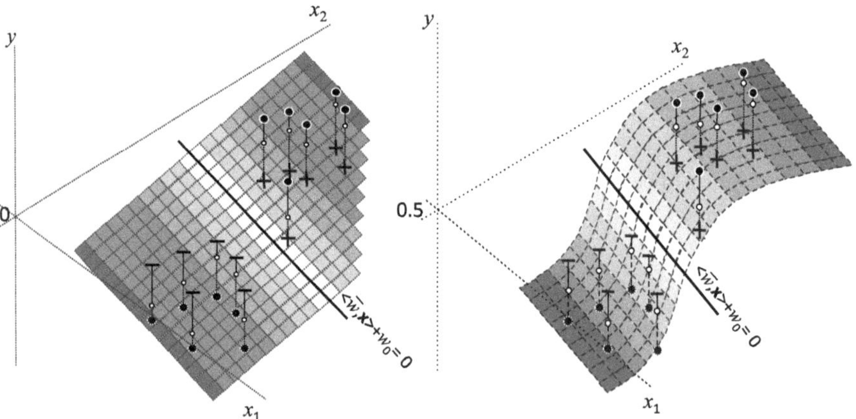

Fig. 4.7 The difference between adaline (left) and logistic regression (right). The output values of the two models are represented by surfaces, curved for logistic regression and flat for Adaline. The outputs predicted by the logistic model are bounded by 0 and 1, while for Adaline they grow, in absolute value, with respect to the distance of the points from the decision boundary

This connection to the ERM principle highlights that logistic regression can be viewed as a method for minimizing the risk of misclassification. The empirical logistic loss is a convex function, which makes it amenable to optimization using gradient-based methods.

The update rule using the stochastic gradient descent algorithm with a learning rate $\eta > 0$ is:

$$\forall (\mathbf{x}, y), \begin{pmatrix} w_0 \\ \bar{\boldsymbol{w}} \end{pmatrix} \leftarrow \begin{pmatrix} w_0 \\ \bar{\boldsymbol{w}} \end{pmatrix} + \eta y (1 - \sigma(h_{\boldsymbol{w}}(\mathbf{x}))) \begin{pmatrix} 1 \\ \mathbf{x} \end{pmatrix}.$$

4.3 Multi-Layer Perceptron

The multi-layer Perceptron (MLP) is a generalization of the Perceptron model. This model is inherently multi-class and represents the association between an observation (or input) and its real output with a multivariate function defined as a composition of several non-linear functions.

4.3.1 Formal Representation

By analogy with the formal models associated with the Perceptron (Sect. 4.1.2), the formal representation of a MLP is a network with successive layers of basic units, where each unit of a layer is linked to the units of the layer succeeding it. The two most widely studied types of network are recurrent networks, where loops exist between the different hidden layers and also between the units of these layers, and loop-free networks (or *feedforward*), which we consider in the following.

Figure 4.8 shows an example of such a network with n hidden layers, where there are d input cells, $(\ell_c)_{1 \leqslant c \leqslant n}$ units on the cth hidden layer, $c = \{1, \dots, n\}$, and K units on the output layer. For a feedforward network, each unit on a hidden layer, as well as those on the output, is linked to all the units on the layer preceding it, as well as to the unit representing the bias with a certain weight that would have to be learned. The network is said to be deep when the number of hidden layers exceeds two (i.e. $n \geqslant 2$). The network weights form a set of matrices $(\mathbf{W}^{(c)})_{1 \leqslant c \leqslant n}$:

$$\forall c \in \{1, \dots, n+1\}, \mathbf{W}^{(c)} = \begin{pmatrix} w_{1,1}^{(c)} & w_{1,2}^{(c)} & \cdots & w_{1,\ell_{c-1}}^{(c)} \\ w_{2,1}^{(c)} & w_{2,2}^{(c)} & \cdots & w_{2,\ell_{c-1}}^{(c)} \\ \vdots & \vdots & \ddots & \vdots \\ w_{\ell_c,1}^{(c)} & w_{\ell_c,2}^{(c)} & \cdots & w_{\ell_c,\ell_{c-1}}^{(c)} \end{pmatrix}.$$

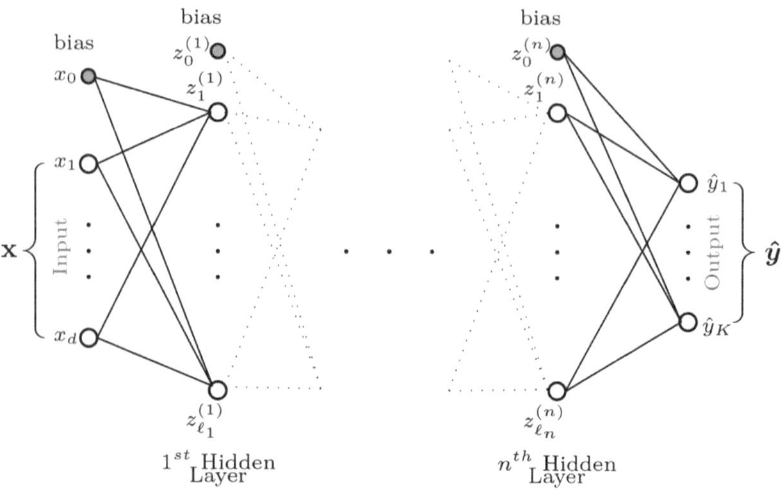

Fig. 4.8 Architecture of a multi-layer Perceptron with n hidden layers. In this example, the bias parameters are introduced by weights linked to two additional units associated with the input layer and each hidden layer with respectively fixed values $z_0^{(c)} = 1, \forall c = \{0, \dots, n\}$. A network with more than two hidden layers is called a deep neural network [66]

where for all $c \in \{1, \dots, n\}$, each row $\boldsymbol{w}_{j.}^{(c)} = (w_{ji}^{(c)})_{i=1,\dots,\ell_{c-1}}; j \in \{1, \dots, \ell_c\}$ in $\mathbf{W}^{(c)}$ represents the weights connecting the feature values of $\mathbf{z}^{(c-1)}$ from units in the previous layer, $c-1$, to the jth unit of layer c. By convention, the index of the input (respectively output) is $c = 0$ (respectively $c = n+1$), i.e. $\mathbf{z}^{(0)} = \mathbf{x}$ and $\mathbf{z}^{(n+1)} = \hat{\mathbf{y}}$.

For this network, the value of the jth unit of the hidden layer c is obtained by a composition:

- of a dot product $a_j^{(c)}$, between the vector representing the unit values of the $c-1$ layer; $\mathbf{z}^{(c-1)} = (z_1^{(c-1)}, \dots, z_{\ell_{c-1}}^{(c-1)})$ and the weight vectors $\boldsymbol{w}_{j.}^{(c)}$ and the bias parameters $w_{j0}^{(c)}$ (taking as bias value $z_0^{(c-1)} = 1$):

$$\forall c \in \{1, \dots, n+1\}, \forall j \in \{1, \dots, \ell_c\}, a_j^{(c)} = \langle \boldsymbol{w}_{j.}^{(c)}, \mathbf{z}^{(c-1)} \rangle + w_{j0}^{(c)} = \sum_{i=0}^{\ell_{c-1}} w_{ji}^{(c)} z_i^{(c-1)}; \tag{4.19}$$

- and an activation function, which is generally differentiable, $\bar{H}(.) : \mathbb{R} \to \mathbb{R}$:

$$\forall c \in \{1, \dots, n+1\}, \forall j \in \{1, \dots, \ell_c\}, z_j^{(c)} = \bar{H}(a_j^{(c)}). \tag{4.20}$$

The predicted output for an observation $\mathbf{x}$, $\hat{\mathbf{y}} = (\hat{y}_1, \dots, \hat{y}_K)^\top$, is a vector where each of its components is a composition of n functions, usually non-linear, of this input using the previous two steps: $\forall \mathbf{x}, \forall j, \hat{y}_j = \text{softmax}(a_j^{(n+1)})$. Where,

$$\forall j, a_j^{(n+1)} = \sum_{i=0}^{\ell_n} w_{ji}^{(n+1)} \bar{H}\left(\dots \bar{H}\left(\sum_{l=0}^{d} w_{kl}^{(1)} x_l\right)\right),$$

and softmax is expressed as follows.

Definition 4.4 (Softmax Function) For a given output unit $j \in \{1, \dots, K\}$ of a MLP with n hidden layers, the softmax function is defined as:

$$\text{softmax}(a_j^{(n+1)}) = \frac{e^{a_j^{(n+1)}}}{\sum_{k=1}^{K} e^{a_k^{(n+1)}}}.$$

4.3.2 Backpropagation Algorithm

Various learning algorithms have been proposed for learning the parameters of a MLP, the best known of which is undoubtedly the back propagation algorithm first proposed by Werbos [181] and later popularized by Parker [132] and Rumelhart, Hinton and Williams [145]. This algorithm is a succession of two phases, known as "information propagation" (or propagation) and "error backpropagation" (or backpropagation).

In the propagation phase, the algorithm estimates the predicted outputs of observations from a training set. In the backpropagation phase, the algorithm minimizes the discrepancy between the predicted and actual outputs of the examples by updating the network parameters in turn from the output layer to the input layer.

The gradient descent algorithm in *batch* mode (Chap. 3, Sect. 3.1) was proposed to minimize the error estimated as the average of the deviations between, $(\hat{\mathbf{y}}_i)_{i=1,\dots,m}$, the predicted outputs for examples $(\mathbf{x}_i)_{i=1,\dots,m}$ of a training set S, and their class indicator vectors $(\mathbf{y}_i)_{i=1,\dots,m}$.

$$\mathcal{L}(\mathbf{W}) = \frac{1}{m} \sum_{i=1}^{m} \ell^{\text{cvx}}(\mathbf{y}_i, \hat{\mathbf{y}}_i), \tag{4.21}$$

where $\ell^{\mathrm{cvx}} : \{0, 1\} \times [0, 1] \rightarrow \mathbb{R}_+$ is a convex loss function, usually the binary cross entropy loss (BCE):

$$\forall(\mathbf{x}, \mathbf{y}), \ell_{BCE}(\mathbf{y}, \hat{\mathbf{y}}) = -\sum_{j=1}^{K} y_j \log(\hat{\mathbf{y}}_j). \tag{4.22}$$

Other alternatives have been proposed for large-scale classification problems, including the stochastic gradient algorithm studied in detail in [20] and which we will outline in the following.

To simplify the presentation, we will first consider any model weight without reference to the hidden layer on which it depends. In this case, for a randomly drawn example $(\mathbf{x}, \mathbf{y})$, each weight w_{ji} of the network is updated by considering the partial derivative of the instantaneous error bound, $\ell_{BCE}(\mathbf{y}, \hat{\mathbf{y}})$, with respect to the current value of w_{ji} and a learning step η following the rule:

$$w_{ji} \leftarrow w_{ji} - \eta \frac{\partial \ell_{BCE}(\mathbf{y}, \hat{\mathbf{y}})}{\partial w_{ji}}, \tag{4.23}$$

which we will denote by $\Delta w_{ji} = -\eta \frac{\partial \ell_{BCE}(\mathbf{y}, \hat{\mathbf{y}})}{\partial w_{ji}}$. Since the instantaneous error depends on the weight w_{ji} via the scalar product a_j estimated for the jth unit in the propagation phase, the partial derivative of the error can be calculated by applying the derivation rule for compound functions, or the chain rule:

$$\frac{\partial \ell_{BCE}(\mathbf{y}, \hat{\mathbf{y}})}{\partial w_{ji}} = \underbrace{\frac{\partial \ell_{BCE}(\mathbf{y}, \hat{\mathbf{y}})}{\partial a_j}}_{=\delta_j} \frac{\partial a_j}{\partial w_{ji}}, \tag{4.24}$$

where $\frac{\partial a_j}{\partial w_{ji}} = z_i$ is the value of the ith neuron on the layer preceding the jth neuron, and w_{ji} is the weight between these two neurons.

For BCE and in the case where the j unit is on the output layer of the network, the partial derivative is written:

$$\delta_j = \frac{\partial \ell_{BCE}(\mathbf{y}, \hat{\mathbf{y}})}{\partial a_j} = \frac{\partial \ell_{BCE}(\mathbf{y}, \hat{\mathbf{y}})}{\partial \hat{y}_j} \frac{\partial \hat{y}_j}{\partial a_j}. \tag{4.25}$$

The derivative of the softmax function is:

$$\frac{\partial \hat{y}_j}{\partial a_j} = \hat{y}_j(1 - \hat{y}_j),$$

and

$$\delta_j = -\frac{y_j}{\hat{y}_j} . \hat{y}_j(1 - \hat{y}_j) = y_j(\hat{y}_j - 1). \tag{4.26}$$

If jth neuron is on a hidden layer $1 \leqslant c \leqslant n$, we have by applying the chain rule again:

$$\delta_j = \frac{\partial \ell_{BCE}(\mathbf{y}, \hat{\mathbf{y}})}{\partial a_j} = \sum_{l \in Af(j)} \frac{\partial \ell_{BCE}(\mathbf{y}, \hat{\mathbf{y}})}{\partial a_l} \frac{\partial a_l}{a_j} = \bar{H}'(a_j) \sum_{l \in Af(j)} \delta_l \times w_{lj}, \tag{4.27}$$

where $Af(j)$ is the set of units on the layer following that of the jth unit.

The last inequality in Eq. (4.27) is due to the fact that, $a_l = \sum_{i=0}^{\ell_{c-1}} w_{li} z_i$, and $z_j = \bar{H}(a_j)$, hence $\frac{\partial a_l}{a_j} = w_{lj} \bar{H}'(a_j)$. The two phases of propagation and back-propagation for the online mode are illustrated in Fig. 4.9.

Following this principle, the weights of the MLP in Fig. 4.8 are updated successively from the output layer to the input according to the following update rule:

$$c \in \{n, \ldots, 0\}, j \in \{1, \ldots, \ell_{c+1}\}, i \in \{1, \ldots, \ell_c\}, \Delta w_{ji}^{(c+1)} = -\eta \times \delta_j^{(c+1)} \times z_i^{(c)},$$

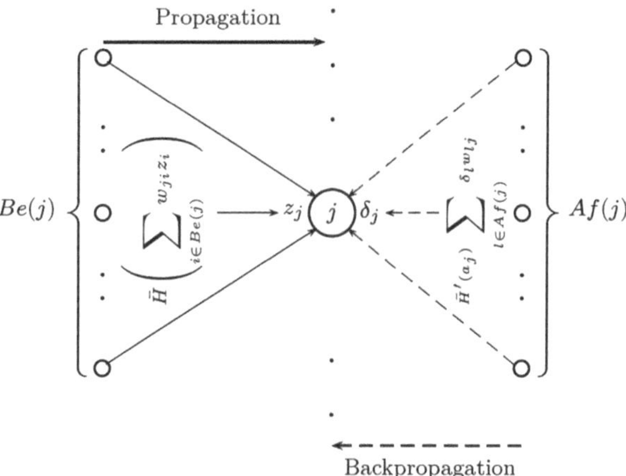

Fig. 4.9 Schematic illustration of the propagation (solid line) and backpropagation (dashed line) phases of the error backpropagation algorithm. For an input example $\mathbf{x}$, the j unit value of a hidden layer is determined on the basis of a transformation $\bar{H} : \mathbb{R} \to \mathbb{R}$ of the dot product between the vector of unit values on the $Be(j)$ layer preceding the j unit layer and the vector of weights linking the j unit to these units: $a_j = \sum_{i \in Be(j)} z_i w_{ji}$ (propagation phase). In the back-propagation phase, the errors of the units located on the layer succeeding the layer of unit j, $Af(j)$, are combined with the weights linking unit j to these units to determine the error δ_j associated with this unit: $\delta_j = \bar{H}'(a_j) \sum_{l \in Af(j)} \delta_l \times w_{lj}$

where the partial derivates $\delta_j^{(c)}; 1 \leqslant c \leqslant n+1, 1 \leqslant j \leqslant \ell_c$ are equal to:

$$\delta_j^{(c)} = \begin{cases} y_j(\hat{y}_j - 1) & \text{if } c = n+1 \\ \bar{H}'(a_j^{(c)}) \times \sum_{l=1}^{\ell_{c+1}} \delta_l^{(c+1)} \times w_{lj}^{(c)} & \text{if } n \geqslant c \geqslant 1 \end{cases}$$

The multi-layer Perceptron algorithm is presented in Algorithm 10.

It is essential to compute the values of $(\delta_j^{(c)})_{1 \leqslant j \leqslant \ell_c}$ for all $c \in \{1, \ldots, n\}$ using the current model weights before applying updates, ensuring accurate gradient calculations. Failure to do so may lead to incorrect weight adjustments and suboptimal performance.

4.3.3 Universal Approximation Theorem

One of the fundamental results in the theory of neural networks is the universal approximation theorem which states that a feedforward network with a single hidden layer containing a finite number of neurons can approximate any continuous function on compact subsets of $\mathbb{R}^n$, under mild assumptions on the activation function. This theorem highlights the expressive power of neural networks, showcasing their versatility in modeling diverse functions, essential for applications in pattern recognition, data analysis, and control systems. Proven by Cybenko [41] and Hornik et al. [85], the theorem leverages functional analysis and approximation theory, focusing on a subalgebra crucial for understanding continuous functions on compact spaces.

Definition 4.5 (Subalgebra) Let $\mathcal{A}$ be a subset of $\mathcal{C}(\mathcal{K})$, where $\mathcal{C}(\mathcal{K})$ is the space of real-valued continuous functions on a compact Hausdorff space $\mathcal{K}$. $\mathcal{A}$ is a subalgebra of $\mathcal{C}(\mathcal{K})$, if it is closed under addition, multiplication and scalar multiplication. That is if $f, g \in \mathcal{A}$ and $c \in \mathbb{R}$, then $f+g \in \mathcal{A}$, $fg \in \mathcal{A}$, and $cf \in \mathcal{A}$.

This idea is central to the Stone-Weierstrass Theorem, which provides a powerful tool for approximating continuous functions [159].

Input :

- A training set $S = ((\mathbf{x}_1, \mathbf{y}_1), \dots, (\mathbf{x}_m, \mathbf{y}_m))$; Precision $0 < \epsilon < 1$; Maximum number of epochs T; Activation function $\bar{H} : \mathbb{R} \rightarrow \mathbb{R}$; Learning rate $\eta > 0$;
- Number of hidden layers n; Number of neurons ℓ_c for each layer $c \in \{1, \dots, n\}$;
  // The input layer is denoted by $c = 0$ and the output layer by $c = n + 1$.

Initialization:

- Randomly initialize network weights;
- $t \leftarrow 0$;
- $old_e \leftarrow -1, new_e \leftarrow 0$;

while $t \leqslant T \wedge |new_e - old_e| > \epsilon$ **do**
    $old_e \leftarrow new_e$;
    $new_e \leftarrow 0$;
    for $l = 1 \dots m$ **do**
        Randomly select an example $(\mathbf{x}, \boldsymbol{y}) \in S$;
        // Propagation phase
        for $c = 1 \dots n$ **do**
            // $\forall i, z_i^{(0)} = x_i$ for input layer cells
            for $j = 1 \dots \ell_c$ **do**
                $a_j^{(c)} \leftarrow \sum_{i=0}^{\ell_{c-1}} w_{ji}^{(c)} z_i^{(c-1)}$;
                $z_j^{(c)} \leftarrow \bar{H}(a_j^{(c)})$; // $\triangleright$ Fig. 4.9 left part
        for $j = 1 \dots K$ **do**
            $a_j^{(n+1)} \leftarrow \sum_{i=0}^{\ell_n} w_{ji}^{(n+1)} z_i^{(n)}$;
            $\hat{y}_j \leftarrow \text{softmax}(a_j^{(n+1)})$;
        // Backpropagation phase
        $s \leftarrow 0$;
        for $k = 1 \dots K$ **do**
            $s \leftarrow s + y_k \ln(\hat{y}_k)$;
            $\delta_k^{(n+1)} \leftarrow y_k(\hat{y}_k - 1)$;
        $new_e \leftarrow new_e - s$;
        for $c = n \dots 1$ **do**
            for $i = 1 \dots \ell_c$ **do**
                $\delta_i^{(c)} = \bar{H}'(a_i^{(c)}) \times \sum_{j=1}^{\ell_{c+1}} \delta_l^{(c+1)} \times w_{ji}^{(c)}$; // $\triangleright$ Fig. 4.9 right part
        // Update of weights
        for $c = n + 1 \dots 1$ **do**
            for $i = 1 \dots \ell_c$ **do**
                for $j = 1 \dots \ell_{c+1}$ **do**
                    $\Delta w_{ji}^{(c)} \leftarrow -\eta \delta_j^{(c)} \times z_i^{(c-1)}$; // $\triangleright$ (4.23)
    $t \leftarrow t + 1$;

Output : The parameters of the MLP, $(\mathbf{W}^{(c)})_{1 \leqslant c \leqslant n}$.

Algorithme 10: Multi-layer perceptron

Theorem 4.2 (Approximation on Locally Compact Spaces [159]) *Consider a compact Hausdorff space $\mathcal{K}$, and let $\mathcal{C}(\mathcal{K})$ denote the space of real-valued continuous functions on $\mathcal{K}$. Assume $\mathcal{A}$ is a subalgebra of $\mathcal{C}(\mathcal{K})$ with the following properties:*

1. *For any two distinct points $\mathbf{x}, \mathbf{y} \in \mathcal{K}$, there exists a function $f \in \mathcal{A}$ such that $f(\mathbf{x}) \neq f(\mathbf{y})$.*
2. *$\mathcal{A}$ contains all constant functions; that is, for any constant $c \in \mathbb{R}$, there exists $f \in \mathcal{A}$ such that for all $\mathbf{x} \in \mathcal{K}$, $f(\mathbf{x}) = c$.*

Under these conditions, $\mathcal{A}$ is dense in $\mathcal{C}(\mathcal{K})$ with respect to the sup-norm (or uniform norm). This implies that for any continuous function, $f \in \mathcal{C}(\mathcal{K})$ and any $\epsilon > 0$, there exists a function $g \in \mathcal{A}$ such that:

$$\|f - g\|_{\infty,\mathcal{K}} = \sup_{\mathbf{x}\in\mathcal{K}} |f(\mathbf{x}) - g(\mathbf{x})| < \epsilon.$$

The universal approximation theorem can be stated as follows.

Theorem 4.3 (Universal Approximation Theorem) *Let $\bar{H} : \mathbb{R} \to \mathbb{R}$ be a continuous non-polynomial activation function. Then, for any continuous function $f : \mathbb{R}^n \to \mathbb{R}$ and any compact subset $\mathcal{K} \subset \mathbb{R}^n$, and any $\epsilon > 0$, there exists a feedforward neural network with a single hidden layer, such that the network function $F : \mathcal{K} \to \mathbb{R}$ satisfies:*

$$\sup_{\mathbf{x}\in\mathcal{K}} |F(\mathbf{x}) - f(\mathbf{x})| < \epsilon.$$

Proof Consider a multi-layer Perceptron with one hidden layer containing N neurones and the set of corresponding functions:

$$\mathcal{A} = \{F : \mathcal{K} \to \mathbb{R} | \forall \mathbf{x} \in \mathcal{K}, F(\mathbf{x}) = \sum_{i=1}^{N} \alpha_i \bar{H}(\langle \boldsymbol{w}_i, \mathbf{x}\rangle + b_i)\},$$

where $\forall i \in \{1, \ldots, N\}$, α_i represent the weights connecting the hidden layer to the output; $\boldsymbol{w}_i$ are the weight vectors, and b_i are the bias term for each neuron i in the hidden layer.

The set $\mathcal{A}$, is closed under addition and scalar multiplication, forming a linear space. From Definition (4.5), $\mathcal{A}$ then forms a subalgebra in $\mathcal{C}(\mathcal{K})$.

Now consider two distinct points $\mathbf{x}_1, \mathbf{x}_2 \in \mathcal{K}$.

- For all $i \in \{1, \ldots, N\}$, we can choose α_i, $\boldsymbol{w}_i$ and b_i such that

$$\alpha_i \bar{H}(\langle \boldsymbol{w}_i, \mathbf{x}_1 \rangle + b_i) \neq \alpha_i \bar{H}(\langle \boldsymbol{w}_i, \mathbf{x}_2 \rangle + b_i),$$

  and since the activation function is non-constant we have $F(\mathbf{x}_1) \neq F(\mathbf{x}_2)$.
- Furthermore, the set $\mathcal{A}$ also includes constant functions by setting $\boldsymbol{w}_i = \mathbf{0}$ and adjusting b_i for all $i \in \{1, \ldots, N\}$.

By the Stone-Weierstrass Theorem, for any continuous function $f \in \mathcal{C}(\mathcal{K})$ and any $\epsilon > 0$, there exists a function F induced by the neural network such that $\sup_{\mathbf{x} \in \mathcal{K}} |F(\mathbf{x}) - f(\mathbf{x})| < \epsilon$. □

4.4 Convolutional and Recurrent Networks

In the following, we will present the architectures of convolutional and recurrent networks.

4.4.1 Convolutional Neural Networks

A convolutional neural network (CNN) is a variant of MLP whose architecture was inspired by the organization of the visual cortex. In the 1960s, Hubel and Wiesel [86] showed that individual neurons respond to stimuli only in a restricted region of the visual field known as the receptive field, with several of these fields overlapping to cover the entire visual area. They had identified two basic types of visual cell in the brain: simple cells, whose output is maximized by characteristic spikes with particular orientations in their receptive field, and complex cells, which have larger receptive fields and whose output is insensitive to the exact position of patterns in the latter.

Inspired by this work, Fukushima [62] proposed a model called neocognitron. This model comprises two types of base layers: convolution layers and subsampling layers. A convolution layer contains units whose receptive fields cover part of the previous layer, and the weight vector is often called a filter or kernel. Subsampling layers contain units whose receiving fields cover parts of previous convolutional layers. Such a unit generally averages the activations of the units in its set. This sub-sampling enables objects to be correctly classified in visual scenes, even when they are moved. On this basis, Waibel [180] introduced the first convolutional network, the so-called time-delay neural network. The weights of this model were shared and learned with the error backpropagation algorithm (see Sect. 4.3.2). Like neocognition, this model also had a pyramidal structure, but the updating of weights was global, rather than local as in neocognition. LeCun [103] also used the error backpropagation algorithm to learn the coefficients of the convolution kernel in conjunction with the weights of the convolutional network directly from images of handwritten numbers. This made learning automatic, outperformed manual design of these coefficients, and was suitable for a wider range of image recognition problems. LeCun [103] showed that his convolutional network was able to successfully capture spatial and temporal dependencies in an image through the application of the learned kernels. The network architecture was better adapted to the image dataset due to the reduced number of parameters involved and the reusability of the weights.

The three types of layers in a convolutional network are convolution, pooling and fully connected layers.

Convolution Layer In the convolution layer, an input (usually a tensor) is modified with a kernel or filter that is learned. A convolution layer thus has three attributes:

- convolution kernels defined by width and height;
- the number of input channels and output channels (hyper-parameter);
- the depth of the convolution filter (input channels), which must be equal to the number of channels (depth) of the input map.

The example below shows the convolution of an image $5 \times 5 \times 1$ (of depth 1) with a kernel $2 \times 2 \times 1$ (of depth 1) to obtain convolved features $4 \times 4 \times 1$:

$$\underbrace{\begin{bmatrix} 1\,1\,0\,1\,1 \\ 1\,0\,1\,1\,0 \\ 0\,0\,1\,1\,0 \\ 1\,1\,0\,1\,1 \\ 0\,0\,0\,1\,1 \end{bmatrix}}_{\text{Image}} + \underbrace{\begin{bmatrix} 0\,1 \\ 1\,1 \end{bmatrix}}_{\text{Kernel}} = \underbrace{\begin{bmatrix} 2\,1\,3\,2 \\ 0\,2\,3\,1 \\ 2\,2\,2\,2 \\ 1\,0\,2\,3 \end{bmatrix}}_{\text{Convoluted image}}$$

In this example, the kernel shifts 9 times from top left to bottom right, each time performing a matrix multiplication operation between the kernel and the part of the image to which it is applied. In the case of multi-channel images (e.g. RGB), the kernel has the same depth as the input image. Matrix multiplication is performed between the stacks of kernels and images, and all the results are added together with the bias to give a single-channel convolved image as output.

The aim of the convolution operation is to extract high-level features. Conventionally, the first convolution layer is responsible for capturing low-level features such as edges, color, gradient orientation and so on. With additional layers, the architecture also accommodates high-level features. The convolution result is of two types: either the convolved feature is reduced in dimensionality relative to the input, which is the general case, or the dimensionality is increased, or remains the same.

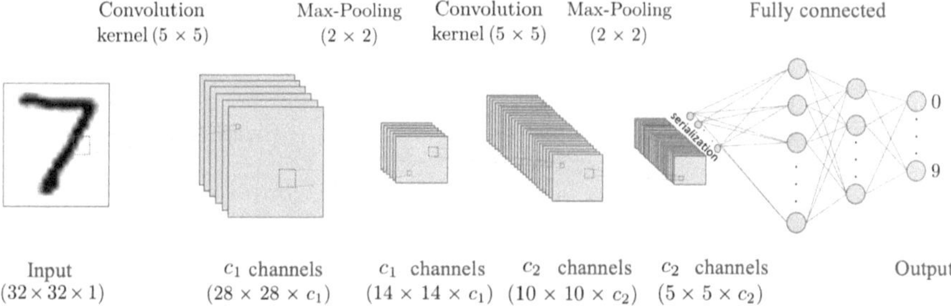

Fig. 4.10 An example of a convolutional network for handwritten digit recognition with two convolution layers and two grouping layers of c_1 and c_2 obtained with a kernel 5 and a max pooling 2

Pooling Layer The pooling layer is responsible for reducing the spatial dimensions of the convolved feature map to decrease the computational power required for data processing through dimensionality reduction. It is also used to extract dominant features that are invariant to rotation and position. There are two common types: *maximum pooling* (more commonly referred to as max pooling) and *average pooling*. Max pooling returns the maximum value from the portion of the image covered by the kernel, while average pooling returns the average of all values in that portion. In the example below, the result of max pooling is:

$$\underbrace{\begin{bmatrix} 2 & 1 & 3 & 2 \\ 0 & 2 & 3 & 1 \\ 2 & 2 & 2 & 2 \\ 1 & 0 & 2 & 3 \end{bmatrix}}_{\text{Convoluted image}} \rightarrow \begin{bmatrix} 2 & 3 \\ 2 & 3 \end{bmatrix}$$

Fully Connected Layer Fully-connected layers, such as those presented in Sect. 4.3 and generally at the end of the network, allow non-linear combinations of the high-level features represented by the output of the convolutional layer to be learned. These are generally few in number, typically one to three. There must be a transition between the last grouping layer and the first fully connected layer. This is achieved simply by serializing the representations produced by this last grouping layer.

Figure 4.10 illustrates an example of a convolutional network for handwritten digit recognition, inspired by the architecture of the LeNet[5] network [103].

[5] http://yann.lecun.com/exdb/lenet/.

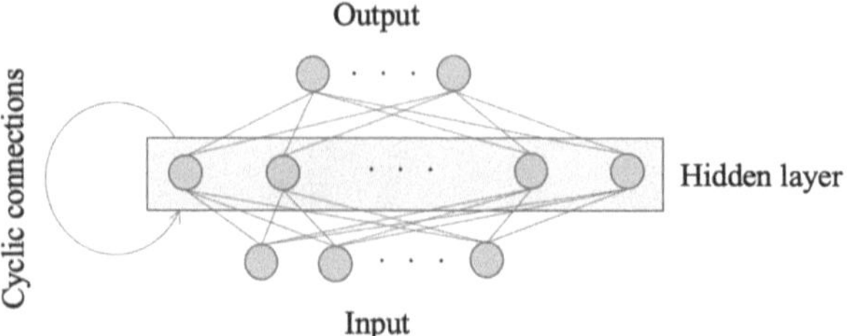

Fig. 4.11 Illustration of a recurrent network

4.4.2 Recurrent Networks

Recurrent networks are neural networks with cyclic connections between units. They have proved very effective for learning prediction tasks on sequential data. Figure 4.11 shows an example. In these networks, the propagation phase is the same as that of a MLP with a single hidden layer, except that inputs arrive at the hidden layer from both the network input and the hidden layer outputs at the previous iteration. Consider $\mathbf{x}$ a sequence of T vectors of size d at the input of a recurrent network with d input units, H hidden units and K output units. Let x_i^t be the ith characteristic of the tth vector of the sequence, the value of the hth unit of the hidden layer is:

$$a_j^t = \sum_{i=1}^{d} w_{ji} x_i^t + \sum_{h=1}^{H} w_{jh} z_h^{t-1}. \tag{4.28}$$

The value of the hth unit of the hidden layer is obtained after a non-linear transformation:

$$z_j^t = \bar{H}(a_j^t). \tag{4.29}$$

The complete sequence of hidden layer values can be calculated by starting at $t = 1$ and recursively applying the Eqs. (4.28) and (4.29), incrementing t at each step. The initial values $(z_h^0)_{1 \leqslant h \leqslant H}$ are typically set to 0.

Weights are updated using the error backpropagation method (Sect. 4.3.2). The two well-known techniques that have been proposed to efficiently estimate partial derivatives of weights for recurrent networks are: real-time recurrent learning (RTRL) proposed by [142] and backpropagation through time (BPTT) proposed by [184]. The latter proved to be both conceptually simpler and more computationally (but not memory) efficient. Like standard backpropagation, BPTT consists of a repeated application of the chain rule. The subtlety is that, for recurrent networks, the objective function depends on the activation of the hidden layer not only through its influence on the output layer, but also through its influence on the hidden layer at the next time step, i.e.:

$$\Delta w_{ji} = -\eta \times \sum_{t=1}^{T} \delta_j^t \times z_i^t, \tag{4.30}$$

where δ_j^t is the partial derivative of the error with respect to the scalar product a_j^t at time t.

$$\delta_j^t = \bar{H}'(a_j^t)\left(\sum_{k=1}^{K} \delta_k^t w_{kj} + \sum_{h=1}^{H} \delta_h^{t+1} w_{hj}\right). \tag{4.31}$$

The complete sequence of terms $(\delta^t)_t$ can be calculated by starting at $t = T$ and recursively applying (4.31), decrementing t at each step. We set $\delta^{T+1} = 0$ since no error is received beyond the end of the sequence. It has been observed that, in some cases, the gradient becomes extremely weak, effectively preventing the weights from changing their values. In the worst case, this can even prevent the neural network from learning altogether. This issue is known as the vanishing gradient problem.

4.5 Practical Considerations

In practice, various tricks have been found to increase the performance of these models or reduce their training time. Here we present some of the most popular tricks.

4.5.1 Activation Functions

The activation function $\bar{H} : \mathbb{R} \to \mathbb{R}$ is generally used to introduce non-linearity at the output of a neural unit. In practice, some nonlinear functions have been shown to be more efficient in terms of learning and convergence speed than other classical examples of activation functions:

- The piecewise linear function.:

$$\bar{H} : x \mapsto \begin{cases} 1 & \text{if } x \geqslant 0.5, \\ x + 0.5 & \text{if } -0.5 < x < 0.5, \\ 0 & \text{otherwise.} \end{cases} \text{ and}$$

$$\bar{H}'(x) = \begin{cases} 1 & \text{if } -0.5 < x < 0.5, \\ 0 & \text{otherwise.} \end{cases}$$

- The hyperbolic tangent function:

$$\bar{H} : x \mapsto \frac{1 - e^{-2x}}{1 + e^{-2x}} \text{ and } \bar{H}'(x) = 1 - \bar{H}^2(x).$$

- The sigmoid function:

$$\bar{H} : x \mapsto \frac{1}{1 + e^{-x}} \text{ and } \bar{H}'(x) = \bar{H}(x)(1 - \bar{H}(x)).$$

  In this case, Richard et al. [140] have shown that learning a MLP amounts to finding the estimator of the a posteriori probabilities of the classes of the examples. With this equivalence, a MLP without a hidden layer gives the same result as the multi-class logistic regression [74, p. 95] whose parameters are learned by maximizing the logarithm of the complete likelihood.
- The rectified linear unit function (ReLU):

$$\bar{H} : x \mapsto x\mathbb{1}_{x\geqslant 0} \text{ and } \bar{H}'(x) = \mathbb{1}_{x\geqslant 0}.$$

  This function proved very efficient, with convergence times far better than those obtained with the sigmoid function, and enabled the training of deep networks. Moreover, it is these functions, in particular, that have enabled the training of very deep and ultra-deep networks, by greatly alleviating the problem of vanishing gradients. A variant of this function is the PReLU defined by:

$$\bar{H} : x \mapsto \max(x, \beta x) \text{ and } \bar{H}'(x) = \begin{cases} \beta & \text{if } \beta x > x, \\ 1 & \text{otherwise.} \end{cases}$$

  where β is a hyperparameter that is either included in the model parameters and learned, or set to fairly small positive values ($\beta \in [0.05, 0.1]$).

4.5.2 Momentum Method

As presented in Chap. 3, Sect. 3.5, the momentum technique involves assigning inertia to the model parameters by applying an update to their *speed* [124, 137]. Hence, instead of updating the weights of the model at iteration t, $\mathbf{W}^{(t)}$, by subtracting them from a multiple of the gradient of the objective function $\mathcal{L}(\mathbf{W}^{(t)})$, this amounts to having the following update:

$$\boldsymbol{u}^{(t)} = \mathbf{W}^{(t)} + \beta_t \left(\mathbf{W}^{(t)} - \mathbf{W}^{(t-1)}\right), \tag{4.32}$$

$$\mathbf{W}^{(t+1)} = \boldsymbol{u}^{(t)} - \eta_t \nabla\mathcal{L}(\boldsymbol{u}^{(t)}), \tag{4.33}$$

where η_t is the learning rate and β_t is the momentum parameter that determines the strength of inertia (the lower the coefficient, the less important inertia). As it has been seen, the choice of hyperparameters β_t and η_t has a direct impact on the speed of convergence for cost function minimization. Sutskever et al. [161] found that weight initialization and the moment method are crucial for efficient deep network learning. They showed that poorly initialized networks cannot be trained with the momentum method, and that well-initialized networks perform significantly worse when the momentum is absent or its parameters are incorrectly set.

4.5.3 *Processing by Mini-Batches*

Another technique for attenuating the noise generated by weight updating using the stochastic gradient descent rule is the so-called *mini-batches* treatment. This technique has also proved to be much more efficient than the gradient descent algorithm, which considers the entire training database for weight updating. According to this principle, the weights are updated by minimizing the $\mathcal{L}_b$ error defined on a subset of the training database, $S_b = (\mathbf{x}_i^b, \mathbf{y}_i^b)_{1\leqslant i\leqslant m_b}$ of size m_b:

$$\mathcal{L}_b(\mathbf{W}) = \frac{1}{m_b}\sum_{i=1}^{m_b} \ell^{\text{cvx}}(\boldsymbol{h}_{\mathbf{W}}(\mathbf{x}_i^b), \mathbf{y}_i^b). \tag{4.34}$$

In practice, the size of the mini-batches $(m_b)_b$ is much smaller than that of the initial training set, and is chosen for values between 32 and 256. During an epoch, each sample must appear once and only once in one of the mini-batches. Experimentally, it has been found that it is better to randomly swap the samples in the mini-batches differently at each epoch. The use of mini-batches facilitates the distribution of computations on parallel architectures such as GPUs.

4.5.4 *Batch Normalization*

Training deep neural networks with numerous hidden layers is challenging, primarily due to their sensitivity to the random initialization of weights. A key reason highlighted in the literature is that the distribution of inputs to the deeper layers of the network can shift after updating the weights for each mini-batch. This phenomenon, known as internal covariate shift (or *internal covariate shift*), can disrupt the learning process. Batch Normalization [87] is a technique designed for training very deep networks, which normalizes the inputs of a layer for each mini-batch. This approach stabilizes the learning process and significantly reduces the number of epochs needed to train these models. Thus, for each hidden layer $c \in \{1, \dots, n\}$ of a network, if we consider $\mathbf{z}_i^{(c)} = (z_{i,1}^{(c)}, \dots, z_{i,\ell_c}^{(c)})$ as the vector of unit values for an input example $\mathbf{x}_i$, each of these values is individually normalized by:

$$\forall j \in \{1, \ldots, \ell_c\}, \tilde{z}_{i,j}^{(c)} = \frac{z_{i,j}^{(c)} - \mu_j^{(c)}}{\sqrt{\sigma_j^{(c)2} + \epsilon}},$$

where ϵ is an arbitrarily small positive value added to the denominator for reasons of numerical stability, and $\mu_j^{(c)}$ and $\sigma_j^{(c)}$ are the mean and variance of its values calculated for examples of the mini-batch S_b of current m_b size:

$$\mu_j^{(c)} = \frac{1}{m_b}\sum_{i=1}^{m_b} z_{i,j}^{(c)} \text{ and } \sigma_j^{(c)2} = \frac{1}{m_b}\sum_{i=1}^{m_b}\left(z_{i,j}^{(c)} - \mu_j^{(c)}\right)^2.$$

The resulting normalized activation has a mean of zero and a variance equal to 1, if ϵ is not taken into account. To restore the network's representational power, a transformation step then follows, as:

$$\forall i, \forall j \in \{1, \ldots, \ell_c\}, \hat{z}_{i,j}^{(c)} = \kappa_j^{(c)} \tilde{z}_{i,j}^{(c)} + \gamma_j^{(c)},$$

where the hyperparameters $(\kappa_j^{(c)})_{1 \leqslant j \leqslant \ell_c; 1 \leqslant c \leqslant n}$ and $(\gamma_j^{(c)})_{1 \leqslant j \leqslant \ell_c; 1 \leqslant c \leqslant n}$ are learned along with the model weights. This technique gives up to 10 times faster convergence, or simply ensures convergence that would not otherwise occur.

4.5.5 *Dropout*

Dropout is a regularization technique designed to mitigate overfitting in neural networks by preventing complex co-adaptations on training data [79, 157]. This method involves the temporary and random omission of units from the hidden layers and input layers. Specifically, it cancels the signal arriving at a connection during the forward pass and similarly nullifies the associated gradient component during the backward pass.

The dropout process can be mathematically described using a Bernoulli distribution. Let $\mathcal{B}(p)$ be the Bernoulli distribution of parameter p thus:

$$r \sim \mathcal{B}(p) \text{ means that } r = \begin{cases} 1 & \text{with probability } p, \\ 0 & \text{with probability } 1 - p. \end{cases}$$

Typically, the parameter p is chosen between 0.2 and 0.5. This parameter controls the probability of retaining a unit in the network; for example, $p = 0.5$ means each unit has a 50% chance of being temporarily removed during training.

During the forward propagation phase on the cth layer of the network, the dropout technique is applied as follows:

$$\forall i \in \{1, \ldots, \ell_{c-1}\}, r_i^{(c-1)} \sim \mathcal{B}(p),$$

$$\forall i \in \{1, \ldots, \ell_{c-1}\}, \tilde{z}_i^{(c-1)} = r_i^{(c-1)} z_i^{(c-1)},$$

$$\forall j \in \{1, \ldots, \ell_c\}, a_j^{(c)} = \langle \mathbf{w}_{j.}^{(c)}, \tilde{\mathbf{z}}^{(c-1)} \rangle + w_{j0}^{(c)},$$

$$\forall j \in \{1, \ldots, \ell_c\}, z_j^{(c)} = \bar{H}(a_j^{(c)}).$$

Here, $r_i^{(c-1)}$ is a Bernoulli random variable that determines whether the ith unit in layer $c-1$ is retained. $\tilde{z}_i^{(c-1)}$ represents the modified output of the unit after applying dropout.

During the test phase, no connections are dropped. Instead, each connection is scaled down by the factor p to maintain the average signal intensity. This scaling ensures that the network behaves similarly during testing as it did during training, preventing any single connection from becoming dominant. While dropout may slow down the convergence process due to the stochastic nature of training, it significantly reduces overfitting by preventing the network from relying too heavily on specific paths. This results in a more robust model with better generalization performance on unseen data.

Dropout is particularly effective in deep neural networks where overfitting is a significant concern due to the large number of parameters. By encouraging the network to learn redundant representations, dropout enhances the model's ability to generalize, making it a valuable tool for training robust and efficient neural networks.

4.5.6 ℓ_1 and ℓ_2 Regularization

ℓ_1 and ℓ_2 regularization are techniques mostly used in deep learning to prevent overfitting (see Chap. 1) and improve the generalization of models. These techniques add a penalty term to the loss function in the form:

$$\mathcal{L}(\mathbf{W}) + \lambda \ell_{reg}(\mathbf{W}),$$

where λ is the regularization parameter controlling the strength of the penalty; and ℓ_{reg} is the regularization term (i.e. ℓ_1 or ℓ_2).

ℓ_1 regularization, also known as Lasso (Least Absolute Shrinkage and Selection Operator) [167], adds a penalty equal to the absolute value of the magnitude of coefficients. The ℓ_1 regularization term is given by:

$$\ell_1(\mathbf{W}) = \sum_i |w_i|,$$

where $(w_i)_i$ are the model coefficients. This can lead to sparse models where some coefficients are exactly zero, effectively performing feature selection by excluding irrelevant features. ℓ_1 regularization is particularly useful when dealing with high-dimensional data where many features may not contribute significantly to the prediction task.

ℓ_2 regularization, also known as Ridge regression [82], adds a penalty equal to the square of the magnitude of coefficients. The ℓ_2 regularization term is given by:

$$\ell_2(\mathbf{W}) = \sum_i w_i^2.$$

Unlike ℓ_1, ℓ_2 regularization does not produce sparse models but rather shrinks the coefficients uniformly. This technique is effective in reducing the complexity of the model and preventing overfitting, especially when the features are correlated. The choice between ℓ_1 and ℓ_2 regularization depends on the specific problem and the nature of the data. Often, a combination of both, known as Elastic Net, is used to leverage the strengths of each method. The objective function with the elastic net method writes then:

$$\mathcal{L}(\mathbf{W}) + \lambda_1 \sum_i |w_i| + \lambda_2 \sum_i w_i^2.$$

4.5.7 Early Stopping

The fundamental idea behind early stopping is to monitor the model's performance on a validation set during training and halt the process when performance stops improving. This approach helps in finding the optimal point where the model generalizes well to unseen data without fitting the noise in the training data.

Implementing early stopping involves dividing the available data into training and validation sets. During training, the model's performance is evaluated on the validation set at regular intervals. If the performance metric (e.g., accuracy or loss) on the validation set does not improve for a specified number of iterations, known as the "patience" parameter, the training process is stopped. Mathematically, early stopping can be described as:

$$\text{Stop training if Validation Loss}_{t+p} \geqslant \text{Validation Loss}_t,$$

where t is the current training iteration, and p is the *patience* parameter. This prevents the model from continuing to learn patterns that do not generalize well. Early stopping is particularly useful in training deep learning models, where the risk of overfitting is high due to the large number of parameters.

4.6 Generalization in Overparameterized Deep Learning Models

Overparameterized deep learning models have demonstrated remarkable success across various domains, such as computer vision, natural language processing, and speech recognition, even when explicit regularization techniques are not employed. This observation challenges the traditional structural risk minimization principle (Chap. 1), which suggests that overparameterization typically leads to overfitting. Traditional theory suggests that models with more parameters than data points should overfit, leading to poor generalization. However, empirical evidence from numerous studies has shown that deep learning models can generalize exceptionally well despite being overparameterized.

This discrepancy between theory and practice has sparked significant interest in the machine learning community, leading to the development of new theoretical frameworks aimed at understanding the underlying mechanisms that enable such generalization. Several factors contribute to the generalization ability of deep learning models:

- The characteristics of the training data play an important role in determining how well a model generalizes. A large and diverse dataset helps the model learn robust features that are applicable to a wide range of scenarios, including those not encountered during training. Balanced data ensures that the model does not become biased towards certain classes or patterns, further enhancing its generalization capability. For instance, a model trained on a diverse set of images is more likely to perform well on new, unseen images compared to one trained on a homogeneous dataset [160].
- The design of the neural network architecture significantly influences its ability to generalize. For example, CNNs are particularly well-suited for image data because they can capture spatial hierarchies through layers of convolutional filters. This hierarchical feature extraction process allows CNNs to learn complex patterns in images, such as edges, textures, and shapes, which are crucial for generalization. Similarly, RNNs are designed to handle sequential data, such as text and time series, by capturing temporal dependencies.
- The choice of optimization algorithm is another critical factor in the generalization of deep learning models. Stochastic gradient descent (SGD) with momentum is a popular choice because it tends to find flat minima in the loss landscape. Flat minima are regions where the loss function changes slowly with respect to the model parameters, making the model less sensitive to small variations in the input data. This insensitivity to variations contributes to better generalization, as the model is more likely to perform well on new data [93].

While explicit regularization techniques like ℓ_1 and ℓ_2 norms are not always used in modern deep learning models, these models often benefit from implicit regularization. Implicit regularization can arise from the optimization process itself, where algorithms like SGD introduce noise that acts as a form of regularization.

This noise helps prevent the model from becoming too specialized to the training data, thereby improving generalization. Additionally, techniques such as dropout (Sect. 4.5.5) can be seen as a form of implicit regularization. Dropout forces the model to learn redundant representations, making it more robust and less likely to overfit [158].

To Sum Up

1. *The update rule for different models based on a formal neuron can be generalized as:*

$$\forall(\mathbf{x}, y), \begin{pmatrix} w_0 \\ \bar{\boldsymbol{w}} \end{pmatrix} \leftarrow \begin{pmatrix} w_0 \\ \bar{\boldsymbol{w}} \end{pmatrix} + \eta\kappa(\mathbf{x}, y) \begin{pmatrix} 1 \\ \mathbf{x} \end{pmatrix}$$

Model	*Perceptron*	*Adaline*	*Logistic regression*
$\kappa(\mathbf{x}, y)$	$y\mathbb{1}_{yh_{\boldsymbol{w}}(\mathbf{x})\leqslant 0}$	$(y - h_{\boldsymbol{w}}(\mathbf{x}))$	$y(1 - \sigma(h_{\boldsymbol{w}}(\mathbf{x})))$

   This generalized update rule highlights the similarities and differences between various linear classification models. The Perceptron updates the weights only when a misclassification occurs, Adaline updates the weights based on the linear prediction error, and Logistic Regression updates the weights based on the gradient of the logistic loss. Each model has its strengths and weaknesses, and the choice of model depends on the specific characteristics of the classification problem at hand.
2. *Feedforward networks with a single hidden layer have the capability to approximate any continuous function, given certain conditions on the activation function used.*
3. *The training of deep neural networks relies heavily on the backpropagation algorithm, which efficiently computes the gradient of the loss function with respect to the network's weights. Gradient descent methods, such as stochastic gradient descent (SGD) and its variants (e.g., Adam, RMSprop), are used to update the weights iteratively to minimize the loss function. These optimization techniques are crucial for training deep networks effectively.*
4. *Regularization methods are essential to prevent overfitting and improve the generalization of deep learning models. Common techniques include ℓ_1 and ℓ_2 regularization, dropout, and early stopping. These methods help in controlling the model complexity and ensuring that the model performs well on unseen data by adding constraints to the learning process or modifying the network architecture.*
5. *Despite their complexity, overparameterized neural networks exhibit excellent generalization capabilities, and further theoretical insights are required to better understand this phenomenon.*

4.7 Exercises

1. Let the following training set of size 4 be:

$$S = \left\{ \left(\begin{pmatrix} 1 \\ 1 \end{pmatrix}, +1 \right); \left(\begin{pmatrix} -1 \\ 1 \end{pmatrix}, +1 \right); \left(\begin{pmatrix} -1 \\ -1 \end{pmatrix}, -1 \right); \left(\begin{pmatrix} 1 \\ -1 \end{pmatrix}, 1 \right) \right\}$$

We consider the Perceptron model to separate the points of class $+1$ from those of class -1. We assume that the initial weight vector is the null vector, that the learning rate is set to 1, and that the bias $w_0 = 0$.

(a) What is in this case the weight vector found by the Perceptron algorithm after 4 updates if we consider that the updates are done in the trigonometric direction (counterclockwise) starting with the point of coordinates: $\begin{pmatrix} 1 \\ 1 \end{pmatrix}$, that is $\begin{pmatrix} -1 \\ 1 \end{pmatrix}$ then $\begin{pmatrix} -1 \\ -1 \end{pmatrix}$ then $\begin{pmatrix} 1 \\ -1 \end{pmatrix}$, etc. ?
(b) What is the equation of the separating line found after two updates?
(b) How far are the points in the training set from this line?

2. Show that wo sets of points in $\mathbb{R}^d$ can be separated by a hyperplane if and only if the intersection of their convex hulls is empty. *Hint: Consider the convex hulls C_A and C_B of two sets A and B in $\mathbb{R}^d$. Define $A - B$ as the set of points whose position vectors are $a - b$ for $a \in A$ and $b \in B$, and similarly for $C_A - C_B$. Show that A and B are linearly separable if and only if $A - B$ is linearly separable from the origin. If $C_A \cap C_B = \emptyset$, then $C_A - C_B$ is convex and does not contain the origin, implying it is linearly separable from the origin. Conversely, if A and B are linearly separable, use a separating hyperplane to show that $C_A \cap C_B = \emptyset$.*
3. Consider the network structure of a multi-layer Perceptron with 9 neurons, as shown in the following figure:

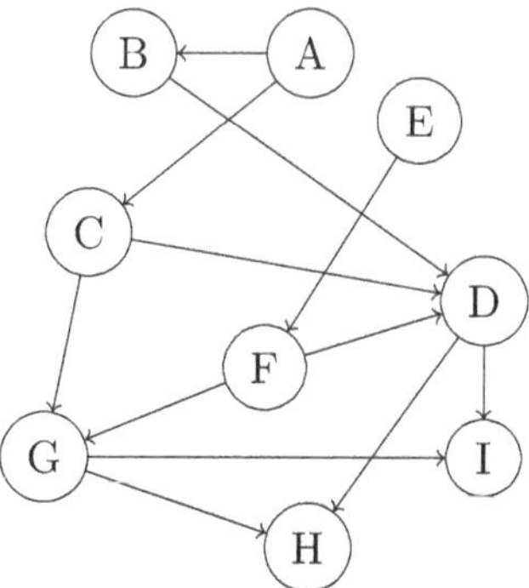

(a) Among the neurons, which are input neurons and which are output neurons?
(b) How many layers does this MLP have? Which neurons belong to which layer?
(c) Provide an order in which the neurons are activated.

4. Consider the following neural network structure:

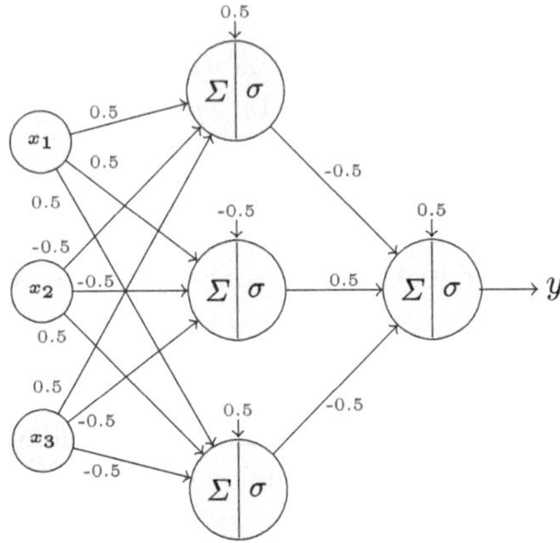

The sigmoid activation function is used for all neurons in the hidden layer and the output layer. The network has three input neurons and one output neuron.

(a) Estimate the values of the hidden layer neurons for an input $\mathbf{x} = (0.5, -0.5, 0.5)^\top$ and an output $y = 1$.
(b) What is the estimated output of this network?
(c) Calculate the Binary Cross-Entropy (BCE) error (4.22).
(d) Update the weights using the backpropagation algorithm.

5. Consider a simple feedforward neural network with one hidden layer using ReLU activations and an output layer with a linear activation. Assume you have a regression problem with mean squared error loss.

(a) Derive the expressions for the gradients of the loss with respect to the weights in both the hidden and output layers using backpropagation.
(b) Explain how the ReLU activation affects gradient flow compared to sigmoid or tanh activations.
(c) Discuss potential issues that can arise during training such as dead ReLUs and how they impact learning.

6. Consider a convolutional neural network (CNN) with multiple convolutional and pooling layers followed by fully connected layers.

(a) Explain how backpropagation works through convolutional and pooling layers.

(b) Discuss the role of weight sharing and local connectivity in CNNs and how they reduce the number of parameters compared to fully connected networks.
(c) Describe how batch normalization can help stabilize and accelerate training in deep CNNs.

7. Consider a neural network with 5 layers, each using the hyperbolic tangent (tanh) activation function. Assume the input to each neuron is such that the tanh function's derivative is always less than 0.2.

(a) Calculate the gradient of the loss with respect to the weights in the first layer during backpropagation.
(b) Show how the gradient diminishes as it propagates backward through the layers.
(c) Explain why gradients vanish even faster with tanh than with sigmoid in deep networks.
(d) Discuss the implications for initializing weights in tanh-based networks.

Modify the network to use parametric ReLU (PReLU) activations, where the slope for negative inputs is a learnable parameter α (initialized to 0.1). Assume inputs are distributed such that 50% of activations are in the negative region.

(e) Calculate the gradient of the loss with respect to the weights in the first layer and the parameter α.
(f) Show how gradients propagate through layers compared to standard ReLU .
(g) Explain why PReLU mitigates the "dying ReLU" problem in deep networks.
(h) Analyze how learnable parameters like α affect optimization dynamics.

8. Residual networks (ResNet) use shortcut connections that skip one or more layers. These connections help mitigate the vanishing gradient problem by ensuring that gradients remain informative as they propagate backward through the network. This is represented as:

$$\mathbf{z}_{l+1} = \mathbf{z}_l + F(\mathbf{z}_l),$$

where $\mathbf{z}_l$ is the input to the layer l and $\mathbf{z}_{l+1}$ is the output of layer l.

(a) For a residual block structure where $\mathbf{z}_L = \mathbf{z}_l + \sum_{i=l}^{L-1} F(\mathbf{z}_i)$, with $L > l$ show that the gradient of the loss function $\mathcal{L}$ is

$$\nabla_{\mathbf{z}_l}\mathcal{L} = \nabla_{\mathbf{z}_L}\mathcal{L} \times \left(1 + \nabla_{\mathbf{z}_l} \sum_{i=l}^{L-1} F(\mathbf{z}_i)\right)$$

(b) Discuss the implications of stable gradient flow on the training of very deep networks.
(c) Explain why residual networks can be effectively trained even with hundreds or thousands of layers.

9. Generative Adversarial Networks (GANs) involve two neural networks—a generator G and a discriminator D—engaged in a competitive game [67]. The generator learns to produce data that resembles real data, while the discriminator learns to distinguish between real and generated data. The minmax objective function for GANs writes:

$$\min_G \max_D V(D, G) = \mathbb{E}_{\mathbf{x} \sim p_{data}}[\ln D(\mathbf{x})] + \mathbb{E}_{\mathbf{z} \sim p_{\mathbf{z}}}[\ln(1 - D(G(\mathbf{z})))],$$

where p_{data} represents the probability distribution of the real data that the GAN aims to mode; and $p_{\mathbf{z}}$ represents the probability distribution of the noise vectors $\mathbf{z}$ that are fed into the generator.

(a) Explain the meaning of each term in the objective function.
(b) Show that the optimal discriminator is given by:

$$D^*(\mathbf{x}) = \frac{p_{data}(\mathbf{x})}{p_{data}(\mathbf{x}) + p_g(\mathbf{x})},$$

where p_g represents the distribution of the data generated by the generator. *Hint: By analyzing the objective function, notice that the discriminator should output the probability that a sample* $\mathbf{x}$ *comes from the real data distribution* p_{data} *rather than the generator's distribution* p_g. *Using Bayes' theorem, show the optimal discriminator* $D^*(\mathbf{x})$ *is the ratio of the probability densities of* $\mathbf{x}$ *under the real data distribution to the sum of the probabilities under both the real and generated distributions.*
(c) Explain the concept of Nash equilibrium in the context of GANs. *Hint: A Nash equilibrium is a state where neither the generator nor the discriminator can unilaterally improve their performance by changing their strategies. Imagine a scenario where the generator has learned to produce data samples that are indistinguishable from real data, and the discriminator is maximally uncertain, assigning a probability of 0.5 to both real and generated samples.*
(d) Describe the conditions under which the global minimum of the objective function is achieved.
(e) Show that the global minimum is attained when the generator's distribution p_g matches the real data distribution p_{data}.
(f) Calculate the value of the objective function at this equilibrium and explain its significance.

10. The Transformer architecture relies on key components such as the self-attention mechanism, which uses queries (Q), keys (K), and values (V) to weigh the importance of input elements relative to each other [179]. Multi-head attention extends this by performing multiple attention functions in parallel, capturing diverse aspects of the input sequence. Positional encodings are added to input embeddings to preserve sequence order, while the encoder-decoder

structure processes input sequences to generate continuous representations and output sequences. Each layer employs layer normalization and residual connections to stabilize training, enabling the model to capture long-range dependencies and process sequences in parallel efficiently.

(a) Given attention scores $A_{ij} = \frac{Q_i K_j^\top}{\sqrt{d_k}}$, derive:

$$\frac{\partial A_{ij}}{\partial Q_i} \text{ and } \frac{\text{softmax}(A_i)}{\partial A_{ik}}.$$

Hint: Use chain rule for softmax derivatives.

(b) Prove that concatenating H attention heads:

$$head_h = \text{softmax}\left(\frac{QW_h^Q(KW_h^K)^\top}{\sqrt{d_k}}\right)VW_h^V,$$

and linearly projecting via W^O preserves dimensionality when $d_{model} = H.d_v$.
Hint: Notice that concatenating the outputs of H attention heads, each producing a vector of dimension d_v, the resulting concatenated vector will have a dimension of $H.d_v$. The linear projection matrix W^O then maps this concatenated vector back to the original model dimension d_{model} which is defined as $H.d_v$.

(c) For positional encoding $PE_{(pos,2i)} = \sin(pos/10000^{2i/d_{model}})$, show it allows the model to attend to relative positions via trigonometric identities.

(d) Prove $PE_{pos+\Delta)}$ can be expressed as linear combination of $PE_{pos.}$.

(e) Given $\tilde{q}_m = q_m e^{im\theta}$ and $\tilde{k}_n = k_n e^{in\theta}$, show that

$$\tilde{q}_m^\top \tilde{k}_n = q_m^\top k_n e^{i(m-n)\theta},$$

and explain how relative positions emerge in attention scores.
Hint: Since we are dealing with complex numbers, the dot product $\tilde{q}_m^\top \tilde{k}_n$ involves multiplying $\tilde{q}_m$ by conjugate of $\tilde{k}_n$. Notice that the attention score between two positions m and n is influenced by the phase shift $e^{i(m-n)\theta}$. This phase shift encodes the relative distance between the positions, rather than their absolute positions.

Chapter 5
Support Vector Machines

In this chapter, we introduce support vector machines (SVMs) , which gained significant popularity in the early 2000s due to their strong theoretical foundations. We will outline the theoretical basis of SVMs, which has led to the development of various optimization techniques proposed in recent years.

In Sect. 5.1, we derive objective functions based on the concept of margin for binary classification. In Sect. 5.2, we explore how the kernel trick allows us to extend the solutions of the underlying optimization problems into higher-dimensional Hilbert spaces. Finally, in Sect. 5.3.2, we discuss the extension of these models to the multi-class classification case.

5.1 The Notion of Margin

SVMs are designed around the notion of margin, which is also the premise of the perceptron algorithm presented in Chap. 4.

5.1.1 Hard Margin

Let us first consider the case of linear functions for a linearly separable classification problem, where we assume that there exists $(\boldsymbol{w}, w_0)$ such that:

$$\forall \mathbf{x}_i \in S_+, \ \langle \bar{\boldsymbol{w}}, \mathbf{x}_i \rangle + w_0 \geqslant 0, \quad \text{and}$$
$$\forall \mathbf{x}_i \in S_-, \ \langle \bar{\boldsymbol{w}}, \mathbf{x}_i \rangle + w_0 \leqslant 0,$$

where S_+ (respectively S_-) is the set of positive (respectively negative) examples of the training set. By normalizing the parameters, we can rewrite these equations in

M.-R. Amini, *Advanced Supervised and Semi-supervised Learning*,
Cognitive Technologies, https://doi.org/10.1007/978-3-031-99928-4_5

the following form, which will be used hereafter:

$$\forall \mathbf{x}_i \in S_+,\ \langle \bar{\boldsymbol{w}}, \mathbf{x}_i \rangle + w_0 \geqslant +1, \quad \text{and} \tag{5.1}$$

$$\forall \mathbf{x}_i \in S_-,\ \langle \bar{\boldsymbol{w}}, \mathbf{x}_i \rangle + w_0 \leqslant -1. \tag{5.2}$$

These two conditions can be combined in a single set of inequalities:

$$\forall (\mathbf{x}, y) \in S,\ y(\langle \bar{\boldsymbol{w}}, \mathbf{x} \rangle + w_0) - 1 \geqslant 0. \tag{5.3}$$

If there are points verifying the equality in (5.1), they will belong to the hyperplane of equation $\langle \bar{\boldsymbol{w}}, \mathbf{x}_i \rangle + w_0 = 1$ and $d_+ = \frac{1}{||\bar{\boldsymbol{w}}||}$ will be the distance between the nearest point of the class of positive examples and the decision boundary. Similarly, the points verifying equality in (5.2) will belong to the hyperplane of equation $\langle \bar{\boldsymbol{w}}, \mathbf{x}_i \rangle + w_0 = -1$ and $d_- = \frac{1}{||\bar{\boldsymbol{w}}||}$ will be the distance between the nearest point of the class of negative examples and the hyperplane. Under these conditions, the margin is defined as $\rho = d_+ = d_- = \frac{1}{||\bar{\boldsymbol{w}}||}$.

Furthermore, among the points of the training set, those that verify equality in (5.3) are called the *support vectors*. Figure 5.1 illustrates these facts for a two-dimensional problem. This formulation is similar to the one we gave for the margin perceptron (Chap. 4, Sect. 4.1.2) with the threshold ρ set to 1 in this case. Furthermore, if we consider the minimization of the global cost function (Chap. 4, Eq. (4.8)), we find the parameters for which the sum of the average distances of the examples in the training set by the two hyperplanes $\langle \bar{\boldsymbol{w}}, \mathbf{x} \rangle + w_0 - y\rho = 0, y \in \{-1, +1\}$ is minimum for a given ρ, whereas in the case of SVM, we try directly to

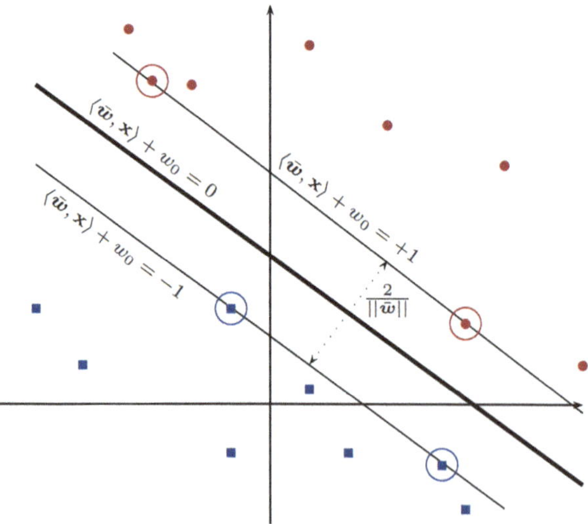

Fig. 5.1 Hyperplanes for a linearly separable classification problem in dimension 2. Support vectors belonging to hyperplanes of equations $\langle \bar{\boldsymbol{w}}, \mathbf{x} \rangle + w_0 = \pm 1$ are circled

maximize the margin for the threshold $\rho = 1$, that is, to minimize the norm of the weights, since in this case the margin is equal to $\rho = \frac{1}{||\bar{\boldsymbol{w}}||}$.

The main idea behind SVMs is that the larger the margin associated with a decision boundary, the more likely the associated prediction function is to generalize well on new data. This idea translates into the search for a hyperplane with the largest margin while respecting the constraints of (5.3), which leads to the solution of the following optimization problem:

$$\begin{aligned} &\min_{\bar{\boldsymbol{w}}\in\mathbb{R}^d, w_0\in\mathbb{R}} \frac{1}{2}||\bar{\boldsymbol{w}}||^2, \\ &\text{subject to } \forall i,\ y_i(\langle\bar{\boldsymbol{w}}, \mathbf{x}_i\rangle + w_0) - 1 \geqslant 0. \end{aligned} \tag{5.4}$$

The optimization problem (5.4) is called a constrained optimization problem in the form of inequalities, and the type of SVM verifying it is called *hard margin* SVM.

Definition 5.1 (Saturated and Unsaturated Constraints) Posing $\boldsymbol{w}^* = (\bar{\boldsymbol{w}}^*, w_0)$ as the solution to (5.4), two situations are conceivable for each constraint $i \in \{1, \ldots, m\}$:

- $y_i(\langle\bar{\boldsymbol{w}}^*, \mathbf{x}_i\rangle + w_0^*) - 1 = 0$, constraint i is said to be saturated.
- $y_i(\langle\bar{\boldsymbol{w}}^*, \mathbf{x}_i\rangle + w_0^*) - 1 > 0$, constraint i is said to be unsaturated.

The Lagrangian formulation of this constrained optimization problem using the Lagrangian multipliers $\boldsymbol{\alpha} = (\alpha_1, \ldots, \alpha_m)$; $\alpha_i \geqslant 0, \forall i \in \{1, \ldots, m\}$ associated with the m constraints of the problem is written as follows:

$$L_p(\bar{\boldsymbol{w}}, w_0, \boldsymbol{\alpha}) = \frac{1}{2}||\bar{\boldsymbol{w}}||^2 - \sum_{i=1}^{m} \alpha_i \left[y_i(\langle\bar{\boldsymbol{w}}, \mathbf{x}_i\rangle + w_0) - 1\right]. \tag{5.5}$$

To be able to use the Lagrangian to solve this optimization problem, it is sufficient for one of the following conditions to be verified:

- The matrix of partial derivatives of the constraints, or Jacobian matrix, of n saturated constraints is of rank n when evaluated at the optimum.
- The constraint functions are all affine, which is the case for the problem (5.4).

The loss function $\mathcal{L} : \bar{\boldsymbol{w}} \mapsto \frac{1}{2}||\bar{\boldsymbol{w}}||^2$ is twice differentiable. Its gradient vector and Hessian matrix at the point $\bar{\boldsymbol{w}}$ are $\nabla\mathcal{L}(\bar{\boldsymbol{w}}) = \bar{\boldsymbol{w}}$ and $\mathbf{H}(\bar{\boldsymbol{w}}) = \mathbf{I}$, the identity matrix. The Hessian $\mathbf{H}(\bar{\boldsymbol{w}})$ thus admits d strictly positive eigenvalues, the loss function $\mathcal{L}$ is strictly convex and the inequality constraints $c_i : (\bar{\boldsymbol{w}}, w_0) \mapsto 1 - y_i(\langle\bar{\boldsymbol{w}}, \mathbf{x}_i\rangle + w_0)$ are affine functions and therefore qualified constraints.

The optimization problem (5.4) thus admits a unique solution. Furthermore, since the objective function is convex and the points satisfying the constraints also form a convex set, solving this problem is a quadratic programming problem, a family of techniques widely studied in optimization (see Exercise 12).

Definition 5.2 The Karush-Kuhn-Tucker (KKT) conditions are the necessary conditions that must be satisfied by the solution $\bar{\boldsymbol{w}}^*$ of the optimization problem. They are derived by setting the gradient of the Lagrangian with respect to the primal variables $\bar{\boldsymbol{w}}^*$ and w_0^* to zero and including the complementarity conditions. The KKT conditions are given by:

$$\nabla L_p(\bar{\boldsymbol{w}}^*) = \bar{\boldsymbol{w}}^* - \sum_{i=1}^{m} \alpha_i y_i \mathbf{x}_i = 0, \text{ thus } \bar{\boldsymbol{w}}^* = \sum_{i=1}^{m} \alpha_i y_i \mathbf{x}_i. \tag{5.6}$$

$$\nabla L_p(w_0^*) = -\sum_{i=1}^{m} \alpha_i y_i = 0, \text{ thus } \sum_{i=1}^{m} \alpha_i y_i = 0. \tag{5.7}$$

$$\forall i, \alpha_i \left[y_i(\langle \bar{\boldsymbol{w}}^*, \mathbf{x}_i \rangle + w_0^*) - 1 \right] = 0, \text{ thus } \begin{cases} \alpha_i = 0, \text{ or} \\ y_i(\langle \bar{\boldsymbol{w}}^*, \mathbf{x}_i \rangle + w_0^*) = 1. \end{cases} \tag{5.8}$$

With (5.6), we can see that the weight vector $\bar{\boldsymbol{w}}^*$ solution to the optimization problem is a linear combination of the training examples. The examples $\mathbf{x}_i$ that appear in this solution are those for which the coefficients $\alpha_i > 0$. These examples are called support vectors, which by complementarity conditions (5.8) verify $y_i(\langle \bar{\boldsymbol{w}}^*, \mathbf{x}_i \rangle + w_0^*) = 1$. The support vectors thus lie on the hyperplanes of equations $\langle \bar{\boldsymbol{w}}^*, \mathbf{x}_i \rangle + w_0^* = \pm 1$ and the other examples for which $\alpha_i = 0$ have no influence on the solution.

Substituting equalities (5.6) and (5.7) into (5.5), we get:

$$\begin{aligned} L_p &= \frac{1}{2} \left\| \sum_{i=1}^{m} \alpha_i y_i \mathbf{x}_i \right\|^2 - \sum_{i=1}^{m} \sum_{j=1}^{m} \alpha_i \alpha_j y_i y_j \langle \mathbf{x}_i, \mathbf{x}_j \rangle - w_0^* \sum_{i=1}^{m} \alpha_i y_i + \sum_{i=1}^{m} \alpha_i \\ &= -\frac{1}{2} \sum_{i=1}^{m} \sum_{j=1}^{m} y_i y_j \alpha_i \alpha_j \langle \mathbf{x}_i, \mathbf{x}_j \rangle + \sum_{i=1}^{m} \alpha_i. \end{aligned} \tag{5.9}$$

Input :

- A training set $S = ((\mathbf{x}_1, y_1), \ldots, (\mathbf{x}_m, y_m))$;

- Find, $\boldsymbol{\alpha}^*$, the solution of the optimization problem:

$$\max_{\boldsymbol{\alpha} \in \mathbb{R}^m} \sum_{i=1}^{m} \alpha_i - \frac{1}{2} \sum_{i=1}^{m} \sum_{j=1}^{m} y_i y_j \alpha_i \alpha_j \langle \mathbf{x}_i, \mathbf{x}_j \rangle,$$

$$\text{subject to} \sum_{i=1}^{m} y_i \alpha_i = 0 \text{ and } \forall i, \alpha_i \geqslant 0.$$

- Choose $i \in \{1, \ldots, m\}$ such that $\alpha_i^* > 0$;

  Let $w_0^* = y_i - \sum_{j=1}^{m} \alpha_j^* y_j \langle \mathbf{x}_j, \mathbf{x}_i \rangle$. // $\triangleright$ (5.13)

- Set $\bar{\boldsymbol{w}}^* = \sum_{i=1}^{m} y_i \alpha_i^* \mathbf{x}_i$. // $\triangleright$ (5.6)

Output : Decision function $\forall \mathbf{x}, f(\mathbf{x}) = \text{sign}\left(\sum_{i=1}^{m} y_i \alpha_i^* \langle \mathbf{x}_i, \mathbf{x} \rangle + w_0^*\right)$.

Algorithm 11: Hard margin (dual formulation)

Hence we obtain the following dual optimization problem, also called the Wolfe dual problem:

$$\max_{(\alpha_1, \ldots, \alpha_m) \in \mathbb{R}^m} \sum_{i=1}^{m} \alpha_i - \frac{1}{2} \sum_{i=1}^{m} \sum_{j=1}^{m} y_i y_j \alpha_i \alpha_j \langle \mathbf{x}_i, \mathbf{x}_j \rangle, \tag{5.10}$$

$$\text{subject to} \sum_{i=1}^{m} y_i \alpha_i = 0 \text{ and } \forall i, \alpha_i \geqslant 0. \tag{5.11}$$

We note that the solution of the dual problem $\boldsymbol{\alpha}$ could be directly used to determine the decision function f (5.6):

$$f(\mathbf{x}') = \text{sign}\left(\sum_{i=1}^{m} y_i \alpha_i \langle \mathbf{x}_i, \mathbf{x}' \rangle + w_0^*\right). \tag{5.12}$$

Moreover, since the support vectors belong to the hyperplanes of equations $\langle \bar{\boldsymbol{w}}^*, \mathbf{x} \rangle + w_0^* = \pm 1$ in the feature space, we can deduce the expression for the intercept by taking any support vector $(\mathbf{x}_i, y_i)$:

$$\langle \bar{\boldsymbol{w}}^*, \mathbf{x}_i \rangle + w_0^* = y_i \Rightarrow w_0^* = y_i - \sum_{j=1}^{m} \alpha_j y_j \langle \mathbf{x}_i, \mathbf{x}_j \rangle. \tag{5.13}$$

With the previous equality, we can obtain a simple expression for the margin. Indeed, as (5.13) is valid for all support vectors with $\alpha_i \neq 0$, multiplying both sides of this equality by $\alpha_i y_i$ and taking the sum over all examples in the training set, we get:

$$\sum_{i=1}^{m} \alpha_i y_i w_0^* = \sum_{i=1}^{m} \alpha_i y_i^2 - \sum_{i=1}^{m}\sum_{j=1}^{m} \alpha_i \alpha_j y_i y_j \langle \mathbf{x}_i, \mathbf{x}_j \rangle.$$

From (5.7) and (5.6), and considering the fact that $\forall i,\ y_i^2 = 1$ and $\alpha_i \geqslant 0$, we have:

$$\sum_{i=1}^{m} \alpha_i - \|\bar{\boldsymbol{w}}^*\|_2^2 = \|\boldsymbol{\alpha}\|_1 - \|\bar{\boldsymbol{w}}^*\|_2^2 = 0.$$

Recall that the margin is equal to:

$$\rho = \frac{1}{\|\bar{\boldsymbol{w}}^*\|_2} = \frac{1}{\sqrt{\|\boldsymbol{\alpha}\|_1}}.$$

With these notations, the hard margin SVM algorithm is given in Algorithm 11. There are a multitude of programs that effectively implement SVMs, the most popular of which are undoubtedly those proposed by Joachims [90] and Fan et al. [51].

5.1.2 Soft Margin

A major drawback of this optimization problem is that when the data are complex or noisy, which is generally the case, the linearly separable condition is no longer respected. One solution to this problem is to relax the constraints on the margin by introducing new ξ_i variables, called slack variables; these variables relax the constraints in (5.1) and (5.2) which, in the vector space, are written:

$$\forall i, \qquad \xi_i \geqslant 0, \tag{5.14}$$

$$\forall \mathbf{x}_i \in S_+,\ \langle \bar{\boldsymbol{w}}, \mathbf{x}_i \rangle + w_0 \geqslant 1 - \xi_i, \tag{5.15}$$

$$\forall \mathbf{x}_i \in S_-,\ \langle \bar{\boldsymbol{w}}, \mathbf{x}_i \rangle + w_0 \leqslant -1 + \xi_i. \tag{5.16}$$

According to these inequalities, when an example $\mathbf{x}_i$ from the training set is misclassified, the corresponding variable ξ_i must be greater than unity, since in this case $1-\xi_i$ or $-1+\xi_i$ changes sign. As a result, the sum $\sum_i \xi_i$ determines an upper bound on the number of classification errors on the training set (Fig. 5.2).

We then seek to minimize the classification errors while maximizing the margin, which amounts to changing the objective function of the primal problem from $\frac{\|\bar{\boldsymbol{w}}\|^2}{2}$

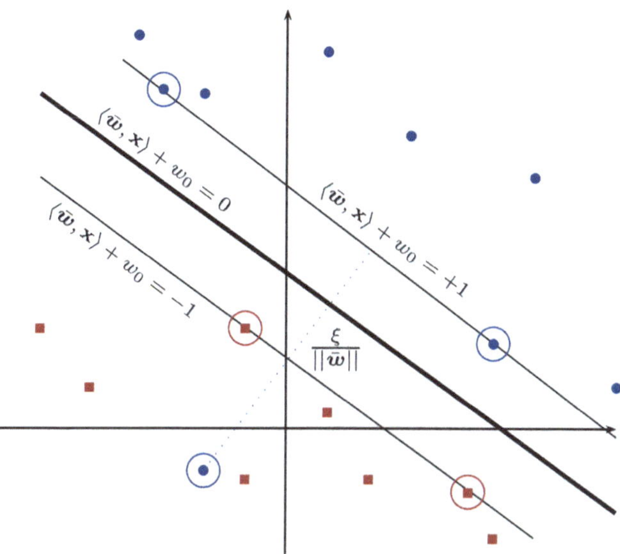

Fig. 5.2 Linear hyperplanes for a nonlinearly separable classification problem. Support vectors are circled. Either these vectors lie on one of the hyperplanes of equation $\langle \bar{\boldsymbol{w}}, \mathbf{x}\rangle + w_0 = \pm 1$, or they are outliers. The distance of an outlier $\mathbf{x}$ from the hyperplane associated with its class is $\frac{\xi}{||\bar{\boldsymbol{w}}||}$

to $\frac{||\bar{\boldsymbol{w}}||^2}{2} + C\sum_{i=1}^{m} \xi_i$, where C is a parameter fixed in advance (the larger C is, the more errors are penalized). The optimization problem then takes the form:

$$\begin{aligned} &\min_{\bar{\boldsymbol{w}}\in\mathbb{H}, w_0\in\mathbb{R}, \xi\in\mathbb{R}^m} \frac{1}{2}||\bar{\boldsymbol{w}}||^2 + C\sum_{i=1}^{m} \xi_i, \\ &\text{subject to } \forall i, \xi_i \geqslant 0 \text{ and } y_i(\langle \bar{\boldsymbol{w}}, \mathbf{x}_i\rangle + w_0) \geqslant 1 - \xi_i. \end{aligned} \tag{5.17}$$

This type of SVM is called *soft-margin SVM*, since the constraints on the margin are relaxed for certain examples.

As in the separable case, the constraints are affine and qualified; as the objective function is also convex, the Karush-Kuhn-Tucker conditions apply to the optimum. Let $\beta_i \geqslant 0, i \in \{1, \ldots, m\}$ be the Lagrange multipliers associated with the non-negativity constraints of the slack variables and $\alpha_i \geqslant 0, i \in \{1, \ldots, m\}$ the variables associated with the set of m following constraints. We denote by $\boldsymbol{\xi}$ the vector $(\xi_1, \ldots, \xi_m)^\top$, by $\boldsymbol{\alpha}$ the vector $(\alpha_1, \ldots, \alpha_m)^\top$, and by $\boldsymbol{\beta}$ the vector $(\beta_1, \ldots, \beta_m)^\top$. The Lagrangian can then be defined for any $\bar{\boldsymbol{w}} \in \mathbb{H}$, $w_0 \in \mathbb{R}$, $\boldsymbol{\xi} \in \mathbb{R}_+^m$, $\boldsymbol{\alpha} \in \mathbb{R}_+^m$ and $\boldsymbol{\beta} \in \mathbb{R}_+^m$ by:

$$L_p(\bar{\boldsymbol{w}}, w_0, \boldsymbol{\xi}, \boldsymbol{\alpha}, \boldsymbol{\beta}) = \frac{1}{2}\|\bar{\boldsymbol{w}}\|^2 + C\sum_{i=1}^{m} \xi_i - \sum_{i=1}^{m} \alpha_i[y_i(\langle \mathbf{x}_i, \bar{\boldsymbol{w}}\rangle + w_0) - 1 + \xi_i] - \sum_{i=1}^{m} \beta_i \xi_i.$$

The KKT conditions (Definition 5.2) are obtained by cancelling the partial derivatives of the Lagrangian with respect to the variables $\bar{\boldsymbol{w}}$, w_0, $\boldsymbol{\xi}$, $\boldsymbol{\alpha}$ and $\boldsymbol{\beta}$ and writing the complementarity conditions:

$$\nabla L_p(\bar{\boldsymbol{w}}) = \bar{\boldsymbol{w}} - \sum_{i=1}^{m} \alpha_i y_i \mathbf{x}_i = 0, \text{ thus } \quad \bar{\boldsymbol{w}} = \sum_{i=1}^{m} \alpha_i y_i \mathbf{x}_i, \tag{5.18}$$

$$\nabla L_p(w_0) = -\sum_{i=1}^{m} \alpha_i y_i = 0, \text{ thus } \sum_{i=1}^{m} \alpha_i y_i = 0, \tag{5.19}$$

$$\nabla L_p(\boldsymbol{\xi}) = C \times \mathbf{1}_m - \boldsymbol{\alpha} - \boldsymbol{\beta} = 0, \text{ thus } \forall i, C = \alpha_i + \beta_i, \tag{5.20}$$

$$\forall i, \alpha_i \left[y_i h_{\boldsymbol{w}}(\mathbf{x}_i) - 1 + \xi_i\right] = 0, \quad \text{thus } \alpha_i = 0 \text{ or } y_i h_{\boldsymbol{w}}(\mathbf{x}_i) = 1 - \xi_i, \tag{5.21}$$

$$\forall i, \beta_i \xi_i = 0, \quad \text{thus } \beta_i = 0 \text{ or } \xi_i = 0, \tag{5.22}$$

where $\mathbf{1}_m$ is the m-dimensional vector with all its components equal to 1, and $h_{\boldsymbol{w}}(\mathbf{x}) = \langle \mathbf{x}, \bar{\boldsymbol{w}} \rangle + w_0$.

As in the separable case, the weight vector $\bar{\boldsymbol{w}}$ that solves the problem is a linear combination of the examples in the training set (5.18). From (5.21), we see that there are, however, two types of support vectors. Indeed, an example $\mathbf{x}_i \in S$ is a support vector if its associated coefficient α_i is non-zero.

In this case, according to equation $y_i(\langle \bar{\boldsymbol{w}}, \mathbf{x}_i \rangle + w_0) = 1 - \xi_i$, if $\xi_i = 0$ then the example $\mathbf{x}_i$ is on one of the hyperplanes $\langle \bar{\boldsymbol{w}}, \mathbf{x} \rangle + w_0 = \pm 1$, and if $\xi_i \neq 0$ the support vector is an outlier (see Fig. 5.2). For an outlier $\mathbf{x}_i$, we have from (5.22) $\beta_i = 0$, which from (5.20) implies that $\alpha_i = C$.

Thus, the support vectors are either outliers whose coefficients α are all equal to C, or points lying on one of the hyperplanes $\langle \bar{\boldsymbol{w}}, \mathbf{x} \rangle + w_0 = \pm 1$. The dual form of the constrained optimization problem is obtained by replacing the expression for the weight vector (5.18) in the Lagrangian and using Eqs. (5.20) and (5.19).

From (5.20), we have:

$$C \sum_{i=1}^{m} \xi_i - \sum_{i=1}^{m} \alpha_i \xi_i - \sum_{i=1}^{m} \beta_i \xi_i = \sum_{i=1}^{m} \xi_i \underbrace{(C - \alpha_i - \beta_i)}_{=0} = 0.$$

Thus

$$\begin{aligned} L_p &= \frac{1}{2} \left\| \sum_{i=1}^{m} \alpha_i y_i \mathbf{x}_i \right\|^2 - \sum_{i=1}^{m} \sum_{j=1}^{m} \alpha_i \alpha_j y_i y_j \langle \mathbf{x}_i, \mathbf{x}_j \rangle - w_0 \underbrace{\sum_{i=1}^{m} \alpha_i y_i}_{=0} + \sum_{i=1}^{m} \alpha_i \\ &= -\frac{1}{2} \sum_{i=1}^{m} \sum_{j=1}^{m} y_i y_j \alpha_i \alpha_j \langle \mathbf{x}_i, \mathbf{x}_j \rangle + \sum_{i=1}^{m} \alpha_i. \end{aligned} \tag{5.23}$$

Input :

- A training set $S = ((\mathbf{x}_1, y_1), \dots, (\mathbf{x}_m, y_m))$;

- Find, $\boldsymbol{\alpha}^*$, the solution of the optimization problem

$$\max_{\boldsymbol{\alpha}\in\mathbb{R}^m} \sum_{i=1}^{m} \alpha_i - \frac{1}{2}\sum_{i=1}^{m}\sum_{j=1}^{m} y_i y_j \alpha_i \alpha_j \langle \mathbf{x}_i, \mathbf{x}_j \rangle,$$

$$\text{subject to } \sum_{i=1}^{m} y_i \alpha_i = 0 \text{ and } \forall i, 0 \leqslant \alpha_i \leqslant C.$$

- Choose $i \in \{1, \dots, m\}$ such that $0 < \alpha_i^* < C$;

  Set $w_0^* = y_i - \sum_{j=1}^{m} \alpha_j^* y_j \langle \mathbf{x}_j, \mathbf{x}_i \rangle$.

- Set $\bar{\boldsymbol{w}}^* = \sum_{i=1}^{m} y_i \alpha_i^* \mathbf{x}_i$. // ▷ (5.18)

Output : Decision function $\forall \mathbf{x}, f(\mathbf{x}) = \text{sign}\left(\sum_{i=1}^{m} y_i \alpha_i^* \langle \mathbf{x}_i, \mathbf{x} \rangle + w_0^*\right)$.

Algorithm 12: Soft-margin SVM (Dual formulation)

This is exactly the form of the objective function obtained in the separable case (5.9) with the only difference being the constraints on the Lagrange multipliers $\beta_i \geqslant 0$, which according to (5.20) imposes an upper bound on the constraints $\alpha_i \leqslant C$. The dual optimization of SVMs in the non-separable case is thus the same as that of SVMs in the separable case with this upper bound on the constraints $(\alpha_i)_{i=1}^m$:

$$\max_{(\alpha_1,\dots,\alpha_m)\in\mathbb{R}^m} \sum_{i=1}^{m} \alpha_i - \frac{1}{2}\sum_{i=1}^{m}\sum_{j=1}^{m} y_i y_j \alpha_i \alpha_j \langle \mathbf{x}_i, \mathbf{x}_j \rangle, \tag{5.24}$$

$$\text{subject to } \sum_{i=1}^{m} y_i \alpha_i = 0 \text{ and } \forall i, 0 \leqslant \alpha_i \leqslant C. \tag{5.25}$$

As in the linearly separable case, we obtain the value of the y-intercept w_0^* by taking a support vector $(\mathbf{x}_i, y_i)$ that is not an outlier (that is $\xi_i = 0$ and $\alpha_i < C$).

For this support vector, we then have $\sum_{j=1}^{m} \alpha_j y_j \langle \mathbf{x}_i, \mathbf{x}_j \rangle + w_0^* = y_i$:

$$w_0^* = y_i - \sum_{j=1}^{m} \alpha_j y_j \langle \mathbf{x}_i, \mathbf{x}_j \rangle; \text{ for all example } \mathbf{x}_i \text{ with } 0 < \alpha_i < C. \tag{5.26}$$

In the dual case, the soft-margin SVM (Algorithm 12) is thus identical to Algorithm 11 with the only difference being the optimization problem (5.24) and the upper bound C on the variables $(\alpha_i)_{i=1}^m$.

Note that the initial problem (5.4) can be solved without using the dual formulation. Equivalently, the problem (5.4) amounts to minimizing the following objective function:

$$\mathcal{L}(\boldsymbol{w}) = \frac{1}{m} \sum_{(\mathbf{x},y)\in S} \max(0, 1 - y h_{\boldsymbol{w}}(\mathbf{x})) + \frac{\lambda}{2}\|\boldsymbol{w}\|^2, \tag{5.27}$$

where λ is a positive constant, $\boldsymbol{w} = (w_0, \bar{\boldsymbol{w}})$ and $h_{\boldsymbol{w}}(\mathbf{x}) = \langle \mathbf{x}, \bar{\boldsymbol{w}} \rangle + w_0$.

We note that the instantaneous loss function $(h(\mathbf{x}), y) \mapsto \max(0, 1 - y h_{\boldsymbol{w}}(\mathbf{x}))$, called the Hinge loss, is a convex upper bound on the classification error $\ell(h(\mathbf{x}), y) = \mathbb{1}_{y \times h(\mathbf{x}) \leqslant 0}$ (Chap. 2, Sect. 2.3.2) and that minimizing Eq. (5.27), comes down to minimizing the regularized empirical error by penalizing models with large weights in the sense of the ℓ_2 norm (Chap. 2, Sect. 2.3.1).

Let A be a convex open set of a normed space, $\mathcal{L} : A \to \mathbb{R}$ a differentiable function and $\lambda > 0$. Let $\boldsymbol{u}$ and $\boldsymbol{v}$ be two weight vectors, we have for all $\alpha \in [0, 1]$:

$$A(S, \alpha, \boldsymbol{u}, \boldsymbol{v}) = \alpha\mathcal{L}(\boldsymbol{u}) + (1-\alpha)\mathcal{L}(\boldsymbol{v}) - \mathcal{L}(\alpha\boldsymbol{u} + (1-\alpha)\boldsymbol{v}) - \frac{\lambda}{2}\alpha(1-\alpha)\|\boldsymbol{u} - \boldsymbol{v}\|^2 =$$

$$\alpha \sum_{(\mathbf{x},y)\in S_u^+} (1 - y h_{\boldsymbol{u}}(\mathbf{x})) + (1-\alpha) \sum_{(\mathbf{x},y)\in S_v^+} (1 - y h_{\boldsymbol{v}}(\mathbf{x})) - \sum_{(\mathbf{x},y)\in S_{u\oplus v}^+} (1 - y h_{\boldsymbol{u}\oplus\boldsymbol{v}}(\mathbf{x})),$$

where $\forall \mathbf{x}$, $h_{\boldsymbol{u}\oplus\boldsymbol{v}}(\mathbf{x}) = \alpha h_{\boldsymbol{u}}(\mathbf{x}) + (1-\alpha) h_{\boldsymbol{v}}$, and for $p \in \{u, v, u \oplus v\}$; $S_p^+ = \{(\mathbf{x}, y) \in S \mid 1 - y h_{\boldsymbol{p}}(\mathbf{x}) > 0\}$.

From the definition of $h_{\boldsymbol{u}\oplus\boldsymbol{v}}$ we have:

$$\forall (\mathbf{x}, y) \in S; \alpha(1 - y h_{\boldsymbol{u}}(\mathbf{x})) + (1 - \alpha)(1 - y h_{\boldsymbol{v}}(\mathbf{x})) = 1 - y h_{\boldsymbol{u}\oplus\boldsymbol{v}}(\mathbf{x}). \tag{5.28}$$

Hence for an example $(\mathbf{x}, y)$ of the training set if $(\mathbf{x}, y) \notin S_u^+$ (that is $1 - y h_{\boldsymbol{u}}(\mathbf{x}) \leqslant 0$) and $(\mathbf{x}, y) \notin S_v^+$ then from (5.28) we obtain $1 - y h_{\boldsymbol{u}\oplus\boldsymbol{v}}(\mathbf{x}) \leqslant 0$ and hence $(\mathbf{x}, y) \notin S_z^+$. Thus:

$$S_{u\oplus v}^+ \subseteq S_u^+ \cup S_v^+ \text{ and } A(S, \alpha, \boldsymbol{u}, \boldsymbol{v}) \geqslant 0.$$

Since there may be examples $(\mathbf{x}, y)$ that belong to $S_{\boldsymbol{u}}^+$ (that is $1 - y h_{\boldsymbol{u}}(\mathbf{x}) > 0$) or S_v^+ (that is $1 - y h_{\boldsymbol{v}}(\mathbf{x}) > 0$) but not to $S_{u\oplus v}^+$. Thus we have $\forall \boldsymbol{u}, \boldsymbol{v}$ and $\alpha \in [0, 1]$:

$$\alpha\mathcal{L}(\boldsymbol{u}) + (1 - \alpha)\mathcal{L}(\boldsymbol{v}) - \mathcal{L}(\alpha\boldsymbol{u} + (1 - \alpha)\boldsymbol{v}) \geqslant \frac{\lambda}{2}\alpha(1 - \alpha)\|\boldsymbol{u} - \boldsymbol{v}\|^2.$$

This verifies one of the properties of λ-strongly convex functions.

Shalev et al. [153] have proposed a highly efficient implementation based on the gradient descent algorithm, presented in Algorithm 13 and called Primal Estimated sub-GrAdient SOlver for SVM (Pegasos), for the minimization of the loss function (5.27). The model weights are iteratively updated according to the

Input :

- A training set $S = ((\mathbf{x}_1, y_1), \ldots, (\mathbf{x}_m, y_m))$;
- Set $\boldsymbol{w}^{(1)} \leftarrow 0$; A constant $\lambda > 0$; A maximum number of iterations T;

for $t = 1, 2, \ldots, T-1$ **do**
  Set $S_t^+ = \{(\mathbf{x}, y) \in S; yh_{\boldsymbol{w}^{(t)}}(\mathbf{x}) < 1\}$;
  Set $\eta_t = \frac{1}{\lambda t}$;
  Update the weights: // ▷ (5.29)

$$\boldsymbol{w}^{(t+1)} \leftarrow (1 - \lambda\eta_t)\boldsymbol{w}^{(t)} + \frac{\eta_t}{m} \sum_{(\mathbf{x},y)\in S_t^+} y\mathbf{x}.$$

Output : $\tilde{\boldsymbol{w}}^* = \frac{1}{T}\sum_{t=1}^{T} \boldsymbol{w}^{(t)}$.

Algorithm 13: Primal Estimated sub-GrAdient SOlver for SVM [153]

following rule:

$$\forall t, \boldsymbol{w}^{(t+1)} \leftarrow \boldsymbol{w}^{(t)} - \left[\lambda\eta_t \boldsymbol{w}^{(t)} - \frac{\eta_t}{m} \sum_{(\mathbf{x},y)\in S_t^+} y\mathbf{x}\right], \tag{5.29}$$

where $S_t^+ = \{(\mathbf{x}, y) \in S | 1 - yh_{\boldsymbol{w}^{(t)}}(\mathbf{x}) > 0\}$ is the set of examples $(\mathbf{x}, y)$ from the training base S that satisfy the condition $1 - yh_{\boldsymbol{w}^{(t)}}(\mathbf{x}) > 0$, and $\eta_t \in \mathbb{R}_+$ is the training step at iteration t. Shalev et al. [153] have shown that for a learning step proportional to the inverse of the iteration number, the update (5.29) converges to the global minimum of the objective function (5.27). This result is stated in the following theorem.

Theorem 5.1 (Convergence of Primal Estimated Sub-gradient Solver for SVM) *Let $S = ((\mathbf{x}_1, y_1), \ldots, (\mathbf{x}_m, y_m))$ be a training base of size m, and $\boldsymbol{w}^*$ the minimizer of the objective function (5.27). Assume that the vectors representing the examples of S are contained in a hypersphere of radius R (that is $\forall(\mathbf{x}, y) \in S, \|\mathbf{x}\| \leqslant R$) and that at each iteration t the norm of the corresponding weight vector is normalized $\|\boldsymbol{w}^{(t)}\| \leqslant \frac{1}{\sqrt{\lambda}}$. For a learning rate $\eta_t = \frac{1}{\lambda t}$; $t \geq 1$, the sequence $(\boldsymbol{w}^{(t)})_{t\in\mathbb{N}^*}$ found by Algorithm 13 converges to $\boldsymbol{w}^*$:*

$$\lim_{T\to\infty} \frac{1}{T} \sum_{t=1}^{T} \boldsymbol{w}^{(t)} = \boldsymbol{w}^*. \tag{5.30}$$

Proof According to the update rule (5.29), for two consecutive weight vectors $\boldsymbol{w}^{(t)}$ and $\boldsymbol{w}^{(t+1)}$ we have:

$$\begin{aligned}\|\boldsymbol{w}^{(t)}-\boldsymbol{w}^*\|^2-\|\boldsymbol{w}^{(t+1)}-\boldsymbol{w}^*\|^2 &= \|\boldsymbol{w}^{(t)}-\boldsymbol{w}^*\|^2-\|\boldsymbol{w}^{(t)}-\eta_t\nabla\mathcal{L}(\boldsymbol{w}^{(t)})-\boldsymbol{w}^*\|^2\\ &= 2\eta_t\langle \boldsymbol{w}^{(t)}-\boldsymbol{w}^*, \nabla\mathcal{L}(\boldsymbol{w}^{(t)})\rangle-\eta_t^2\|\nabla\mathcal{L}(\boldsymbol{w}^{(t)})\|^2.\end{aligned} \tag{5.31}$$

Since the objective function (5.27) is λ-strongly convex , the following relationship is verified for each weight vector $\boldsymbol{w}^{(t)}$ obtained with Algorithm 13 [24]:

$$\forall \boldsymbol{u}, \langle \boldsymbol{w}^{(t)}-\boldsymbol{u}, \nabla\mathcal{L}(\boldsymbol{w}^{(t)})\rangle \geqslant \mathcal{L}(\boldsymbol{w}^{(t)})-\mathcal{L}(\boldsymbol{u})+\frac{\lambda}{2}\|\boldsymbol{w}^{(t)}-\boldsymbol{u}\|^2.$$

Specifically, when $\boldsymbol{u}$ is equal to $\boldsymbol{w}^*$, which is the minimizer of the objective function, the inequality derived from (5.31) after T updates becomes:

$$\sum_{t=1}^{T}\Big(\mathcal{L}(\boldsymbol{w}^{(t)})-\mathcal{L}(\boldsymbol{w}^*)\Big) \leqslant$$
$$\sum_{t=1}^{T}\left(\frac{\|\boldsymbol{w}^{(t)}-\boldsymbol{w}^*\|^2-\|\boldsymbol{w}^{(t+1)}-\boldsymbol{w}^*\|^2}{2\eta_t}-\frac{\lambda}{2}\|\boldsymbol{w}^{(t)}-\boldsymbol{w}^*\|^2\right)+\frac{1}{2}\sum_{t=1}^{T}\eta_t\|\nabla\mathcal{L}(\boldsymbol{w}^{(t)})\|^2.$$

The terms of the first sum after the inequality simplify two by two, giving:

$$\begin{aligned}\sum_{t=1}^{T}\Big(\mathcal{L}(\boldsymbol{w}^{(t)})-\mathcal{L}(\boldsymbol{w}^*)\Big) &\leqslant \frac{-\lambda T}{2}\|\boldsymbol{w}^{(T+1)}-\boldsymbol{w}^*\|^2+\frac{1}{2}\sum_{t=1}^{T}\eta_t\|\nabla\mathcal{L}(\boldsymbol{w}^{(t)})\|^2\\ &\leqslant \frac{1}{2}\sum_{t=1}^{T}\eta_t\|\nabla\mathcal{L}(\boldsymbol{w}^{(t)})\|^2.\end{aligned} \tag{5.32}$$

Assuming that the examples in the training base are contained in a hypersphere of radius R and that the weight vectors are normalized so that $\forall t, \|\boldsymbol{w}^{(t)}\| \leqslant \frac{1}{\sqrt{\lambda}}$, we have from the gradient expression (5.29):

$$\|\nabla\mathcal{L}(\boldsymbol{w}^{(t)})\| \leqslant \sqrt{\lambda}+R.$$

For a learning rate $\eta_t = \frac{1}{\lambda\times t}, t \geq 1$ and the previous inequality, (5.32) becomes;

$$\sum_{t=1}^{T}\mathcal{L}(\boldsymbol{w}^{(t)}) \leqslant T\mathcal{L}(\boldsymbol{w}^*)+\frac{c}{2\lambda}\sum_{t=1}^{T}\frac{1}{t}, \tag{5.33}$$

where $c = (\sqrt{\lambda} + R)^2$. Using the bound $\sum_{t=1}^{T} \frac{1}{t} \leqslant 1 + \int_1^T \frac{dt}{t}$, it then follows that for $T \geqslant 3$:

$$\frac{1}{T}\sum_{t=1}^{T} \mathcal{L}(\boldsymbol{w}^{(t)}) \leqslant \mathcal{L}(\boldsymbol{w}^{*}) + \frac{c(1+\ln(T))}{2\lambda T}. \tag{5.34}$$

Moreover, since the objective function is convex and $\boldsymbol{w}^*$ is the global minimizer, we have according to Jensen's inequality:

$$\mathcal{L}(\boldsymbol{w}^*) \leqslant \mathcal{L}\left(\frac{1}{T}\sum_{t=1}^{T} \boldsymbol{w}^{(t)}\right) \leqslant \frac{1}{T}\sum_{t=1}^{T} \mathcal{L}(\boldsymbol{w}^{(t)}) \leqslant \mathcal{L}(\boldsymbol{w}^*) + \frac{c(1+\ln(T))}{2\lambda T}.$$

The result follows by applying the squeeze theorem. □

5.2 Kernel Trick

The main advantage of using the dual problem is that the dimension of the input space no longer plays a role in solving the optimization problem, which now depends solely on the dot products between the pairs of examples in the training set.

This observation, known as the kernel trick, has motivated the use of *kernels* for SVMs, which learn nonlinear classifiers. This use follows work which has shown that, in some cases, it is possible to render a non-linearly separable problem into a linearly separable one by immersing the initial data in a Hilbert space, $\mathbb{H}$; called the feature space, of higher dimension than the initial vector space.

Definition 5.3 A *kernel* is a function $\kappa : \mathcal{X} \times \mathcal{X} \to \mathbb{R}$ that computes the inner product of the images of all pairs of data in the feature space. It is defined as:

$$\kappa(\mathbf{x}, \mathbf{x}') = \langle \phi(\mathbf{x}), \phi(\mathbf{x}') \rangle,$$

where $\phi : \mathcal{X} \to \mathbb{H}$ is the projection function that maps the input space to the feature space.

5.2.1 Kernel Functions

The kernel function κ often interpreted as a measure of similarity between observations in the feature space. The advantage of using a kernel is that if its underlying projection function exists, then it is not necessary to know its expression in order to estimate it. What is more, this projection function can even, in certain cases, transform the initial space into a feature space of infinite dimension, meaning that its expression is impossible to calculate. We say that the kernel κ is symmetric if for all pairs $(\mathbf{x}, \mathbf{x}') \in \mathcal{X}^2$, we have $\kappa(\mathbf{x}, \mathbf{x}') = \kappa(\mathbf{x}', \mathbf{x})$,. Moreover, for any finite sequence of points $(\mathbf{x}_i)_{1\leqslant i\leqslant n}$ and for all real values $(\alpha_i)_{1\leqslant i\leqslant n}$, the kernel function κ satisfies:

$$\sum_{i=1}^{n}\sum_{j=1}^{n}\alpha_i\alpha_j\kappa(\mathbf{x}_i, \mathbf{x}_j) \geqslant 0,$$

then κ is said to be a positive kernel.

The existence of a projection function is assured if its associated kernel function verifies the following condition, known as the Mercer's condition:

Theorem 5.2 (Mercer's Theorem [116]) *Let κ be a positive definite symmetric kernel function. The integral operator $T_\kappa : L_2(\mathfrak{X}) \to L_2(\mathfrak{X})$ defined on the space of square-integrable functions $L_2(\mathfrak{X})$ over a compact set $\mathfrak{X}$ by:*

$$\forall f \in L_2(\mathfrak{X}); [T_\kappa f](\mathbf{x}) = \int_{\mathfrak{X}^2} \kappa(\mathbf{x}, \mathbf{s}) f(\mathbf{s}) ds$$

has positive eigenvalues $(\lambda_j)_j$ and corresponding eigenvectors $(v_j)_j$ that form an orthonormal basis in $L_2(\mathfrak{X})$. The kernel κ can be represented as:

$$\kappa(\mathbf{x}, \mathbf{x}') = \sum_{j=1}^{N_{\mathfrak{X}}} \lambda_j\, v_j(\mathbf{x}) v_j(\mathbf{x}'),$$

where $N_{\mathfrak{X}}$ is the number of eigenvalues and eigenvectors, which can be infinite. The convergence of this series to the kernel function is absolute and uniform.

The theorem establishes that any positive definite kernel can be decomposed into a series involving its eigenvalues and eigenfunctions, which form a basis for the space of square-integrable functions. This decomposition is crucial in the theory of reproducing kernel Hilbert spaces and has wide-ranging applications in machine learning and functional analysis.

5.2.2 Positive Definite Symmetric Kernel

An equivalent condition to the mercer condition is that the kernel κ is symmetric positive definite. For any set $(\mathbf{x}_i)_{1\leqslant i\leqslant m} \in \mathcal{X}^m$, it is the same that the matrix $\mathbf{K} = [\kappa(\mathbf{x}_i, \mathbf{x}_j)]_{ij} \in \mathbb{R}^{m\times m}$ called Gram matrix is symmetric positive semidefinite, that is, all eigenvalues $\mathbf{K}$ are positive and for $\boldsymbol{\alpha} = (\alpha_1, \ldots, \alpha_m)^\top \in \mathbb{R}^m$ we have:

$$\boldsymbol{\alpha}^\top \mathbf{K}\boldsymbol{\alpha} = \sum_{i=1}^{m}\sum_{j=1}^{m} \alpha_i\alpha_j\kappa(\mathbf{x}_i, \mathbf{x}_j) \geqslant 0.$$

The main symmetric positive definite polynomial kernels used in the literature are:

$$\kappa(\mathbf{x}, \mathbf{x}') = (\langle \mathbf{x}, \mathbf{x}'\rangle + a)^r, \forall r \in \mathbb{N}, a \in \mathbb{R}_+, \qquad \textbf{(polynomial kernel)}$$

$$\kappa(\mathbf{x}, \mathbf{x}') = e^{-\frac{\|\mathbf{x}-\mathbf{x}'\|^2}{2\sigma^2}}, \sigma \in \mathbb{R}_+^*, \qquad \textbf{(RBF Gaussian kernel)}$$

$$\kappa(\mathbf{x}, \mathbf{x}') = \tanh(\alpha\langle \mathbf{x}, \mathbf{x}'\rangle + \beta), (\alpha, \beta) \in \mathbb{R}_+^2. \qquad \textbf{(sigmod kernel)}$$

An important property of symmetric positive definite kernels is that they induce a scalar product in their associated Hilbert space. This property is formulated in the following theorem:

Theorem 5.3 (Reproducing Kernel Hilbert Space) *Let $\kappa : \mathcal{X} \times \mathcal{X} \to \mathbb{R}$ be a symmetric positive definite kernel. Then there exists a Hilbert space $\mathbb{H}$ and a feature map $\phi : \mathcal{X} \to \mathbb{H}$ such that:*

$$\forall(\mathbf{x}, \mathbf{x}') \in \mathcal{X} \times \mathcal{X}, \kappa(\mathbf{x}, \mathbf{x}') = \langle \phi(\mathbf{x}), \phi(\mathbf{x}')\rangle.$$

Furthermore, the Hilbert space $\mathbb{H}$ verifies the reproducing property:

$$\forall f \in \mathbb{H}, \forall \mathbf{x} \in \mathcal{X}, f(\mathbf{x}) = \langle f, \kappa(\mathbf{x}, .)\rangle.$$

The feature space $\mathbb{H}$ is called a *Reproducing kernel Hilbert space* (RKSH) associated with the kernel κ. The previous theorem implies that symmetric positive-definite kernels can be used to implicitly define this feature space. In this case, for a fixed input space $\mathcal{X}$, the problem of finding useful features to represent the observations is replaced by that of finding suitable positive-definite symmetric kernels.

This transformation is significant because it allows us to work with potentially high-dimensional or even infinite-dimensional feature spaces without explicitly computing the feature mapping. Instead, we rely on the kernel function κ to compute the inner products in this feature space. This approach is particularly advantageous in scenarios where the feature mapping ϕ is complex or infeasible to compute directly.

5.2.3 SVM with Symmetric Positive Definite Kernels

Since symmetric positive definite kernels implicitly define a scalar product in the feature space, it is possible to extend the optimization problem of SVMs by replacing the scalar product $\langle \mathbf{x}, \mathbf{x}' \rangle$ with such a kernel $\kappa(\mathbf{x}, \mathbf{x}')$. This leads to the following generalized dual optimization problem:

$$\max_{(\alpha_1,\ldots,\alpha_m)\in\mathbb{R}^m} \sum_{i=1}^{m} \alpha_i - \frac{1}{2}\sum_{i=1}^{m}\sum_{j=1}^{m} y_i y_j \alpha_i \alpha_j \kappa(\mathbf{x}_i, \mathbf{x}_j), \tag{5.35}$$

$$\text{subject to } \sum_{i=1}^{m} y_i \alpha_i = 0 \text{ and } \forall i, 0 \leqslant \alpha_i \leqslant C.$$

In this case, the expression of the prediction function is written according to Eqs. (5.18) and (5.26):

$$\forall \mathbf{x},\ f(\mathbf{x}) = \text{sign}\left(\sum_{i=1}^{m} y_i \alpha_i^* \kappa(\mathbf{x}_i, \mathbf{x}) + w_0^*\right),$$

where $(\alpha_i^*)_{1\leqslant i\leqslant m}$ are the solutions of (5.35) and $w_0^* = y_i \sum_{j=1}^{m} \alpha_j^* y_j \kappa(\mathbf{x}_i, \mathbf{x}_j)$, for any example $\mathbf{x}_i$ with $0 < \alpha_i^* < C$.

5.3 Theoretical Study and Multi-class Case

A remarkable result, justifying the surge of interest for SVMs in the early 2000, is that the generalization bound of these classifiers depends mainly on the margin and not on the dimension of the input space.

5.3.1 Margin Based Generalization Bound

We will explain this bound using the Rademacher complexity presented in Chap. 2, Sect. 2.1. A larger margin results in a tighter bound on the error, and since SVMs directly maximize this margin, we can often expect very accurate estimates of the true risk for these classifiers. This bound is based on the $1/\rho$-lipschitz loss function $\mathfrak{h}_\rho : \mathbb{R} \to [0, 1]$, with $\rho > 0$, dominating the classification error presented in Chap. 2.

$$\mathfrak{h}_\rho(z) = \begin{cases} 1, & \text{if } z \leqslant 0; \\ 1 - z/\rho, & \text{if } 0 < z \leqslant \rho; \\ 0, & \text{otherwise.} \end{cases}$$

From this loss, we define the margin-based empirical error on a training set $S = ((\mathbf{x}_i, y_i))_{i=1}^m$ of size m, as the average number of misclassified examples but with a confidence (an unsigned prediction) less than or equal to ρ by the current classifier h:

$$\mathcal{L}_\rho(h, S) = \frac{1}{m}\sum_{i=1}^m \mathfrak{h}_\rho(y_i h(\mathbf{x}_i)). \tag{5.36}$$

We note that if h is a linear classifier and if the norm of the normal vector equals 1, then the unsigned prediction of h for an example $(\mathbf{x}, y)$, $yh(\mathbf{x})$, is the example margin. The following theorem gives a generalization bound for linear classifiers in a feature space, such as SVMs, which depends exclusively on the margin of these classifiers.

Theorem 5.4 (Margin-dependent Generalization Bound for Linear Classifiers) *Let $\mathcal{H}_B = \{\mathbf{x} \mapsto \langle \boldsymbol{w}, \phi(\mathbf{x})\rangle \mid ||\boldsymbol{w}||| \leqslant B\}$ be the class of linear functions, defined on a feature space $\phi : \mathcal{X} \to \mathbb{H}$, and of bounded norm. Suppose the examples are contained in a hypersphere of radius R in this space, $\forall \mathbf{x} \in \mathcal{X}, \langle \phi(\mathbf{x}), \phi(\mathbf{x})\rangle \leqslant R^2$. Let $\rho > 0$ and let $S = ((\mathbf{x}_i, y_i))_{i=1}^m$ be a training set of size m identically and independently distributed according to a probability distribution $\mathcal{D}$. For any $\delta \in (0, 1)$ and any function $h \in \mathcal{H}_B$, we have with probability at least $1 - \delta$:*

$$\mathcal{L}(h) \leqslant \mathcal{L}_\rho(h, S) + \frac{4B}{\sqrt{m}} \times \frac{R}{\rho} + 3\sqrt{\frac{\ln(2/\delta)}{2m}}. \tag{5.37}$$

Proof Let $\mathfrak{P} = \{z = (\mathbf{x}, y) \mapsto yh(\mathbf{x}) \mid h \in \mathcal{H}_B\}$. Let be the class of $\mathcal{X} \times \mathcal{Y}$ functions in $[0, 1]$ defined by:

$$\mathcal{G}_\rho = \{\mathfrak{h}_\rho \circ g, g \in \mathfrak{P}\}.$$

According to Theorem 2.3, we then have, for any $\delta \in (0, 1)$ and any function $g \in \mathfrak{P}$, the following inequality which holds with probability at least $1 - \delta$:

$$\mathbb{E}_{(\mathbf{x},y)}[\mathfrak{h}_\rho(g(\mathbf{x}, y))-1] \leqslant \frac{1}{m}\sum_{i=1}^{m}(\mathfrak{h}_\rho(g(\mathbf{x}_i, y_i))-1)+\hat{\mathfrak{R}}_S((\mathfrak{h}_\rho-1)\circ\mathfrak{P})+3\sqrt{\frac{\ln(2/\delta)}{2m}}.$$

where $\hat{\mathfrak{R}}_S((\mathfrak{h}_\rho-1)\circ\mathfrak{P})$ is the empirical Rademacher complexity of the function class $(\mathfrak{h}_\rho - 1)\circ\mathfrak{P}$. The function $z \mapsto (\mathfrak{h}_\rho - 1)(z)$ is $1/\rho$-lipschitzian and $(\mathfrak{h}_\rho - 1)(0) = 0$, that is according to Talagrand's lemma:

$$\hat{\mathfrak{R}}_S((\mathfrak{h}_\rho - 1)\circ\mathfrak{P}) \leqslant \frac{2}{\rho}\hat{\mathfrak{R}}_S(\mathfrak{P}), \tag{5.38}$$

with:

$$\begin{aligned}\hat{\mathfrak{R}}_S(\mathfrak{P}) &= \mathbb{E}_\sigma\left[\sup_{g\in\mathfrak{P}}\left|\frac{2}{m}\sum_{i=1}^{m}\sigma_i g(\mathbf{x}_i, y_i)\right|\right] = \mathbb{E}_\sigma\left[\sup_{h\in\mathcal{H}_B}\left|\frac{2}{m}\sum_{i=1}^{m}\sigma_i y_i h(\mathbf{x}_i)\right|\right]\\ &= \frac{2}{m}\mathbb{E}_\sigma\left[\sup_{||\boldsymbol{w}||\leqslant B}\left|\left\langle \boldsymbol{w}, \sum_{i=1}^{m}\sigma_i\phi(\mathbf{x}_i)\right\rangle\right|\right].\end{aligned}$$

The last equality is based on the bilinearity of the scalar product and the fact that $\forall i,\ y_i\sigma_i \in \{-1, +1\}$, and therefore $y_i\sigma_i$ and σ_i have the same expectation. Using the Cauchy-Schwarz inequality, we obtain:

$$\hat{\mathfrak{R}}_S(\mathfrak{P}) \leqslant \frac{2B}{m}\mathbb{E}_\sigma\left[\left\|\sum_{i=1}^{m}\sigma_i\phi(\mathbf{x}_i)\right\|\right] = \frac{2B}{m}\mathbb{E}_\sigma\left[\left(\left\|\sum_{i=1}^{m}\sigma_i\phi(\mathbf{x}_i)\right\|^2\right)^{1/2}\right].$$

According to Jensen's inequality and the concavity of the square root function, this gives:

$$\begin{aligned}\hat{\mathfrak{R}}_S(\mathfrak{P}) &\leqslant \frac{2B}{m}\left(\mathbb{E}_\sigma\left[\left\|\sum_{i=1}^{m}\sigma_i\phi(\mathbf{x}_i)\right\|^2\right]\right)^{1/2}\\ &= \frac{2B}{m}\left(\mathbb{E}_\sigma\left[\left\langle\sum_{i=1}^{m}\sigma_i\phi(\mathbf{x}_i), \sum_{j=1}^{m}\sigma_j\phi(\mathbf{x}_j)\right\rangle\right]\right)^{1/2},\end{aligned}$$

thus:

$$\hat{\mathfrak{R}}_S(\mathfrak{P}) \leqslant \frac{2B}{m}\left(\mathbb{E}_\sigma\left[\sum_{i=1}^{m}\sum_{j=1}^{m}\sigma_i\sigma_j\langle\phi(\mathbf{x}_i),\phi(\mathbf{x}_j)\rangle\right]\right)^{1/2}.$$

Since the Rademacher variables $\forall i, \forall j, \sigma_i\sigma_j$ take values in $\{-1,+1\}$ with equal probability, there are four possible values of these variables when $i \neq j$, with two of these pairs of values yielding products of opposite signs:

$$\hat{\mathfrak{R}}_S(\mathfrak{P}) \leqslant \frac{2B}{m}\left(\sum_{i=1}^{m}\langle\phi(\mathbf{x}_i),\phi(\mathbf{x}_j)\rangle\right)^{1/2} \leqslant \frac{2B}{m}\sqrt{(mR^2)}. \tag{5.39}$$

Finally, as the margin-based cost upper-bounds the classification error and according to Eqs. (5.38) and (5.39), we can write:

$$\begin{aligned}\mathbb{E}_{(\mathbf{x},y)\sim\mathcal{D}}[\mathbb{1}_{yh(\mathbf{x})\leqslant 0} - 1] &\leqslant \mathbb{E}_{(\mathbf{x},y)\sim\mathcal{D}}[\mathfrak{h}_\rho(g(\mathbf{x},y)) - 1]\\ &\leqslant \frac{1}{m}\sum_{i=1}^{m}(\mathfrak{h}_\rho(g(\mathbf{x}_i,y_i)) - 1) + \frac{4B}{\sqrt{m}}\times\frac{R}{\rho} + 3\sqrt{\frac{\ln(2/\delta)}{2m}}.\end{aligned}$$

□

The complexity term in the previous result depends on the bound on the weight norm, B, the number of examples in the training base, m, and above all on the ratio R/ρ. So, if for a fixed margin value ρ we have a training set of size m satisfying $\frac{BR}{\rho} << \sqrt{m}$, the previous result guarantees that the generalization error of the linear classifier in the feature space will be very close to its empirical margin-based loss.

We note that the R/ρ term also plays a central role in Novikoff's theorem (4.1), as it affects the number of iterations required by the perceptron algorithm for a linearly separable classification problem.

5.3.2 Multi-class SVM

Several works have extended SVMs to the multi-class classification case, of which the most cited ones are [39, 71, 104, 182]. In the following, we will outline the extension proposed by [39], which was not based directly on the theoretic guarantees of the bound (2.36), but whose result is a direct consequence.

Consider the function class $\mathcal{H}_B$ (2.35) defined in Chap. 2:

$$\mathcal{H}_B = \{h : (\mathbf{x}, y) \in \mathcal{X}\times\mathcal{Y} \mapsto \langle\mathbf{x}, \boldsymbol{w}_y\rangle \mid \boldsymbol{W} = (\boldsymbol{w}_1,\ldots,\boldsymbol{w}_K), \|\boldsymbol{W}\|_{\mathbb{H}} \leqslant B\},$$

the corresponding classification function $f_{\boldsymbol{h}}$ (2.25):

$$\forall \mathbf{x},\ f_{\boldsymbol{h}}(\mathbf{x}) = \underset{k\in\{1,\dots,K\}}{\operatorname{argmax}}\ h(\mathbf{x}, k),$$

and the 1-lipschitzian instantaneous error loss:

$$\forall (\mathbf{x}_i, y_i) \in S, \forall h \in \mathcal{H}_B, \xi_i = \ell(f_{\boldsymbol{h}}(\mathbf{x}_i), y_i) = \min(1, \max(\mu_h(\mathbf{x}_i, y_i), 0)),$$

where $\mu_h(\mathbf{x}, y)$ is the margin of the function h on the example $(\mathbf{x}, y)$, (2.31). For any function $h \in \mathcal{H}_B$, the generalization error of the function $f_{\boldsymbol{h}}$ (2.25) is bounded, according to Theorem 2.6:

$$\forall \delta \in]0, 1[, \mathbb{P}\left(\mathcal{L}(f_{\boldsymbol{h}}) \leqslant \frac{1}{m}\sum_{i=1}^{m} \xi_i + \frac{8BR}{\sqrt{m}} K(K-1) + \sqrt{\frac{\ln(2/\delta)}{2m}}\right) \geqslant 1 - \delta. \tag{5.40}$$

Proposition 5.1 (Crammer and Singer Formulation [39]) Minimizing the bound of $\mathcal{L}(f_{\boldsymbol{h}})$ is thus equivalent to the joint minimization of the empirical error of $f_{\boldsymbol{h}}$ on a training set S; $\frac{1}{m}\sum_{i=1}^{m}\xi_i$ and the norm of $\boldsymbol{W}$, or $\sum_{k=1}^{K}\|\mathbf{w}_k\|^2$. This boils down to the following optimization problem leading to the multi-class SVM algorithm:

$$\min_{\boldsymbol{W},\boldsymbol{\xi}\in\mathbb{R}^m} \frac{1}{2}\sum_{k=1}^{K}\|\mathbf{w}_k\|^2 + C\sum_{i=1}^{m}\xi_i,$$

$$\text{s.t.} \begin{cases} \xi_i \geqslant 0 & \forall i \in \{1,\dots,m\}; \\ \langle \mathbf{w}_{y_i}, \phi(\mathbf{x}_i)\rangle \geqslant \langle \mathbf{w}_k, \phi(\mathbf{x}_i)\rangle + 1 - \xi_i & \forall i \in \{1,\dots,m\}, \forall k \in \mathcal{Y}\setminus\{y_i\}. \end{cases} \tag{5.41}$$

Proof By introducing $(\mathbf{y}_i)_{i\in\{1,\dots,m\}}$ the indicator vectors per class of examples (2.20) and setting $\forall i, \boldsymbol{b}_i = C\mathbf{y}_i - \boldsymbol{\alpha}_i$, the dual problem of problem (5.41) is written:

$$\max_{\boldsymbol{b}\in\mathbb{R}^{m\times K}} \mathcal{Q}(\boldsymbol{b}) = \sum_{i=1}^{m}\boldsymbol{b}_i^\top\mathbf{y}_i - \frac{1}{2}\sum_{i=1}^{m}\sum_{j=1}^{m}\boldsymbol{b}_i^\top\boldsymbol{b}_j\kappa(\mathbf{x}_i, \mathbf{x}_j),$$

$$\text{s.t.} \begin{cases} \boldsymbol{b}_i \leqslant C\mathbf{y}_i & \forall i \in \{1,\dots,m\}, \\ \boldsymbol{b}_i^\top\mathbf{1}_K = 0 & \forall i \in \{1,\dots,m\}, \end{cases} \tag{5.42}$$

where $\delta_{k,\ell} = \begin{cases} 1, & \text{if } \ell = k \\ 0, & \text{otherwise} \end{cases}$ is the Kronecker symbol, $\boldsymbol{b}_i \leqslant C\mathbf{y}_i$ is equivalent to a term-by-term comparison, $b_{ik} \leqslant C\delta_{y_i,k}; \forall i \in \{1,\dots,m\}, \forall k \in \{1,\dots,K\}$, and $\mathbf{1}_K$ is a vector of dimension K having all its components equal to 1.

The use of the Kronecker symbol makes it possible to have a single type of constraint in (5.41):

$$\begin{aligned} &\min_{W,\boldsymbol{\xi}\in\mathbb{R}^m} \frac{1}{2}\sum_{k=1}^{K} \|\mathbf{w}_k\|^2 + C\sum_{i=1}^{m} \xi_i, \\ &\quad \text{s.t. } \langle \mathbf{w}_{y_i} - \mathbf{w}_k, \phi(\mathbf{x}_i)\rangle + \delta_{y_i,k} \geqslant 1 - \xi_i, \forall i \in \{1,\dots,m\}, \forall k \in \mathcal{Y}. \end{aligned} \tag{5.43}$$

As in the binary case, the constraints are affine and qualified, and since the objective function is also convex, the Karush-Kuhn-Tucker conditions apply to the optimum. Let $\boldsymbol{\alpha} \in \mathbb{R}_+^{m\times K}$ be the matrix each i row of which, $\boldsymbol{\alpha_i}^\top = (\alpha_{i1},\dots,\alpha_{iK})$, corresponds to the set of K constraints associated with example i. The LagrangianindexLagrangian can then be defined by:

$$\begin{aligned} L_p(\mathbf{W},\boldsymbol{\xi},\boldsymbol{\alpha}) =& \frac{1}{2}\sum_{k=1}^{K}\|\mathbf{w}_k\|^2 + C\sum_{i=1}^{m}\xi_i \\ &- \sum_{i=1}^{m}\sum_{k=1}^{K}\alpha_{ik}\left[\langle \mathbf{w}_{y_i} - \mathbf{w}_k, \phi(\mathbf{x}_i)\rangle - 1 + \xi_i + \delta_{y_i,k}\right]. \end{aligned}$$

The partial derivative of the Lagrangian with respect to $\boldsymbol{\xi}$ gives the condition:

$$\forall i, C = \sum_{k=1}^{K} \alpha_{ik}. \tag{5.44}$$

Moreover, by canceling the partial derivatives of the Lagrangian with respect to the vectors $\mathbf{w}_k,\ k \in \mathcal{Y}$, we obtain:

$$\begin{aligned} \forall k \in \mathcal{Y}, \mathbf{w}_k &= \sum_{i:y_i=k} \underbrace{\left[\sum_{\ell\neq k}\alpha_{i\ell}\right]}_{=C-\alpha_{ik}} \phi(\mathbf{x}_i) - \sum_{i:y_i\neq k}\alpha_{ik}\phi(\mathbf{x}_i) \\ &= \sum_{i=1}^{m}\left(C\delta_{y_i,k} - \alpha_{ik}\right)\phi(\mathbf{x}_i). \end{aligned} \tag{5.45}$$

The dual form of the constrained optimization problem (5.43) is obtained by replacing the expression for the weight vectors (5.45) in the Lagrangian and using (5.44). From the latter and the definition of the class indicator vector (2.20), we get:

$$L_p(\boldsymbol{W},\boldsymbol{\xi},\boldsymbol{\alpha}) = \underbrace{\frac{1}{2}\sum_{k=1}^{K}\|\mathbf{w}_k\|^2}_{=S_1} + \sum_{i=1}^{m}\xi_i\underbrace{\left(C-\sum_{k}\alpha_{ik}\right)}_{=0} - \underbrace{\sum_{i,k}\alpha_{ik}\langle\mathbf{w}_{y_i},\phi(\mathbf{x}_i)\rangle}_{=S_2}$$
$$+\underbrace{\sum_{i,k}\alpha_{ik}\langle\mathbf{w}_k,\phi(\mathbf{x}_i)\rangle}_{=S_3} + \underbrace{\sum_{i,k}\alpha_{ik}}_{=\sum_{i=1}^{m}C} - \sum_{i=1}^{m}\boldsymbol{\alpha}_i^\top\mathbf{y}_i. \tag{5.46}$$

By replacing the expressions (5.45) of the vectors $(\mathbf{w}_k)_{k\in\{1,\dots,K\}}$ in the first three non-zero sums, S_1, S_2, S_3 of the previous equality, we get:

$$S_1 = \frac{1}{2}\sum_{k=1}^{K}\|\mathbf{w}_k\|^2 = \frac{1}{2}\sum_{k=1}^{K}\left\langle\sum_{i=1}^{m}(C\delta_{y_i,k}-\alpha_{ik})\phi(\mathbf{x}_i),\sum_{j=1}^{m}(C\delta_{y_j,k}-\alpha_{jk})\phi(\mathbf{x}_j)\right\rangle$$
$$= \frac{1}{2}\sum_{i=1}^{m}\sum_{j=1}^{m}\langle\phi(\mathbf{x}_i),\phi(\mathbf{x}_j)\rangle\sum_{k=1}^{K}\left(C\delta_{y_i,k}-\alpha_{ik}\right)\left(C\delta_{y_j,k}-\alpha_{jk}\right)$$
$$= \frac{1}{2}\sum_{i=1}^{m}\sum_{j=1}^{m}\kappa(\mathbf{x}_i,\mathbf{x}_j)(C\mathbf{y}_i-\boldsymbol{\alpha}_i)^\top(C\mathbf{y}_j-\boldsymbol{\alpha}_j).$$

$$S_2 = \sum_{i=1}^{m}\sum_{k=1}^{K}\alpha_{ik}\langle\mathbf{w}_{y_i},\phi(\mathbf{x}_i)\rangle = \sum_{i=1}^{m}\sum_{k=1}^{K}\alpha_{ik}\left\langle\sum_{j=1}^{m}\left(C\delta_{y_i,y_j}-\alpha_{jy_i}\right)\phi(\mathbf{x}_j),\phi(\mathbf{x}_i)\right\rangle$$
$$= \sum_{i=1}^{m}\sum_{j=1}^{m}\kappa(\mathbf{x}_i,\mathbf{x}_j)(C\delta_{y_i,y_j}-\alpha_{jy_i})\underbrace{\sum_{k=1}^{K}\alpha_{ik}}_{=C}$$
$$= \sum_{i=1}^{m}\sum_{j=1}^{m}\kappa(\mathbf{x}_i,\mathbf{x}_j)\sum_{k=1}^{K}\left(C\delta_{k,y_j}-\alpha_{jk}\right)C\delta y_i,k.$$

$$S_3 = \sum_{i=1}^{m}\sum_{k=1}^{K}\alpha_{ik}\langle\mathbf{w}_k,\phi(\mathbf{x}_i)\rangle = \sum_{i=1}^{m}\sum_{k=1}^{K}\alpha_{ik}\left\langle\sum_{j=1}^{m}\left(C\delta_{k,y_j}-\alpha_{jk}\right)\phi(\mathbf{x}_j),\phi(\mathbf{x}_i)\right\rangle$$
$$= \sum_{i=1}^{m}\sum_{j=1}^{m}\kappa(\mathbf{x}_i,\mathbf{x}_j)\sum_{k=1}^{K}\left(C\delta_{k,y_j}-\alpha_{jk}\right)\alpha_{ik}.$$

The difference $S_2 - S_3$ is thus:

$$\begin{aligned}
S_2 - S_3 &= \sum_{i=1}^{m}\sum_{j=1}^{m}\kappa(\mathbf{x}_i, \mathbf{x}_j)\left[\sum_{k=1}^{K}\left(C\delta_{k,y_j} - \alpha_{jk}\right)C\delta y_i, k - \sum_{k=1}^{K}\left(C\delta_{k,y_j} - \alpha_{jk}\right)\alpha_{ik}\right] \\
&= \sum_{i=1}^{m}\sum_{j=1}^{m}\kappa(\mathbf{x}_i, \mathbf{x}_j)(C\mathbf{y}_i - \boldsymbol{\alpha}_i)^\top(C\mathbf{y}_j - \boldsymbol{\alpha}_j).
\end{aligned}$$

Replacing the expressions for S_1, $S_2 - S_3$ in (5.46) and noting that the class indicator vectors are orthonormal, $\forall i, C = C\mathbf{y}_i^\top \mathbf{y}_i$, we obtain:

$$L_p(\boldsymbol{\alpha}) = -\frac{1}{2}\sum_{i=1}^{m}\sum_{j=1}^{m}(C\mathbf{y}_i - \boldsymbol{\alpha}_i)^\top(C\mathbf{y}_j - \boldsymbol{\alpha}_j)\kappa(\mathbf{x}_i, \mathbf{x}_j) + \sum_{i=1}^{m}(C\mathbf{y}_i - \boldsymbol{\alpha}_i)^\top \mathbf{y}_i, \tag{5.47}$$

with $\forall i \in \{1, \ldots, m\}$ and $\forall k \in \{1, \ldots, K\}, \alpha_{ik} \geqslant 0$ and $\forall i \in \{1, \ldots, m\}$, $\sum_{k=1}^{K}\alpha_{ik} = C$. By posing $\forall i \in \{1, \ldots, m\}$; $\boldsymbol{b}_i = C\mathbf{y}_i - \boldsymbol{\alpha}_i$, we obtain the announced dual form (5.42).

The dual problem (5.42) can be solved using standard quadratic programming techniques. The difficulty, however, is the number of variables to be manipulated by the optimization algorithm, which is of the order of $m \times K$, which for large-scale problems quickly becomes a handicap.

Crammer and Singer [39] have proposed an efficient decomposition method for learning these parameters based on separating the constraints of (5.42) into m disjoint sets $\{\boldsymbol{b}_i | \langle \boldsymbol{b}_i, \mathbf{1}_K\rangle = 0, \boldsymbol{b}_i \leqslant C\mathbf{y}_i\}_{i=1,\ldots,m}$. The proposed algorithm works in a loop and chooses an example $(\mathbf{x}_t, y_t) \in S$ at each loop by optimizing the value of the objective function (5.42) and updating the variables $\boldsymbol{b}_t$ under the set of constraints $\boldsymbol{b}_t^\top \mathbf{1}_K = 0$ and $\boldsymbol{b}_t \leqslant C\mathbf{y}_t$. By isolating the contribution of the example $(\mathbf{x}_t, y_t)$ in the objective function (5.42), we obtain:

$$\mathcal{Q}^{(t)}(\boldsymbol{b}_t) = -\frac{1}{2}(\boldsymbol{b}_t^\top \boldsymbol{b}_t)A_t - \boldsymbol{b}_t^\top B_t + C_t, \tag{5.48}$$

where:

$$A_t = \kappa(\mathbf{x}_t, \mathbf{x}_t) > 0, \tag{5.49}$$

$$B_t = -\mathbf{y}_t + \sum_{i=1, i\neq t}^{m} \kappa(\mathbf{x}_i, \mathbf{x}_t)\boldsymbol{b}_t, \tag{5.50}$$

$$C_t = \sum_{i=1, i\neq t}^{m} \boldsymbol{b}_i^\top \mathbf{y}_i - \frac{1}{2}\sum_{i=1, i\neq t}^{m}\sum_{j=1, j\neq t}^{m} \boldsymbol{b}_i^\top \boldsymbol{b}_j \kappa(\mathbf{x}_i, \mathbf{x}_j). \tag{5.51}$$

Input :

- A training set $S = ((\mathbf{x}_1, y_1), \ldots, (\mathbf{x}_m, y_m))$;

Initialization:

- Maximum number of iterations T;
- Initialize $\forall i \in \{1, \ldots, m\}$, $\boldsymbol{b}_i \leftarrow 0$.

for $t = 1, \ldots, T$ **do**

- Choose an example$(\mathbf{x}_t, y_t)$;
- Estimate the variables A_t and B_t corresponding to: // ▷ (5.50), (5.51)

$$A_t \leftarrow \kappa(\mathbf{x}_t, \mathbf{x}_t),$$

$$B_t \leftarrow -\mathbf{y}_t + \sum_{i=1, i \neq t}^{m} \kappa(\mathbf{x}_i, \mathbf{x}_t)\boldsymbol{b}_t.$$

- Resolve the reduced optimization problem // ▷ (5.52)

$$\min_{\boldsymbol{b}_t \in \mathbb{R}^K} \frac{1}{2}(\boldsymbol{b}_t^\top \boldsymbol{b}_t)A_t + \boldsymbol{b}_t^\top B_t,$$

$$\text{s.t. } \boldsymbol{b}_t^\top \mathbf{1}_K = 0 \text{ and } \boldsymbol{b}_t \leqslant C\mathbf{y}_t.$$

Output : The classifier $\mathbf{x} \mapsto \operatorname{argmax}_{k \in \mathcal{Y}} \sum_{i=1}^{m} b_{ik}\kappa(\mathbf{x}, \mathbf{x}_i)$.

Algorithm 14: Multi-class SVM

The function C_t (5.51) does not depend on $\boldsymbol{b}_t$. Maximizing $\mathcal{Q}^{(t)}$ (5.48) is then equivalent to minimizing the following problem, with K variables and $K + 1$ constraints:

$$\begin{aligned} &\min_{\boldsymbol{b}_t \in \mathbb{R}^K} \frac{1}{2}(\boldsymbol{b}_t^\top \boldsymbol{b}_t)A_t + \boldsymbol{b}_t^\top B_t, \\ &\text{s.t. } \boldsymbol{b}_t^\top \mathbf{1}_K = 0 \text{ and } \boldsymbol{b}_t \leqslant C\mathbf{y}_t. \end{aligned} \tag{5.52}$$

□

The pseudo-code for the multi-class SVM algorithm, as proposed by Crammer and Singer in their 2001 publication [39], is presented in Algorithm 14. This algorithm extends the traditional binary SVM to handle multiple classes by formulating the problem as a single optimization task. The authors introduce a method that maximizes the margin between all classes simultaneously, ensuring robust classification performance.

In their publication, Crammer and Singer discuss various aspects crucial for the practical implementation of this algorithm. They present different stopping criteria that can be employed to determine when the optimization process should terminate. These criteria are essential for balancing the trade-off between achieving an optimal solution and computational efficiency.

Additionally, the authors provide insights into several tricks and techniques for an efficient implementation of the multi-class SVM algorithm. These include strategies for managing the computational complexity, which can be significant, especially for large datasets with many classes. They also explore methods for improving convergence speed and ensuring numerical stability during the optimization process.

To Sum Up

1. *SVMs implement the principle of structural risk minimization by balancing the empirical risk (classification error on training data) and the model complexity. This is achieved through regularization, where the margin maximization acts as a form of regularization that prevents overfitting by controlling the capacity of the model.*
2. *The concept of the margin is central to both SVMs and the perceptron algorithm, particularly in scenarios where classes are linearly separable. The margin represents the distance between the hyperplane and the nearest data points from either class, influencing the model's robustness and generalization performance.*
3. *The original SVM formulation, called the hard margin, seeks a hyperplane that perfectly separates data points with the maximum possible margin, applicable when data is linearly separable.*
4. *An extension of the hard margin introduces slack variables to handle non-separable data, allowing for some misclassifications and making the model more robust to noise.*
5. *The kernel trick, is a technique that uses kernel functions to implicitly map data into higher-dimensional spaces, enabling SVMs to handle non-linear classification by finding a linear separator in the transformed space.*
6. *In the context of SVMs, the Rademacher complexity term for a class of linear classifiers in the feature space with bounded norm, where observations lie within a hypersphere of radius R, is $O(\frac{R}{\rho\sqrt{m}})$, where ρ is the margin and m is the size of the training set. This term highlights the importance of the margin in controlling the model's complexity and generalization ability, especially in cases where classes are linearly separable.*
7. *While traditional SVMs are designed for binary classification, they can be extended to handle multi-class problems using combined strategies such as one-vs-one, one-vs-rest, or more sophisticated uncombined methods like the Crammer-Singer formulation.*

5.4 Exercises

1. Consider the following points in $\mathbb{R}^2$:
   - Class 1: (2, 2), (4, 4);
   - Class 2: (2, −2), (4, −4).

   Find the equation of the maximum margin hyperplane that separates these points. Calculate the margin.
2. Consider the polynomial kernel function of degree 3:

$$\kappa : \mathbb{R}^2 \times \mathbb{R}^2 \to \mathbb{R}$$
$$(\mathbf{x}, \mathbf{x}') \mapsto (\mathbf{x}^\top \mathbf{x}' + 1)^3.$$

   Furthermore, let $\phi : \mathbb{R}^2 \to \mathbb{R}^{10}$ be a feature map defined as:

$$\phi(\mathbf{x}) = (1, \sqrt{3}x_1, \sqrt{3}x_2, \sqrt{3}x_1^2, \sqrt{3}x_2^2, \sqrt{3}x_1x_2, x_1^3, x_2^3, x_1^2x_2, x_1x_2^2).$$

   Show that $\kappa(\mathbf{x}, \mathbf{x}') = \langle \phi(\mathbf{x}), \phi(\mathbf{x}') \rangle$.
3. Show that the following functions are Mercer kernels:
   (a) $\kappa_1(\mathbf{x}, \mathbf{x}') = \exp(\gamma \|\mathbf{x} - \mathbf{x}'\|^2)$ for $\gamma > 0$.
   (b) $\kappa_2(\mathbf{x}, \mathbf{x}') = (\alpha \mathbf{x}^\top \mathbf{x}' + \beta)^2$ with $\alpha, \beta > 0$.
   (c) $\kappa_3(\mathbf{x}, \mathbf{x}') = \min(\mathbf{x}^\top \mathbf{x}', c)$ for some constant $c > 0$.
4. Show that if κ_1 and κ_2 are two kernels, then:
   (a) $c\kappa_1$ is a kernel for some constant $c > 0$.
   (b) $\kappa_1 + \kappa_2$ is a kernel.
   (c) $\kappa_1\kappa_2$ is a kernel.
5. For a d-dimensional space $\mathcal{X} \subseteq \mathbb{R}^d$ and a homogeneous polynomial kernel of degree p, defined as:

$$\kappa : \mathcal{X} \times \mathcal{X} \to \mathbb{R}$$
$$(\mathbf{x}, \mathbf{x}') \mapsto (\mathbf{x}^\top \mathbf{x}')^p.$$

   Prove that the dimension of the minimal embedding space is given by the binomial coefficient: $\binom{p+d-1}{p}$.
   Hint: Consider the feature map ϕ associated with the polynomial kernel and analyze the structure of the resulting feature space. Focus on the combinatorial nature of the monomials formed by the kernel and how they relate to the binomial coefficient.
6. Let κ be a positive definite kernel corresponding to a minimal embedding space $\mathbb{H}$. Show that the VC dimension of the SVM associated with this kernel, where the error penalty C can take any value, is given by $dim(\mathbb{H}) + 1$.

Hint: Consider the minimal embedding space $\mathbb{H}$ induced by the positive definite kernel κ. Identify $dim(\mathbb{H})$ points in the feature space whose position vectors are linearly independent. Use the fact that any set of $dim(\mathbb{H})$ linearly independent points in a $dim(\mathbb{H})$-dimensional space can be shattered by hyperplanes. Show that an SVM with kernel κ can realize these hyperplanes by adjusting the error penalty C to ensure separation, and conclude.

7. Consider the class of Mercer kernels $\kappa(\mathbf{x}_1, \mathbf{x}_2)$ that satisfy the following properties:

   - The kernel function $\kappa(\mathbf{x}_1, \mathbf{x}_2)$ approaches zero, as the distance between $\mathbf{x}_1$ and $\mathbf{x}_2$ approaches infinity,
   - The kernel evaluated at the same point $\mathbf{x}$ is bounded by a constant.

   If the data points can be chosen arbitrarily from $\mathbb{R}^d$, then show the family of SVM classifiers using these kernels, with the error penalty C allowed to take any value, has an infinite VC dimension.
   Hint: Consider the Gram matrix $\mathbf{K} = [\kappa_{ij} = \kappa(\mathbf{x}_i, \mathbf{x}_j)]_{ij}$ in the feature space $\mathbb{H}$. By choosing training data such that the off-diagonal elements $\kappa_{i \neq j}$ are arbitrarily small and ensuring the diagonal elements $\kappa_{i=i}$ are bounded by a constant, the matrix $\mathbf{K}$ becomes full rank. This implies that the set of vectors corresponding to these kernel evaluations are linearly independent in $\mathbb{H}$. Using the fact that any set of linearly independent points in a feature space can be shattered by hyperplanes, and knowing that SVMs with sufficiently large error penalties can realize these hyperplanes, conclude that the SVMs can shatter any finite set of points.
8. Apply the Karush-Kuhn-Tucker (KKT) conditions to the problem:

$$\min_{\mathbf{w} \in \mathbb{R}^2} w_1^2 + w_2^2 \quad \text{subject to} \quad w_1 + w_2 \geqslant 1,$$

   and verify the optimal solution.
9. Slater's condition is a constraint qualification that guarantees strong duality in convex optimization problems. Informally, it requires the existence of a strictly feasible point-one that satisfies all equality constraints exactly and strictly satisfies all inequality constraints (i.e., inequalities hold with strict inequality).

   Formally, consider a convex problem with inequality constraints $g_i(\boldsymbol{w}) \leqslant 0$ and equality constraints $h_j(\boldsymbol{w}) = 0$. Slater's condition holds if there exists some $\boldsymbol{w}_0$ such that

$$g_i(\boldsymbol{w}_0) < 0 \quad \text{for all } i, \quad \text{and} \quad h_j(\boldsymbol{w}_0) = 0 \quad \text{for all } j.$$

   For the linear programming above, where the inequality is $\boldsymbol{w} \succeq 0$ (component-wise non-negativity), Slater's condition requires the existence of some $\boldsymbol{w}_0 > 0$ (all components strictly positive) satisfying $\mathbf{A}\boldsymbol{w}_0 = \mathbf{b}$.

   When Slater's condition holds, the Karush-Kuhn-Tucker (KKT) conditions are both necessary and sufficient for optimality, and the duality gap-the difference between the primal and dual optimal values-is zero. This means the primal and dual problems have the same optimal value, allowing one to solve either problem equivalently.

Consider the primal linear optimization problem

$$\min_{w} \mathbf{c}^\top w \quad \text{subject to} \quad \mathbf{A}w = \mathbf{b}, \quad w \succeq 0,$$

where $\mathbf{c} \in \mathbb{R}^d$, $\mathbf{b} \in \mathbb{R}^m$, and $\mathbf{A} \in \mathbb{R}^{m \times d}$.

(a) Derive the dual problem associated with this primal problem.
(b) Define the duality gap and explain its significance in convex optimization.
(c) State Slater's condition for convex optimization problems and explain its role in ensuring strong duality.
(d) For the given problem, specify what Slater's condition means in terms of the existence of a strictly feasible point.
(e) Show that if Slater's condition holds, then the duality gap is zero; that is, strong duality holds between the primal and dual problems.

Hints: Use the Lagrangian function to derive the dual problem. Recall that weak duality always holds: the dual optimal value is a lower bound on the primal optimal value. Slater's condition ensures no duality gap and that the dual optimum is attained. For linear programs, strict feasibility means there exists a feasible point in the interior of the nonnegative orthant satisfying the linear equality constraints.

10. Let $\mathbf{u} \in \mathbb{R}^d$ be a vector, and define $\mathbf{w}$ as the vector with coordinates $w_i = \max(0, u_i)$. For any $p \in [1, +\infty)$, the L_p norm of a vector $\mathbf{v} \in \mathbb{R}^d$ is defined as:

$$\|\mathbf{v}\|_p = \left(\sum_{i=1}^{d} |v_i|^p \right)^{1/p}.$$

The dual norm p_* is defined as:

$$p_* = \frac{p}{p-1}.$$

Prove that:

$$\|\mathbf{w}\|_{p_*} = \max_{\mathbf{z} \in B_p^+} \mathbf{z}^\top \mathbf{u},$$

where B_p^+ denotes the non-negative part of the unit ball in the L_p norm, i.e., the set of vectors $\mathbf{z}$ such that $\mathbf{z} \geqslant 0$ (component-wise) and $\|\mathbf{z}\|_p \leqslant 1$.

Hint: Consider the vector $\mathbf{z} \in B_p^+$ *with the components* $z_i = \frac{w_i^{p_*-1}}{\left(\sum_j w_j^{p_*}\right)^{1/p}}$. *Show that* $\|\mathbf{w}\|_{p_*} \geqslant \|\mathbf{z}\|_p \|\mathbf{w}\|_{p_*} \geqslant \mathbf{z}^\top \mathbf{w} \geqslant \mathbf{z}^\top \mathbf{u}$ *and conclude by taking into account that* $p(p_* - 1) = p_*$.

11. Demonstrate that a SVM with a RBF kernel, defined as:

$$\kappa : \mathcal{X} \times \mathcal{X} \to \mathbb{R}$$
$$(\mathbf{x}, \mathbf{x}') \mapsto \exp\left(-\frac{\|\mathbf{x} - \mathbf{x}'\|^2}{2\sigma^2}\right),$$

can approximate any continuous function on a compact subset of the input space.
Hint: Show that the RBF kernel separates points (i.e. $\forall(\mathbf{x}, \mathbf{x}') \in \mathcal{X}, \kappa(\mathbf{x}, \mathbf{x}) \neq \kappa(\mathbf{x}, \mathbf{x}')$*). Verify that the RKHS* $\mathbb{H}$ *associated with the RBF kernel contains the constant functions. Apply the Stone-Weierstrass theorem (Chap. 4, Theorem 4.2) and deduce that* $\mathbb{H}$ *is dense in the space of continuous functions* $C(\mathcal{X})$ *with respect to the supremum norm* $\|.\|_\infty$. *From the fact that a SVM with a RBF kernel can represent and function in* $\mathbb{H}$ *by choosing appropriate support vectors and weights, conclude by using the triangle inequality.*

12. Sequential Minimal Optimization (SMO) is an efficient algorithm for solving the quadratic programming (QP) problem that arises in training SVMs. It breaks the large QP problem into a series of smallest possible subproblems involving only two Lagrange multipliers, which can be solved analytically.

    Consider a binary classification dataset $\{(\mathbf{x}_i, y_i)\}_{i=1}^m$ with inputs $\mathbf{x}_i \in \mathbb{R}^d$ and labels $y_i \in \{-1, +1\}$. Consider the dual form of the SVM optimization problem (5.35):

$$\max_{\boldsymbol{\alpha}} \quad \sum_{i=1}^m \alpha_i - \frac{1}{2} \sum_{i,j=1}^m \alpha_i \alpha_j y_i y_j \kappa(\mathbf{x}_i, \mathbf{x}_j)$$

subject to

$$0 \leqslant \alpha_i \leqslant C, \quad \text{for } i = 1, \ldots, m, \quad \text{and} \quad \sum_{i=1}^m \alpha_i y_i = 0,$$

where $\kappa(\mathbf{x}_i, \mathbf{x}_j)$ is the kernel function and $C > 0$ is the regularization parameter.

    (a) Explain why SMO optimizes two Lagrange multipliers α_i and α_j at each iteration instead of one, and why this is the smallest possible optimization subproblem.

(b) Describe how the equality constraint $\sum_i \alpha_i y_i = 0$ restricts the feasible region of α_i and α_j, and how this leads to the feasible region being a line segment in the (α_i, α_j) plane.
(c) Derive the update formulas for α_i and α_j when optimizing these two variables jointly, assuming all other α_k are fixed.
(d) Explain the role of the box constraints $0 \leqslant \alpha_i, \alpha_j \leqslant C$ and how the updated α_j is clipped to lie within the feasible bounds.
(e) Discuss the heuristics used in SMO for selecting the pair (α_i, α_j) to optimize at each step, including the use of the KKT conditions and error terms E_i.
(f) Explain how SMO uses the KKT conditions to check for convergence and why satisfying these conditions implies that the solution is optimal.
(g) Discuss the computational advantages of SMO compared to solving the full QP problem directly, especially in terms of memory and speed.

Hints: The equality constraint $\sum_i \alpha_i y_i = 0$ means that when one multiplier α_i changes, at least one other α_j must change to maintain feasibility. The two-dimensional subproblem can be solved analytically by minimizing a quadratic function along the feasible line segment defined by the constraints. The error term $E_i = f(\mathbf{x}_i) - y_i$, where $f(\mathbf{x}_i)$ is the current prediction, is used to select the second multiplier to maximize the step size and improve convergence speed. The KKT conditions for each α_i specify when it is optimal to stop updating that multiplier.

Chapter 6
Boosting

Ensemble methods are a general family of learning algorithms that have their origins in the pioneering work of Valiant [175], who founded the principle of the boosting algorithm. The aim of this algorithm is to combine classifiers known as *weak learners* in order to build a final, higher-performance classifier known as *strong learner* defined as follows:

Definition 6.1 A learner is said to be strong, if for any probability distribution $\mathcal{D}$ on the data, for any precision $\epsilon \in]0, \frac{1}{2}[$ and for any $\delta > 0$; finds a function from a class of functions $\mathcal{F}$ and a polynomial number of training samples generated i.i.d. following $\mathcal{D}$; that will have a generalization error less than ϵ with a probability of at least $1 - \delta$.

Similarly, the definition of a γ-weak learner is:

Definition 6.2 A learner is said to be weak, if for any probability distribution $\mathcal{D}$ on the data, for any accuracy $\epsilon \in]0, \frac{1}{2}[$ and for any $\delta > 0$; finds a function from a class of functions $\mathcal{F}$ and a polynomial number of training examples generated i.i.d. following $\mathcal{D}$; which will have a generalization error greater than $\epsilon - \frac{1}{2}$ with a probability of at least $1 - \delta$.

On this basis Kearns and Valiant [92], have posed the fundamental question in the following boosting theory:

Can weak learnability imply strong learnability?

M.-R. Amini, *Advanced Supervised and Semi-supervised Learning*,
Cognitive Technologies, https://doi.org/10.1007/978-3-031-99928-4_6

Schapire [148] gave an affirmative answer to this question by proposing the first boosting algorithm. Subsequently, other boosting algorithms were proposed. In this chapter, we will present the Adaptive Boosting (AdaBoost) algorithm [58, 149], which, among all these methods, has enjoyed great popularity due in particular to its performance and theoretical justifications.

In the following, Sect. 6.1 provides a detailed explanation of the principles behind the AdaBoost algorithm. We will explore how this algorithm corresponds to the concept of empirical risk minimization. In Sect. 6.2, we will set a generalization bound for this model using the concept of Rademacher complexity. Our analysis will show that the complexity of the function class, created by linearly combining weak learners, aligns with the complexity of the original function class. The adaptation of this model for multi-class classification is covered in Sect. 6.3.

6.1 AdaBoost

This algorithm generates a set of weak learners and combines them using a vote-based method. The final classifier is thus called the voting classifier. Weak classifiers are trained sequentially; the tth classifier generated takes into account the errors of the previous classifier already built. This is achieved by assigning a weight $D(i)$ to each example i in the training set: an example misclassified by the $(t-1)$th classifier will be assigned a higher weight than a well-classified example. In this way, at step t, the new weak classifier will focus on classifying examples deemed difficult by the classifier at step $t-1$.

Algorithm 15 describes the first variant of the AdaBoost algorithm proposed by Freund and Schapire [58] for a two-class classification problem.

At initialization, the distribution of weights over the examples is a uniform distribution ($\forall i \in \{1, \ldots, m\}, D^{(1)}(i) = \frac{1}{m}$) and we choose the number T of classifiers involved in the final vote. At each iteration t, a binary classifier, belonging to the class $\mathcal{F} = \{f : \mathbb{R}^d \to \{-1, +1\}\}$, is learned by minimizing the empirical classification error and considering the weights $D^{(t)}$ over the examples ($f_t = \underset{f \in \mathcal{F}}{\operatorname{argmin}} \sum_{i: f(\mathbf{x}_i) \neq y_i} D^{(t)}(i)$). From the empirical error ϵ_t of the classifier learned on the training set:

$$\epsilon_t = \sum_{i: f_t(\mathbf{x}_i) \neq y_i} D^{(t)}(i),$$

the learned classifier is assigned a weight a_t that is all the greater as the error ϵ_t is large. The weight distribution is then updated according to the following rule:

$$\forall i \in \{1, \ldots, m\}, D^{(t+1)}(i) = \frac{D^{(t)}(i) e^{-a_t y_i f_t(\mathbf{x}_i)}}{Z^{(t)}}, \quad (6.1)$$

Input :

- A training set $S = ((\mathbf{x}_1, y_1), \dots, (\mathbf{x}_m, y_m))$.

Initialization:

- Maximum number of iterations T;
- Initalize the distribution of weights $\forall i \in \{1, \dots, m\}, D^{(1)}(i) = \frac{1}{m}$.

for $t = 1, \dots, T$ **do**

- Train a classifier $f_t : \mathbb{R}^d \to \{-1, +1\}$ with the distribution $D^{(t)}$;
- Set $\epsilon_t = \sum_{i: f_t(\mathbf{x}_i) \neq y_i} D^{(t)}(i)$;
- Choose $a_t = \frac{1}{2} \ln \frac{1-\epsilon_t}{\epsilon_t}$;
- Update weight distribution on examples: // ▷ (6.1)

$$\forall i \in \{1, \dots, m\}, D^{(t+1)}(i) = \frac{D^{(t)}(i) e^{-a_t y_i f_t(\mathbf{x}_i)}}{Z^{(t)}},$$

where $Z^{(t)} = \sum_{i=1}^{m} D^{(t)}(i) e^{-a_t y_i f_t(\mathbf{x}_i)}$ is a normalization factor so that $D^{(t+1)}$ remains a distribution.

Output : The voting classifier $\forall \mathbf{x}, F(\mathbf{x}) = \text{sign}\left(\sum_{t=1}^{T} a_t f_t(\mathbf{x})\right)$.

Algorithme 15: AdaBoost

where $Z^{(t)} = \sum_{i=1}^{m} D^{(t)}(i) e^{-a_t y_i f_t(\mathbf{x}_i)}$ is a normalization factor so that $D^{(t+1)}$ remains a distribution.

If the current classifier f_t gives a bad class prediction on a training example $(\mathbf{x}_i, y_i)$ (i.e. $y_i f_t(\mathbf{x}_i) = -1$), the weight associated with this example, $D^{(t+1)}(i)$, will be higher than its previous weight, $D^{(t)}(i)$.

The final classifier F is determined by taking the sign of the weighted linear combination of the classifiers $(f_t)_{t=1}^{T}$ with their corresponding weights $(a_t)_{t=1}^{T}$. For a new example $\mathbf{x}$, the class prediction is made based on a weighted majority vote among these classifiers.

6.1.1 Link with the ERM Principle

Schapire [149] showed that Algorithm 15 minimizes the exponential convex bound of the empirical error of the final classifier $F(\mathbf{x}) = \text{sign}(h(\mathbf{x}))$ where $h(\mathbf{x}) = \sum_{t=1}^{T} a_t f_t(\mathbf{x})$ (Chap. 2, Fig. 2.1):

$$\frac{1}{m} \sum_{i=1}^{m} \mathbb{1}_{y_i \neq F(\mathbf{x}_i)} \leqslant \frac{1}{m} \sum_{i=1}^{m} e^{-y_i h(\mathbf{x}_i)}. \tag{6.2}$$

Theorem 6.1 (Exponential Convergence of AdaBoost [149]) *Given the AdaBoost Algorithm 15, which minimizes the exponential loss (6.2), the empirical error of the final classifier learned by the algorithm converges exponentially to zero as the number of iterations T increases, provided that each weak classifier used within the algorithm performs better than random guessing.*

Proof Consider the exponential bound of the previous inequality. We have:

$$\begin{aligned}\frac{1}{m}\sum_{i=1}^{m} e^{-y_i h(\mathbf{x}_i)} &= \frac{1}{m}\sum_{i=1}^{m} e^{-y_i \sum_{t=1}^{T} a_t f_t(\mathbf{x}_i)} \\ &= \sum_{i=1}^{m} \frac{1}{m} e^{-y_i a_1 f_1(\mathbf{x}_i)} \prod_{t>1} e^{-y_i a_t f_t(\mathbf{x}_i)} \\ &= \sum_{i=1}^{m} D^{(1)}(i) e^{-y_i a_1 f_1(\mathbf{x}_i)} \prod_{t>1} e^{-y_i a_t f_t(\mathbf{x}_i)}.\end{aligned}$$

According to the definition of the distribution of examples at iteration 2, $D^{(2)}$ (6.1), this gives:

$$\begin{aligned}\frac{1}{m}\sum_{i=1}^{m} e^{-y_i h(\mathbf{x}_i)} &= \sum_{i=1}^{m} D^{(2)}(i) Z^{(1)} \prod_{t>1} e^{-y_i a_t f_t(\mathbf{x}_i)} \\ &= Z^{(1)} \sum_{i=1}^{m} D^{(2)}(i) e^{-y_i a_2 f_2(\mathbf{x}_i)} \prod_{t>2} e^{-y_i a_t f_t(\mathbf{x}_i)}.\end{aligned}$$

By recursion, we arrive at:

$$\frac{1}{m}\sum_{i=1}^{m} e^{-y_i h(\mathbf{x}_i)} = \prod_{t=1}^{T} Z^{(t)} \underbrace{\sum_{i=1}^{m} D^{(T+1)}(i)}_{=1}. \tag{6.3}$$

Since the predictions of the weak classifier, f_t, are in $\{-1, +1\}$, and writing $Z^{(t)}$ as two sums over the well-classified and misclassified examples according to f_t, we have:

$$Z^{(t)} = \sum_{i:f_t(\mathbf{x}_i)\neq y_i} D^{(t)}(i) e^{a_t} + \sum_{i:f_t(\mathbf{x}_i)=y_i} D^{(t)}(i) e^{-a_t},$$

hence:

$$Z^{(t)} = \epsilon_t e^{a_t} + (1 - \epsilon_t)e^{-a_t}, \tag{6.4}$$

where $\epsilon_t = \sum_{i:f_t(\mathbf{x}_i) \neq y_i} D^{(t)}(i)$.

The coefficient a_t at iteration t is chosen so that the normalization factor $Z^{(t)}$ is minimal, which by canceling the derivative of $Z^{(t)}$ with respect to a_t gives the following analytical expression for this coefficient:

$$a_t = \frac{1}{2} \ln \frac{1 - \epsilon_t}{\epsilon_t}. \tag{6.5}$$

Replacing this expression in the definition of $Z^{(t)}$ (6.4), gives:

$$Z^{(t)} = 2\sqrt{\epsilon_t(1 - \epsilon_t)}. \tag{6.6}$$

Now using the inequality $\forall z \in [0, \frac{1}{2}], \sqrt{1 - 4z^2} \leqslant e^{-2z^2}$ and in the case where $\epsilon_t < \frac{1}{2}$ we obtain:

$$Z^{(t)} \leqslant e^{-2\gamma_t^2}, \tag{6.7}$$

where $\gamma_t = \frac{1}{2} - \epsilon_t > 0$. In the case where $\forall t, \epsilon_t < \frac{1}{2}$, we then obtain the following exponential bound on the empirical error of the classifier learned by Algorithm 15:

$$\begin{aligned} \frac{1}{m}\sum_{i=1}^{m} \mathbb{1}_{y_i \neq F(\mathbf{x}_i)} &\leqslant \frac{1}{m}\sum_{i=1}^{m} e^{-y_i h(\mathbf{x}_i)} = \prod_{t=1}^{T} Z^{(t)} \\ &\leqslant \underbrace{\prod_{t=1}^{T} e^{-2\gamma_t^2}}_{=e^{-2\sum_{t=1}^{T}\gamma_t^2}}. \end{aligned}$$

□

Thus, if weak classifiers give better predictions than random prediction ($\forall t, \epsilon_t < \frac{1}{2}$), the exponential upper bound of the empirical error of the final classifier F decreases exponentially as a function of the number of iterations.

6.1.2 Sampling with Rejection

A simple yet effective way to sample the training examples at iteration t from the distribution $D^{(t)}$ is to employ the rejection sampling technique. This method

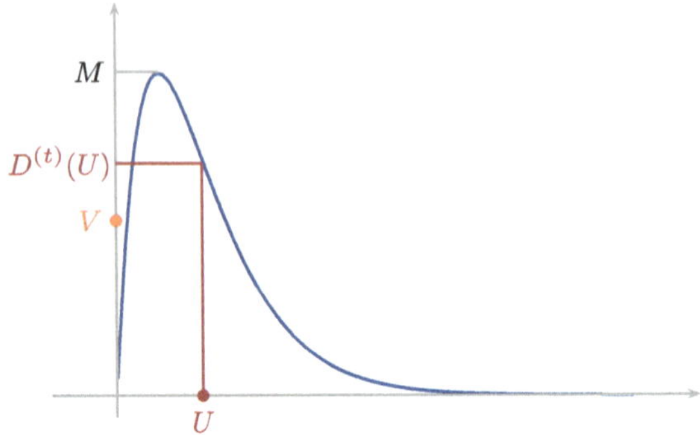

Fig. 6.1 Illustration of the rejection sampling technique

is particularly useful when dealing with weighted distributions of examples, as is the case in the AdaBoost algorithm, where the weights of the examples are adjusted based on their classification difficulty.

The rejection sampling technique involves two main steps:

1. Random Drawing:
   - Simultaneously draw a random index U from the set $\{1, \ldots, m\}$, where m is the total number of training examples;
   - Draw a random real number V from the interval $[0, M]$, where M is an upper bound of the distribution $D^{(t)}$. This upper bound is determined as $M = \max_{i \in \{1,\ldots,m\}} D^{(t)}(i)$, ensuring that M is the maximum weight assigned to any example in the distribution at iteration t.
2. Acceptance or Rejection:
   - Compare the drawn value V with the weight $D^{(t)}(U)$ of the example indexed by U. If $D^{(t)}(U) > V$, the example with index U is accepted and included in the sample for the current iteration. otherwise, the example is rejected, and the process is repeated.

The rejection sampling technique, illustrated in Fig. 6.1, ensures that examples are sampled according to their weights in the distribution $D^{(t)}$, with higher-weighted examples having a greater probability of being selected. The process is efficient and straightforward to implement (Algorithm 16), making it a popular choice for algorithms like AdaBoost that require adaptive sampling strategies.

6.2 Theoretical Study

We will now present a margin-based generalization bound of classifiers learned with the AdaBoost algorithm [150]. This presentation follows the analysis given in Sect. 5.3.1 and allows us to examine the generalization properties of Algorithm 15.

The notion of margin of the voting classifier considered in this section is based on the ℓ_1 norm and it differs somewhat from the notion of margin used so far (4.1). If we consider the weight vector $\boldsymbol{a} = (a_1, \ldots, a_T)^\top$ obtained after T iterations of Algorithm 15, the linear combination of weak learners $h : \mathbf{x} \mapsto \sum_{t=1}^{T} a_t f_t$ can be seen as the scalar product between the weight vector $\boldsymbol{a} = (a_1, \ldots, a_T)^\top$ and the decision vector of weak classifiers for a given $\mathbf{x}$ example $\boldsymbol{f}(\mathbf{x}) = (f_1(\mathbf{x}), \ldots, f_T(\mathbf{x}))$; that is, $h(\mathbf{x}) = \langle \boldsymbol{a}, \boldsymbol{f}(\mathbf{x})\rangle$. Compared to SVMs, the vector $\boldsymbol{f}(\mathbf{x})$ can be seen as the new representation of $\mathbf{x}$ in the feature space defined by the weak learners $(f_t)_{t=1}^{T}$ and the vector $\boldsymbol{a} = (a_1, \ldots, a_T)^\top$ plays the same role as the weight vector $\boldsymbol{w}$ in the case of SVMs with the difference that the components of the vector $\boldsymbol{a}$ are this time all strictly positive. The ℓ_1 margin of an example $(\mathbf{x}, y) \in \mathcal{X} \times \{-1, +1\}$ with respect to the voting classifier is then defined by:

$$\rho_1(\mathbf{x}, y) = \frac{y\sum_{t=1}^{T} a_t f_t(\mathbf{x})}{\sum_{t=1}^{T} a_t} = \frac{y\langle \boldsymbol{a}, \boldsymbol{f}(\mathbf{x})\rangle}{||\boldsymbol{a}||_1}. \tag{6.8}$$

Similarly, we define the ℓ_1 margin of a voting classifier on a training set $S = (\mathbf{x}_i, y_i)_{i=1}^{m}$ as:

$$\rho = \max_{\boldsymbol{a}} \min_{i\in\{1,\ldots,m\}} \rho_1(\mathbf{x}_i, y_i) = \max_{\boldsymbol{a}} \min_{i\in\{1,\ldots,m\}} \frac{y_i\langle \boldsymbol{a}, \boldsymbol{f}(\mathbf{x}_i)\rangle}{||\boldsymbol{a}||_1}. \tag{6.9}$$

Input :

- A training set $S = ((\mathbf{x}_1, y_1), \ldots, (\mathbf{x}_m, y_m))$;
- A probability distribution $D^{(t)}(i), i \in \{1, \ldots, m\}$;

Initialization:

- $S_t \leftarrow \emptyset$; $M = \max_{i\in\{1,\ldots,m\}} D^{(t)}(i)$; $i \leftarrow 1$.

while $i \leqslant m$ **do**
  Randomly select an index $U \in \{1, \ldots, m\}$;
  Randomly select a number $V \in [0, M]$;
  if $D^{(t)}(U) > V$ **then**
    $S_t \leftarrow S_t \cup \{(\mathbf{x}_U, y_U)\}$;
    $i \leftarrow i + 1$;

Output : Sampled training set S_t.

Algorithme 16: Sampling with rejection

6.2.1 Margin Based Bound on the Empirical Error

Theorem 6.1 can be extended to consider the margin-based empirical error of the voting classifier learned by the AdaBoost algorithm. This extension is defined for a given margin $\rho > 0$, which serves as a threshold to evaluate the confidence of the classifier's predictions. The margin-based empirical error is a measure that takes into account not only whether the predictions are correct but also the confidence level of these predictions, and it is expressed as follows:

$$\hat{\mathcal{L}}_m\left(\frac{h}{||\boldsymbol{a}||_1}, S\right) = \sum_{i=1}^{m} \mathbb{1}_{y_i \frac{h(\mathbf{x}_i)}{||\boldsymbol{a}||_1} \leqslant \rho}$$

$$= \sum_{i=1}^{m} \mathbb{1}_{y_i h(\mathbf{x}_i) - \rho ||\boldsymbol{a}||_1 \leqslant 0}. \tag{6.10}$$

Theorem 6.2 (Exponential Convergence of Margin-based Empirical Error in AdaBoost) *For the AdaBoost algorithm, the margin-based empirical error of the voting classifier* (6.10) *converges exponentially to zero as the number of iterations T increases. This convergence is guaranteed under the condition that the weak classifiers used in the AdaBoost algorithm perform better than random guessing.*

Proof Let $\rho > 0$, From the inequality $\forall z \in \mathbb{R}, \mathbb{1}_{z \leqslant 0} \leqslant e^{-z}$, we can write:

$$\hat{\mathcal{L}}_m\left(\frac{h}{||\boldsymbol{a}||_1}, S\right) \leqslant \sum_{i=1}^{m} e^{-y_i h(\mathbf{x}_i) + \rho ||\boldsymbol{a}||_1}. \tag{6.11}$$

Hence from (6.3):

$$\hat{\mathcal{L}}_m\left(\frac{h}{||\boldsymbol{a}||_1}, S\right) \leqslant e^{\rho ||\boldsymbol{a}||_1} \prod_{t=1}^{T} Z^{(t)}$$

$$= e^{\rho \sum_{t=1}^{T} a_t} \prod_{t=1}^{T} Z^{(t)}.$$

Using the expressions of the coefficients $a_t, t \geqslant 1$ (6.5) and the normalizing coefficients $Z^{(t)}, t \geqslant 1$ (6.7) found by Algorithm 15, we get:

$$\hat{\mathcal{L}}_m\left(\frac{h}{||\boldsymbol{a}||_1}, S\right) \leqslant 2^T \prod_{t=1}^{T} \left[(1 - \epsilon_t)^{1+\rho} \epsilon_t^{1-\rho}\right]^{1/2}. \tag{6.12}$$

For some values of the margin ρ, the upper bound of the previous margin-based empirical error will tend towards 0 depending on the number of iterations t. Indeed, the function $z \mapsto \sqrt{(1-z)^{1+\rho} z^{1-\rho}}$ is strictly increasing positive on the interval $[0, \frac{1}{2} - \frac{\rho}{2}]$. By setting $\varsigma = \left(\min_{t \in \{1,\dots T\}} \left[\frac{1}{2} - \epsilon_t \right] \right) > 0$ for ρ, checking the inequality $0 < \rho \leqslant 2\varsigma$, we have $\forall \epsilon_t \in \left[0, \frac{1}{2} - \varsigma\right] \subseteq \left[0, \frac{1}{2} - \frac{\rho}{2}\right]$:

$$\begin{aligned}\sqrt{(1-\epsilon_t)^{1+\rho}\epsilon_t^{1-\rho}} &\leqslant \sqrt{\left(1 - \frac{1}{2} + \varsigma\right)^{1+\rho} \left(\frac{1}{2} - \varsigma\right)^{1-\rho}} \\ &= \frac{1}{2}\sqrt{(1+2\varsigma)^{1+\rho}(1-2\varsigma)^{1-\rho}}.\end{aligned}$$

Thus:

$$\hat{\mathcal{L}}_m\left(\frac{h}{||\boldsymbol{a}||_1}, S\right) \leqslant \left[(1+2\varsigma)^{1+\rho}(1-2\varsigma)^{1-\rho}\right]^{T/2}. \tag{6.13}$$

Finally, as $\varsigma > 0$, by rewriting $(1+2\varsigma)^{1+\rho}(1-2\varsigma)^{1-\rho} = (1-4\varsigma^2)\left(\frac{1+2\varsigma}{1-2\varsigma}\right)^{\rho}$, we have $\left(\frac{1+2\varsigma}{1-2\varsigma}\right) > 1$ and the function $z \mapsto (1-4\varsigma^2)\left(\frac{1+2\varsigma}{1-2\varsigma}\right)^{z}$ which is strictly increasing. Hence, by choosing $0 < \rho < \varsigma < \frac{1}{2}$, we have:

$$(1+2\varsigma)^{1+\rho}(1-2\varsigma)^{1-\rho} < (1+2\varsigma)^{1+\varsigma}(1-2\varsigma)^{1-\varsigma}. \tag{6.14}$$

The function $z \mapsto (1+2z)^{1+z}(1-2z)^{1-z}$ is strictly decreasing on the interval $]0, \frac{1}{2}[$ bounded by 1 (Fig. 6.2), i.e. for $0 < \varsigma < \frac{1}{2}$, $(1+2\varsigma)^{1+\varsigma}(1-2\varsigma)^{1-\varsigma} < 1$. The upper bound of the empirical margin-based error thus decreases exponentially as a function of the maximum number of iterations T. □

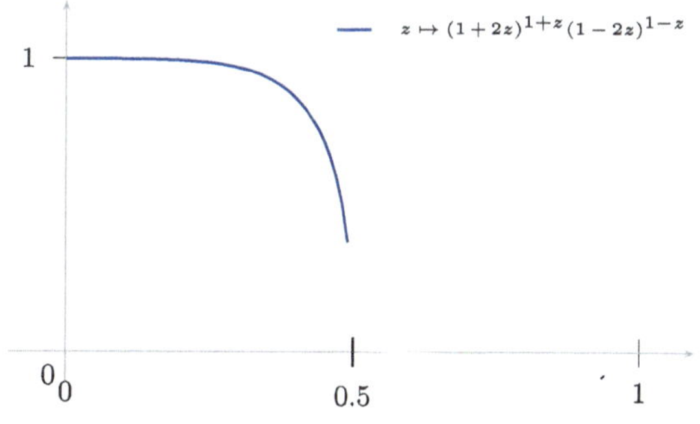

Fig. 6.2 Representative curve of the function $z \mapsto (1+2z)^{1+z}(1-2z)^{1-z}$ on the interval $]0, \frac{1}{2}[$

6.2.2 Margin Based Generalization Error of the Vote Classifier

We will now present a margin-based generalization bound of the voting classifier learned by Algorithm 15. We begin our presentation by showing the link between the convex hull of the class of weak functions $\mathcal{F}$ and the class of functions learned by the AdaBoost algorithm $\mathcal{H} = \{\sum_{t=1}^{T} a_t f_t \mid f_t \in \mathcal{F}\}$. This convex hull is defined by:

$$conv(\mathcal{F}) = \left\{\sum_{t=1}^{T} \lambda_t f_t \mid \forall t \in \{1, \dots, T\}, f_t \in \mathcal{F}, \lambda_t \geqslant 0 \wedge \sum_{t=1}^{T} \lambda_t = 1\right\}.$$

According to Theorem 5.4, we have seen that for a fixed margin $\rho > 0$ and for any $\delta \in (0, 1)$, the generalization error of any function h of a set of functions $\mathcal{H}$ is bounded by its margin-based empirical error on a training set S with a probability of at least $1 - \delta$ by:

$$\mathcal{L}(h) \leqslant \hat{\mathcal{L}}_m(h, S) + \frac{2}{\rho}\hat{\mathfrak{R}}_S(\mathcal{H}) + 3\sqrt{\frac{\ln \frac{2}{\delta}}{2m}}. \tag{6.15}$$

We are now able to state the following theorem which gives a generalization bound of the classifier learned by the AdaBoost algorithm (Algorithm 15).

Theorem 6.3 (Data-dependent Generalization Bound for AdaBoost) *Let $\mathcal{F}$ be the class of weak functions used in Algorithm 15 and T be the maximum number of iterations of the algorithm. Let $\varsigma = \left(\min_{t\in\{1,\dots,T\}} \left[\frac{1}{2} - \epsilon_t\right]\right) > 0$ and set $0 < \rho < \varsigma$. Let $S = ((\mathbf{x}_i, y_i))_{i=1}^{m}$ be a training set of size m identically and independently distributed according to a probability distribution $\mathcal{D}$. Let $\boldsymbol{a} = (a_1, \dots, a_T)$ be the coefficients of the linear combination, and $\mathcal{H} = \left\{\sum_{t=1}^{T} a_t f_t \mid f_t \in \mathcal{F}\right\}$ be the class of functions learned by the AdaBoost algorithm. For any $\delta \in (0, 1)$ and any function $h \in \mathcal{H}$, we have with probability at least $1 - \delta$:*

$$\mathcal{L}(h) \leqslant \left[(1+2\varsigma)^{1+\rho}(1-2\varsigma)^{1-\rho}\right]^{T/2} + \frac{2}{\rho}\hat{\mathfrak{R}}_S(\mathcal{F}) + 3\sqrt{\frac{\ln(2/\delta)}{2m}}. \tag{6.16}$$

Proof Since all the combination coefficients obtained with the AdaBoost algorithm are positive, by normalizing these coefficients $\boldsymbol{a} = (a_1, \dots, a_T)$ by their

ℓ_1 norm, we have $\mathcal{H}' = \left\{\sum_{t=1}^{T} \frac{a_t}{||\boldsymbol{a}||_1} f_t \mid f_t \in \mathcal{F}\right\} \subseteq conv(\mathcal{F})$. Let $\varsigma = \left(\min_{t\in\{1,\dots,T\}} \left[\frac{1}{2} - \epsilon_t\right]\right) > 0$ and set $0 < \rho < \varsigma$. From (6.15), for $\delta \in (0,1)$ we have:

$$\forall h \in \mathcal{H}, \frac{h}{||\boldsymbol{a}||_1} \in \mathcal{H}'; \mathcal{L}\left(\frac{h}{||\boldsymbol{a}||_1}\right) \leqslant \hat{\mathcal{L}}_m\left(\frac{h}{||\boldsymbol{a}||_1}, S\right) + \frac{2}{\rho}\hat{\mathfrak{R}}_S(\mathcal{H}') + 3\sqrt{\frac{\ln\frac{2}{\delta}}{2m}}.$$

Furthermore, $\mathcal{H}' \subseteq conv(\mathcal{F})$, which according to properties 2 and 4 of empirical Rademacher complexity (Theorem 2.5) gives:

$$\hat{\mathfrak{R}}_S(\mathcal{H}') \leqslant \hat{\mathfrak{R}}_S(conv(\mathcal{F})) = \hat{\mathfrak{R}}_S(\mathcal{F}).$$

Finally, we obtain the result by using the inequality (6.13) and noticing that the functions:

$$\mathbf{x} \mapsto \sum_{t=1}^{T} a_t f_t(\mathbf{x}) \text{ and } h : \mathbf{x} \mapsto \sum_{t=1}^{T} \frac{a_t}{||\boldsymbol{a}||_1} f_t(\mathbf{x})$$

have the same generalization errors. □

With this result, we have the guarantee that, for a well-chosen margin value, the margin-based empirical error decreases exponentially towards 0 and that the complexity of the class of voting functions is equal to the complexity of the class of weak learners and it does not depend on the number of iterations T.

6.3 Multiclass AdaBoost

Several extensions of the AdaBoost algorithm to the multi-class case have been proposed. We will present in the following the AdaBoost.M2 algorithm described in [58]. This algorithm is based on the outputs of the weak learners $h \in [0,1]^{\mathcal{X}\times\mathcal{Y}}$ assuming that for any pair $(\mathbf{x}, y) \in \mathcal{X} \times \mathcal{Y}$, the latter estimates the probability that y is the correct class label of the example $\mathbf{x}$. With this assumption, [58] define the binary random variable $b(\mathbf{x}, y)$ which is equal to 1 with probability $h(\mathbf{x}, y)$ and 0 otherwise. They further assume that if the weak learner gives the same confidence score to a pair $(\mathbf{x}_i, y_i)$ where y_i is the correct class of $\mathbf{x}_i$ as to another pair $(\mathbf{x}_i, y)$, then one of the labels among y and y_i is chosen as the true class label of example

$\mathbf{x}_i$ uniformly. The probability that the weak learner h is wrong about the class of example $\mathbf{x}_i$ is then:

$$\begin{aligned}\mathbb{P}\left[b(\mathbf{x}_i, y_i) = 0 \wedge b(\mathbf{x}_i, y) = 1\right] + \frac{1}{2}\mathbb{P}\left[b(\mathbf{x}_i, y_i) = b(\mathbf{x}_i, y)\right] \\ = \frac{1}{2}(1 - h(\mathbf{x}_i, y_i) + h(\mathbf{x}_i, y)),\end{aligned}$$

since for $(u, v) \in \{0, 1\}^2$, the events $b(\mathbf{x}_i, y_i) = u$ and $b(\mathbf{x}_i, y_i) = v$ are independent and the equality $b(\mathbf{x}_i, y_i) = b(\mathbf{x}_i, y)$ is equivalent to the event:

$$(b(\mathbf{x}_i, y_i) = 1 \wedge b(\mathbf{x}_i, y) = 1) \vee (b(\mathbf{x}_i, y_i) = 0 \wedge b(\mathbf{x}_i, y) = 0).$$

6.3.1 Pseudo-misclassification

The classification pseudo-error of the h classifier on an example $\mathbf{x}_i \in S$ with respect to a probability distribution $q : \{1, \ldots, m\} \times \mathcal{Y} \rightarrow [0, 1]$ which weights

---

Input :

- A training set $S = ((\mathbf{x}_1, y_1), \ldots, (\mathbf{x}_m, y_m))$;
- Two probability distributions over the examples $D(i), i \in \{1, \ldots, m\}$ and the classes $q(i, y), i \in \{1, \ldots, m\}, y \in \mathcal{Y} \setminus \{y_i\}$;

Initialization:

- $\tilde{S} \leftarrow \emptyset$;
- $M_1 \leftarrow \max_{i \in \{1,\ldots,m\}} D(i)$;
- $\ell \leftarrow 1$;

```
while ℓ ⩽ m do
    Randomly draw an index U_1 ∈ {1, ..., m};
    Randomly draw a number V_1 ∈ [0, M_1];
    if D(U_1) > V_1 then
        M_2 ← max_{y≠y_{U_1}} q(U_1, y);
        Randomly draw an index U_2 ∈ {1, ..., K − 1};
        Randomly draw a number V_2 ∈ [0, M_2];
        if q(U_1, U_2) > V_2 then
            S̃ ← S̃ ∪ {(x_{U_1}, y_{U_1})};
            ℓ ← ℓ + 1;
```

Output : A sampled training set S_t.

Algorithme 17: Rejection sampling technique with two distributions

the prediction errors with respect to $K - 1$ classes other than the y_i class of the example, and verifying $\forall i \in \{1, \ldots, m\}; \sum_{y \neq y_i} q(i, y) = 1$, is defined by:

$$\begin{aligned} \text{ploss}_q(h, i) &= \sum_{y \neq y_i} q(i, y) \left[\frac{1}{2} (1 - h(\mathbf{x}_i, y_i) + h(\mathbf{x}_i, y)) \right] \\ &= \frac{1}{2} \left(1 - h(\mathbf{x}_i, y_i) + \sum_{y \neq y_i} q(i, y) h(\mathbf{x}_i, y) \right). \end{aligned} \tag{6.17}$$

The weak learner is then learned by minimizing the expectation of this pseudo-error for a distribution D on the examples:

$$\begin{aligned} \epsilon &= \mathbb{E}_{i \sim D} \left[\text{ploss}_q(h, i) \right] \\ &= \frac{1}{2} \sum_{i=1}^{m} D(i) \left(1 - h(\mathbf{x}_i, y_i) + \sum_{y \neq y_i} q(i, y) h(\mathbf{x}_i, y) \right). \end{aligned} \tag{6.18}$$

As in the two-class classification case, the D distribution is iteratively updated so that examples that are difficult to classify have a higher weighting than other examples. This notion of "difficult classification" is defined for each example $\mathbf{x}_i \in S$ of the training set thanks to the distribution $q(i, y)$ on the classes $y \in \mathcal{Y} \backslash \{y_i\}$ which is maintained with weights $w_{i,y}$:

$$\forall i \in \{1, \ldots, m\}, \forall y \neq y_i; q(i, y) = \frac{w_{i,y}}{W_i}, \tag{6.19}$$

where $\forall i$, $W_i = \sum_{y \neq y_i} w_{i,y}$. For an example, these weights are all the higher for certain classes as it is difficult for the weak learner h to distinguish these classes from the true class of the example. For each example $\mathbf{x}_i$, the sum of the weights W_i then gives an indication of how difficult the example is to classify by the classifier h and the distribution D is thus defined by:

$$\forall i \in \{1, \ldots, m\}, D(i) = \frac{W_i}{\sum_{i=1}^{m} W_i}. \tag{6.20}$$

6.3.2 *Sampling by Rejection Following Two Distributions*

The current distributions q and D are then used to sample the examples from the training set to learn a new classifier $h : \mathcal{X} \times \mathcal{Y} \rightarrow [0, 1]$. This double sampling

Input :

- A training set $S = ((\mathbf{x}_1, y_1), \dots, (\mathbf{x}_m, y_m))$;
- A distribution $D^{(1)}$ on the training examples;
- A learning algorithm producing a classifier $h : \mathcal{X} \times \mathcal{Y} \to [0, 1]$ that assigns class labels to examples with some confidence.

Initialization:

- Maximum number of iterations T;
- Initialize the weights $w_{i,y}^{(1)} = \frac{D^{(1)}(i)}{K-1}$ for all $i \in \{1, \dots, m\}, y \in \mathcal{Y} \setminus \{y_i\}$.

for $t = 1, \dots, T$ **do**

- Set $W_i^{(t)} = \sum_{y \neq y_i} w_{i,y}^{(t)}, \forall y \neq y_i; q^{(t)}(i, y) = \frac{w_{i,y}^{(t)}}{W_i^{(t)}}$ and // ▷ (6.19), (6.20)

$$\forall i \in \{1, \dots, m\};\ D^{(t)}(i) = \frac{W_i^{(t)}}{\sum_{i=1}^m W_i^{(t)}};$$

- Train a classifier $h_t : \mathcal{X} \times \mathcal{Y} \to [0, 1]$ on the training set sampled with the distributions $D^{(t)}$ and $q^{(t)}$ (Algorithm 17);
- Estimate the pseudo-loss of the classifier h_t: // ▷ (6.18)

$$\epsilon_t = \frac{1}{2} \sum_{i=1}^{m} D^{(t)}(i) \left(1 - h_t(\mathbf{x}_i, y_i) + \sum_{y \neq y_i} q^{(t)}(i, y) h_t(\mathbf{x}_i, y) \right);$$

- Set $\beta^{(t)} = \frac{\epsilon_t}{1-\epsilon_t}$;
- Update weight on examples

$$\forall i \in \{1, \dots, m\}, y \in \mathcal{Y} \setminus \{y_i\};\ w_{i,y}^{(t+1)} = w_{i,y}^{(t)} \beta_t^{\frac{1}{2}[1+h_t(\mathbf{x}_i, y_i) - h_t(\mathbf{x}_i, y)]}$$

Output : The voting classifier $\forall \mathbf{x}, H(\mathbf{x}) = \underset{y \in \mathcal{Y}}{\operatorname{argmax}} \sum_{t=1}^{T} \left(\ln \frac{1}{\beta^{(t)}} \right) h_t(\mathbf{x}, y)$.

Algorithme 18: AdaBoost multi-class M2

allows to identify examples that are difficult to classify, thanks to the D distribution, and also classes that are difficult to separate, thanks to the q distribution. This process can be easily implemented using the rejection sampling technique presented in Sect. 6.1.2 and summarized in Algorithm 17.

Minimizing the classification error on the sampled base is then equivalent to minimizing the expectation of the pseudo-error, ϵ, defined in (6.18). By setting $\beta = \frac{\epsilon}{1-\epsilon}$, the weights on the classes, $w_{i,y}, \forall i \in \{1, \dots, m\}, y \in \mathcal{Y} \setminus \{y_i\}$ are weighted with the coefficients $\beta^{\frac{1}{2}(1+h(\mathbf{x}_i, y_i)+h(\mathbf{x}_i, y))}$, which will be higher for the misclassified examples.

This weighting ensures that the algorithm focuses more on the examples that are incorrectly classified, thereby improving the overall performance. As in the two-class case, the final rule is then a linear combination of the weak learners, where each weak learner contributes to the final decision based on its performance. The pseudo-code of the AdaBoost M2 algorithm is presented in Algorithm 18.

To Sum Up

1. *In boosting, weak learners are classifiers that perform only slightly better than random guessing. The key idea behind boosting is to combine these weak learners in a way that their combined performance is significantly better than any individual weak learner.*
2. *AdaBoost (Adaptive Boosting) updates the weights of the training examples in each iteration based on the performance of the weak learner. Examples that are correctly classified receive lower weights, while misclassified examples receive higher weights. This ensures that the algorithm focuses more on difficult examples in subsequent iterations, leading to improved performance.*
3. *AdaBoost has strong theoretical convergence properties. Under the condition that each weak learner performs better than random guessing, the training error of the combined classifier converges exponentially to zero as the number of iterations increases. This means that the error rate decreases very rapidly with each additional iteration.*
4. *Implementing boosting algorithms requires careful consideration of several practical aspects. These include choosing the appropriate weak learners, setting the number of iterations, and handling overfitting. Overfitting can be a concern, especially with noisy data, and techniques such as early stopping or regularization can be used to mitigate this issue.*
5. *Several extensions of AdaBoost have been developed to handle different types of data and problems. For example, AdaBoost.M2 is an extensions for multi-class classification problems.*
6. *The generalization error of the voting classifier in AdaBoost can be analyzed using margin theory. The margin of a classifier is defined as the difference between the confidence score of the correct class and the highest confidence score of any incorrect class. Margin theory states that a classifier with a larger margin is more likely to generalize well to unseen data.*

6.4 Exercises

1. Consider a training set S of size 10, where each example is a pair $(\mathbf{x}_i, y_i)$ with $\mathbf{x}_i$ belonging to the input space $\mathcal{X}$ and y_i belonging to the label set $\{-1, +1\}$:

$$S = \{(\mathbf{x}_i, y_i); i \in \{1, \dots, 10\}\} \in (\mathcal{X} \times \{-1, +1\})^{10}.$$

   (a) At the first step, each example in the training set is assigned a uniform weight $D_1(i) = \frac{1}{10}$ for all i. Suppose that after the training phase, the first classifier $f_1 : \mathcal{X} \to \{-1, +1\}$ misclassifies 3 examples of S. Calculate the error $\epsilon_1 = \sum_{i:f_1(\mathbf{x}_i)\neq y_i} D_1(i)$ and deduce the weights a_1 associated to h_1 as found by the AdaBoost algorithm.
   (b) Estimate new weights D_2 for both the misclassified and correctly classified examples by f_1.

2. Let $\Psi : \mathbb{R} \to \mathbb{R}_+^*$ be a strictly increasing, convex, differentiable function satisfying $\forall z < 0, \Psi(z) > 0; \forall z \geqslant 0, \Psi(z) \geqslant 1$.

   Consider the objective function defined over a training set of size m, $S = \{(\mathbf{x}_i, y_i); i \in \{1, \dots, m\}\} \in (\mathcal{X}, \{-1, +1\})^m$:

$$\mathcal{L}_T(\boldsymbol{a}) = \sum_{i=1}^{m} \Psi(-y_i h_T(\mathbf{x}_i)), \tag{6.21}$$

   where $h_T(\mathbf{x}) = \sum_{t=1}^{T} a_t f_t(\mathbf{x})$ is a linear combination of T weak classifiers $f_t \in \mathcal{F}; t \in \{1, \dots, T\}$ with respect to the weights $\boldsymbol{a} = (a_1, \dots, a_T)$.

   (a) At iteration t, we are looking for a weak classifier f_t and a weight a_t which minimize the objective function $\mathcal{L}_t(\boldsymbol{a}) = \sum_{i=1}^{m} \Psi(-y_i h_t(\mathbf{x}_i))$, where $h_t = h_{t-1} + a_t f_t$ and h_{t-1} is the current model.

   A common approach for selecting the classifier and the weight is to proceed in two steps at each iteration.

   - *Weak classifier selection:* For each candidate weak classifier $f \in \mathcal{F}$, compute the gradient of the loss with respect to a at $a = 0$ (i.e., at the current model h_{t-1}). Select the weak classifier f_t that optimizes this quantity. This step corresponds to fitting the weak classifier to the pseudo-residuals.
   - *Weight optimization:* Once f_t is selected, find the optimal weight a_t by minimizing the loss along the direction of f_t:

$$a_t = \underset{a\in\mathbb{R}}{\operatorname{argmin}} \sum_{i=1}^{m} \Psi\left(-y_i\left[h_{t-1}(\mathbf{x}_i) + a f_t(\mathbf{x}_i)\right]\right). \tag{6.22}$$

   This is typically a one-dimensional convex optimization problem, which can often be solved efficiently (sometimes even in closed form, depending on Ψ).

(b) Consider the following candidate functions for Ψ:

- Zero-one loss: $\Psi(u) = \mathbf{1}_{u \leqslant 0}$,
- Least squared loss: $\Psi(u) = (1-u)^2$,
- SVM (hinge) loss: $\Psi(u) = \max(0, 1-u)$,
- Logistic loss: $\Psi(u) = \frac{1}{\log(2)} \log(1+e^u)$.

Determine which satisfy the assumptions (strictly increasing, convex, differentiable, $\Psi(z) > 0$ for $z < 0$ and $\Psi(z) \geqslant 1$ for $z \geqslant 1$).

(c) For each valid Ψ, compare the boosting algorithm in (a) with AdaBoost.

3. Let $\ell_\gamma : \mathbb{R} \to \mathbb{R}_+$ be defined for some parameter $\gamma > 0$ by

$$\ell_\gamma(z) = \begin{cases} \log(1+e^{\gamma z}), & \text{if } z \leqslant 0, \\ \gamma z + \log(1+e^{-\gamma z}), & \text{if } z > 0. \end{cases} \tag{6.23}$$

Consider the following objective function over a training set $S = \{(\mathbf{x}_i, y_i); i \in \{1, \ldots, m\}\} \in (\mathcal{X}, \{-1, +1\})^m$:

$$\mathcal{L}(\boldsymbol{a}) = \sum_{i=1}^{m} \ell_\gamma(-y_i h(\mathbf{x}_i)), \tag{6.24}$$

where $h(\mathbf{x}) = \sum_{t=1}^{T} a_t f_t(\mathbf{x})$ is a linear combination of T weak classifiers $f_t \in \mathcal{F}$.

(a) Show that ℓ_γ (6.23) is convex and differentiable.

(b) Using $\mathcal{L}$ and a gradient descent strategy, derive an iterative boosting algorithm. Clearly state how the weak classifier f_t and its weight a_t are chosen at each iteration.

(c) Discuss how the use of ℓ_γ affects the algorithm's sensitivity to noisy labels. Compare this approach to AdaBoost.

4. Let $\mathcal{X}$ be a finite input space and $\mathcal{Y} = 1, \ldots, K$ a finite label set. Suppose you are given a training set $\mathcal{S} = (\mathbf{x}_i, y_i)i = 1^m$ and a set of feature functions $f_j(\mathbf{x}, y)_{j=1}^d$.

(a) The conditional log-linear (maximum entropy) model is defined as:

$$p_{\boldsymbol{\theta}}(y \mid \mathbf{x}) = \frac{\exp(\sum_{j=1}^{d} \theta_j f_j(\mathbf{x}, y))}{Z_{\boldsymbol{\theta}}}, \tag{6.25}$$

where $Z_{\boldsymbol{\theta}}(\mathbf{x}) = \sum_{y' \in \mathcal{Y}} \exp\left(\sum_{j=1}^{d} \theta_j f_j(\mathbf{x}, y')\right)$.

Explain the intuition behind this model and why it is called a "maximum entropy" model.

(b) Show that maximizing the conditional likelihood of the data under this model is equivalent to finding parameters $\boldsymbol{\theta}$ such that the expected value of each feature under the model matches its empirical expectation:

$$\frac{1}{m}\sum_{i=1}^{m} f_j(\mathbf{x}_i, y_i) = \frac{1}{m}\sum_{i=1}^{m}\sum_{y\in\mathcal{Y}} p_{\boldsymbol{\theta}}(y \mid \mathbf{x}_i) f_j(\mathbf{x}_i, y), \quad \forall j. \tag{6.26}$$

(c) The Generalized Iterative Scaling (GIS) algorithm is a classical method for fitting log-linear models. Derive the update rule that GIS uses to iteratively adjust each parameter θ_j so that the feature expectations under the model match the empirical feature expectations from the data. Clearly state any assumptions required for the update (e.g., the sum of features is constant for all $\mathbf{x}$), and define any constants used in your formula.

5. Let $S = (\mathbf{x}i, y_i)i = 1^m$ be a training set with $y_i \in -1, +1$ of size m, and let $\mathcal{F}$ be a set of weak classifiers $f : \mathcal{X} \to -1, +1$. Consider the additive model

$$h_T(\mathbf{x}) = \sum_{t=1}^{T} a_t f_t(\mathbf{x}), \tag{6.27}$$

and the exponential loss

$$\mathcal{L}(\boldsymbol{a}) = \sum_{i=1}^{m} \exp(-y_i h_T(\mathbf{x}_i)). \tag{6.28}$$

(a) Show that minimizing $\mathcal{L}(\boldsymbol{a})$ with respect to $\boldsymbol{a}$ is equivalent to fitting a log-linear model to the data.
(b) Describe how a coordinate descent version of GIS can be used to minimize $\mathcal{L}(\boldsymbol{a})$, and show that this leads to an algorithm similar to AdaBoost.
(c) Explain the analogy between the weight update in AdaBoost and the parameter update in iterative scaling. Under what conditions do the two algorithms coincide?
(d) Discuss how this perspective unifies boosting and maximum entropy/logistic regression.

6. Consider the optimization problem derived from AdaBoost (6.9). The optimization can be written as:

$$\max_{\boldsymbol{a}} \quad \rho$$
$$\text{subject to} \quad \frac{y_i \sum_{t=1}^{T} a_t f_t(\mathbf{x}_i)}{||\boldsymbol{a}||_1} \geqslant \rho, \forall i\{1, \ldots, m\}, \tag{6.29}$$

where $(f_t : \mathcal{X} \to \{-1, +1\})_t$ are weak classifiers, $(y_i \in \{-1, +1\})_i$ are labels, and $\boldsymbol{a} = (a_t)_t$ are classifier weights.

(a) **LP Structure**
By rewriting the objective function (6.29) as:

$$\max_{\rho} \quad \rho$$

$$\text{subject to} \quad y_i\left(\sum_{t=1}^{T} a_t f_t(\mathbf{x}_i)\right) \geqslant \rho, \qquad \forall i \in \{1, \ldots, m\}, \tag{6.30}$$

$$\sum_{t=1}^{T} a_t = 1, \tag{6.31}$$

$$a_t \geqslant 0, \qquad \forall t \in \{1, \ldots, T\}. \tag{6.32}$$

Identify the following components of this linear program (LP):

i. Decision variables.
ii. Objective function.
iii. Inequality constraints (specify how many).
iv. Equality constraints (specify how many).
v. Non-negativity constraints (specify how many).

(b) **Dual Problem Derivation**
Derive the dual of this linear program. Follow these steps:

i. Introduce Lagrange multipliers $\lambda_i \geqslant 0$ for the margin constraints (6.30) and $\nu \in \mathbb{R}$ for the equality constraint (6.31).
ii. Write the Lagrangian function $\mathcal{L}(a, \rho, \lambda, \nu)$.
iii. Take derivatives of $\mathcal{L}$ with respect to ρ and a_t, set them to zero, and simplify.
iv. Show that the dual problem is:

$$\min_{\nu} \quad \nu$$

$$\text{subject to} \quad \sum_{i=1}^{m} \lambda_i y_i f_t(\mathbf{x}_i) \leqslant \nu, \qquad \forall t \in \{1, \ldots, T\}, \tag{6.33}$$

$$\sum_{i=1}^{m} \lambda_i = 1, \tag{6.34}$$

$$\lambda_i \geqslant 0, \qquad \forall i \in \{1, \ldots, m\}. \tag{6.35}$$

(c) **Interpretation of Dual Variables**

i. Interpret the dual variables λ_i in the context of boosting. How do they relate to the weights AdaBoost assigns to training examples?

ii. What does the dual objective ν represent in terms of the original margin ρ?

(d) **Algorithmic Implications**

i. Explain why solving the dual problem might be computationally advantageous if $T \gg m$ (i.e., the number of weak classifiers is much larger than the number of training examples).

ii. How does the dual problem relate to the "weak learner assumption" in AdaBoost?

(e) **Geometric Connection**

Sketch the feasible region of the primal problem in $\mathbb{R}^T$ (for fixed ρ). How does maximizing ρ correspond to finding the largest-margin hyperplane in this space? Relate this to the margin theory of boosting.

7. Fix $\epsilon \in (0, 1/2)$. Consider a training set of m points in $\mathbb{R}^2$ distributed as follows:

- $\frac{m}{4}$ negative points at $(1, 1)$,
- $\frac{m}{4}$ negative points at $(-1, -1)$,
- $\frac{m(1-\epsilon)}{4}$ positive points at $(1, -1)$,
- $\frac{m(1+\epsilon)}{4}$ positive points at $(-1, 1)$.

AdaBoost is run on this dataset using decision stumps (i.e., classifiers of the form $h(x) = \text{sign}(x_j - \theta)$ for $j = 1, 2$ and some threshold θ).

(a) For the first iteration of AdaBoost, describe which stump(s) will be selected and compute the corresponding weighted error.

(b) Describe how the weights on the training examples evolve after each iteration. Which points become most influential in subsequent rounds?

(c) After T rounds, what is the form of the final AdaBoost classifier? Is it able to perfectly classify all training points? Explain why or why not.

(d) How does the parameter ϵ affect the behavior and the margin of the final classifier?

8. Consider a variant of AdaBoost where the weight assigned to each weak learner is fixed to a constant $\alpha > 0$ throughout the boosting process, regardless of the weak learner's error. Assume the weak learners satisfy a uniform edge condition: for all rounds t, $\epsilon_t \leqslant \frac{1}{2} - \gamma$, for some $\gamma > 0$.

(a) Determine the value of α that minimizes the exponential loss on the training data at each iteration, assuming the edge γ is known and fixed. Explain your reasoning.

(b) Analyze the evolution of the distribution over training examples induced by this fixed-weight AdaBoost. Specifically, compare the total weight assigned to correctly classified versus misclassified examples after each iteration.

(c) Derive an upper bound on the training error after T rounds in terms of γ and T.

(d) Show that the training error can be driven to zero after a sufficient number of rounds T, and provide an explicit expression for such T in terms of m and γ.
(e) Let s denote the VC dimension of the base hypothesis class. Using known VC dimension bounds for linear combinations of base classifiers, provide a uniform convergence bound on the generalization error of the final boosted classifier after T rounds.
(f) Suppose $\gamma = \gamma(m)$ decreases with the sample size m at the rate $\gamma(m) = O(\sqrt{\frac{\log m}{m}})$. Discuss the impact of this rate on the consistency and generalization guarantees of the fixed-weight AdaBoost algorithm.

9. Let $\mathcal{F}$ be a class of weak learners with VC dimension d. AdaBoost constructs its final hypothesis as a weighted majority vote over T weak classifiers from $\mathcal{F}$, i.e., hypotheses of the form:

$$\mathcal{H} = \left\{ \mathbf{x} \mapsto \text{sign}\left(\sum_{t=1}^{T} a_t f_t \right) : f_t \in \mathcal{F}, a \in \mathbb{R}_+ \right\}. \tag{6.36}$$

(a) Show that the final hypothesis of AdaBoost after T rounds can be represented as a linear threshold function in T-dimensional space, where each dimension corresponds to the output of a base classifier f_t.
(b) Prove that the VC dimension of the set of all such linear threshold functions in T dimensions is at most T.
(c) Suppose that $\mathcal{F}$ is finite of size $N_{\mathcal{F}}$. Using Sauer's lemma and the fact that the number of possible sequences of T weak classifiers is at most $(N_{\mathcal{F}})^T$, show that the VC dimension of the AdaBoost hypothesis class $\mathcal{H}$ is bounded by:

$$vc_{\mathcal{H}} \leqslant T(d+1)(3\ln(T(d+1)) + 2). \tag{6.37}$$

(e) Explain why, as the number of boosting rounds T increases, the VC dimension of AdaBoost's hypothesis set grows roughly linearly with T, and discuss the implications for overfitting and generalization.

Chapter 7
Learning-to-Rank

Advancements in information technologies have led to the development of new machine learning frameworks in recent years. One example is the paradigm of learning to rank functions, which focuses on developing automated methods for ranking observations within large datasets. This task is particularly relevant in the field of Information Retrieval (IR), with search engines being a notable example.

Another significant application is document routing, where the goal is to rank incoming documents based on specific information needs. Each new document is compared to those not yet read by the user and then inserted into an ordered list that prioritizes relevant unread documents. This chapter will explore the foundations of learning ranking functions, using the domain of IR as the primary application area.

7.1 Formalism

Formally, learning a ranking function involves developing a real-valued function that takes an element from a set to be ordered as input. The order is then determined by ranking these elements based on their increasing or decreasing scores. Unlike classification, the focus is not on the absolute value of the predicted score for a given input, but rather on the relative scores between elements, which facilitate sorting. To effectively learn score functions, it is essential to define new error criteria, develop algorithms to optimize these errors, and establish a theoretical framework to ensure that the learned function will perform well on future data.

Recent research has concentrated on formulating various ranking methods. These studies have proposed algorithms and developed theoretical frameworks for predicting total or partial orders on examples. This section will present the error functions and discuss the two mentioned ranking tasks.

M.-R. Amini, *Advanced Supervised and Semi-supervised Learning*,
Cognitive Technologies, https://doi.org/10.1007/978-3-031-99928-4_7

7.1.1 Learning-to-Rank with Score Functions

For a given information need or query, q, we will assume that the desired ranking on a set of inputs or examples $\Im = (\mathbf{x}_1, \ldots, \mathbf{x}_n)$ is translated into associated real variables $\mathbf{y} = (y_1, \ldots, y_n) \in \mathbb{R}^n$. Thus, for query q, the ranking $y_i > y_j$ reflects the preference of example $\mathbf{x}_i \in \Im$ over example $\mathbf{x}_j \in \Im$ (denoted by $\mathbf{x}_i \succ \mathbf{x}_j$) over the latter. By analogy with Information Retrieval, we will generally use the term relevance judgment to denote the utility values associated with examples.

In the considered examples of IR, the relevance judgments are generally expressed as a binary value $\{-1, +1\}$: if for q, $\mathbf{x}_i$ is relevant, $y_i = +1$, otherwise, $y_i = -1$. The score function $h : \mathcal{X} \to \mathbb{R}$ that we seek to learn induces an order on a set of inputs by real outputs that it assigns to the examples for a given query q, and the ranking error functions measure the agreement between the induced order and the relevance judgments associated with the examples with respect to q. For a set of n examples X, these error functions are of the form:

$$\ell_o : \mathbb{R}^n \times \mathbb{R}^n \to \mathbb{R}^+. \tag{7.1}$$

Precision and Recall

When the relevance judgments are binary (i.e. $y_i \in \{-1, 1\}$), the two usual error measures are precision and recall at rank k, for $k \leqslant n$, defined on the ordered list according to descending scores given by f on $\Im$ for query q, denoted by $h(\Im, q) = (h(\mathbf{x}_{i,q}))_{i=1}^n$ [1] (rank 1 is therefore the rank of the entry with the highest score):

$$\ell_{r@k}(h(\Im, q), \mathbf{y}) = \frac{1}{n_+^q} \sum_{i:y_i=1} \mathbb{1}_{rg_i^q \leqslant k}, \tag{7.2}$$

$$\ell_{p@k}(h(\Im, q), \mathbf{y}) = \frac{1}{k} \sum_{i:y_i=1} \mathbb{1}_{rg_i^q \leqslant k}, \tag{7.3}$$

where $n_+^q = \sum_{i=1}^m \mathbb{1}_{y_i=1}$ is the total number of relevant entries for query q, and rg_i^q is the rank of entry $\mathbf{x}_i$ in the ordered list induced by the scores given by f for query q. Thus, the recall at rank k, $\ell_{r@k}$, measures the proportion of relevant entries that are in the first k elements, and the precision at rank k, $\ell_{p@k}$, measures the proportion of the first k elements in the list that are relevant entries.

Average Precision

The average precision of a score function f for a given query q, denoted by $\ell_{pm}(h(\Im, q), \mathbf{y})$, is the average of the precision values of the relevant documents

[1] In Sect. 7.1.3, we will present the common vector representation $\mathbf{x}_q \in \mathcal{X}$ used to represent each pair of example and query, (x, q).

with respect to q in the ordered list induced by the scores given by f:

$$\ell_{pm}(h(\Im, q), \mathbf{y}) = \frac{1}{n_+^q} \sum_{i:y_i=1} \ell_{p@rg_i^q}(h(\Im, q), \mathbf{y}). \tag{7.4}$$

In cases where we have a set of queries $Q = \{q_1, \ldots, q_{|Q|}\}$, we can extend the previous evaluation by computing the Mean Average Precision (MAP) over all of these queries:

$$MAP(h) = \frac{1}{|Q|} \sum_{j=1}^{|Q|} \ell_{pm}(h(\Im, q_j), \mathbf{y}) = \frac{1}{|Q|} \sum_{j=1}^{|Q|} \frac{1}{n_+^{q_j}} \sum_{i:y_i=1} \ell_{p@rg_i^{q_j}}(h(\Im, q_j), \mathbf{y}). \tag{7.5}$$

In the various evaluation campaigns, the MAP measure is widely used for its discriminatory capacity and its stability.

Crucial Pairwise Classification Error

Cohen et al. [35] introduced the pairwise ranking error of inputs $(\mathbf{x}_i, \mathbf{x}_j)$ such that $\mathbf{x}_i \succ \mathbf{x}_j$ (or $y_i > y_j$) for a given query q, called crucial pairs and which played a central role in the development of training-based ranking functions. For the considered query q, this measure is the proportion of crucial pairs on which the predicted order is not the desired order:

$$\ell_{pc}(h(\Im, q), \mathbf{y}) = \frac{1}{\sum_{j,\ell} \mathbb{1}_{y_j > y_\ell}} \sum_{j,\ell:y_j > y_\ell} \mathbb{1}_{h(\mathbf{x}_{j,q}) \leqslant h(\mathbf{x}_{\ell,q})}. \tag{7.6}$$

In the case of binary relevance judgments, this error has been considered in many studies on learning ranking functions [1, 144] and its motivation comes from an interpretation of ranking in terms of reproducing a binary partial order relation [37]. Indeed, the strict partial order induced by relevance judgments $y_i \in \{-1, +1\}$ is a binary relation, defined on the considered input set $\Im$.

The score function, too, allows to define a binary relation on $\Im$. It is then natural to measure the error of the score function by comparing on the one hand the desired binary relation, and on the other hand the relation induced by the score function. This function also allows to reduce some ranking problems to the binary classification of crucial pairs and is of particular interest in the theoretical study of these problems. We will return to this point in the next section.

Area Under the *ROC* Curve

The Receiver Operating Characteristics (ROC) curve indicates the ability of a scoring function to order relevant documents with respect to a query q before irrelevant documents. Starting from an ordered list of documents, this curve is constructed by estimating the proportion of relevant documents (or recall) at each rank k on this list, $\ell_{r@k}(h(\Im, q), \mathbf{y})$, as a function of the proportion of irrelevant documents ordered before this rank. The area under this ROC curve (AUC) is an indicator of the ability of the function f to order relevant documents above irrelevant documents with respect to a query q and it can be defined as:

$$\ell_{auc}(h(\Im, q), \mathbf{y}) = \frac{1}{n_+^q n_-^q} \sum_{i:y_i=1} \sum_{j:y_j=-1} \left(\mathbb{1}_{rg_i^q < rg_j^q} + \frac{1}{2} \mathbb{1}_{rg_i^q = rg_j^q} \right), \tag{7.7}$$

where n_-^q (respectively n_+^q) is the number of irrelevant (respectively relevant) documents relative to q.

We note that in the case of binary relevance judgments, $\ell_{pc}(h(\Im, q), \mathbf{y})$ and $\ell_{auc}(h(\Im, q), \mathbf{y})$ are complementary:

$$\begin{aligned}
\ell_{pc}(h(\Im, q), \mathbf{y}) &= \frac{1}{n_+^q n_-^q} \sum_{i:y_i=1} \sum_{j:y_j=-1} \mathbb{1}_{h(\mathbf{x}_{i,q}) \leqslant h(\mathbf{x}_{j,q})} \\
&= \frac{1}{n_+^q n_-^q} \sum_{i:y_i=1} \sum_{j:y_j=-1} \mathbb{1}_{rg_i^q \geqslant rg_j^q} \\
&= 1 - \ell_{auc}(h(\Im, q), \mathbf{y}).
\end{aligned}$$

Figure 7.1 illustrates the calculation of the previous measurements on a toy example.

k	$h(.)$	y	$\ell_{p@k}$	$\ell_{r@k}$	τ_k
1	1.0	1	1	1/4	0
2	0.9	0	1/2	1/4	1/6
3	0.8	0	1/3	1/4	1/3
4	0.7	1	1/2	1/2	1/3
5	0.6	1	3/5	3/4	1/3
6	0.5	0	1/2	3/4	1/2
7	0.4	0	3/7	3/4	2/3
8	0.3	1	1/2	1	2/3
9	0.2	0	4/9	1	5/6
10	0.1	0	2/5	1	1

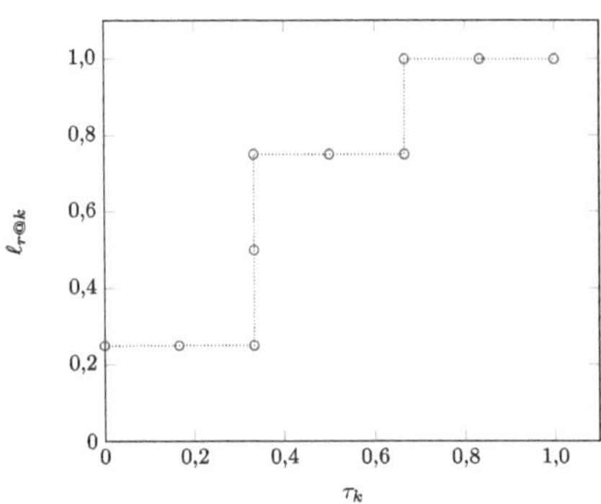

Fig. 7.1 Different error measures computed for a score function f and a given query q on a toy example, as well as the corresponding *ROC* curve (right). τ_k is the rate of irrelevant documents ordered before rank k. The average Precision is in this case, $\ell_{pm}(h(\Im, q), \mathbf{y}) = \frac{1}{4}(1+\frac{1}{2}+\frac{3}{5}+\frac{1}{2}) = \frac{13}{20}$, and the area under the ROC curve is $\ell_{auc}(h(\Im, q), \mathbf{y}) = 1 - \frac{8}{4\times 6} = \frac{2}{3}$. We note that it is the rank of the relevant examples in the ordered list induced by the scores, and not their predicted values, which is involved in the calculation of these measures

Discounted Cumulative Gain

The discounted cumulative gain (DCG) measures the efficiency of a scoring function in the case where the relevance judgments on the documents are no longer binary but take values in a subset of discrete values of $\mathbb{N}$. This measure evaluates the utility, or gain, of a document according to its relevance score. This utility is summed for each document from the head of the list of returned documents, and this up to the desired position. However, since getting all relevant documents at the top of the list is more interesting than finding them at the tail of the list, the gain is weighted by the position at which each document appears (it is in fact the $\log_2$ of the position that is used so as not to give too much importance to this factor). As for the precision at the top of the list, $\ell_{p@k}$, the DCG measure for a given query q is evaluated at a rank k (k generally takes the same values as for the precision at the top of the list: 5, 10, 20 or 25).

$$\ell_{dcg@k}(h(\Im, q), \mathbf{y}) = \sum_i \mathbb{1}_{rg_i^q \leqslant k} \left[\frac{2^{y_i} - 1}{\log_2(1 + rg_i^q)} \right]. \tag{7.8}$$

In practice, it is appropriate to normalize the previous measure by multiplying it by an adequate normalizing factor in order to obtain a gain equal to 1. This is achieved by dividing the reduced cumulative gain by that obtained with an ideal, perfect system, which would place at the top of the list the most relevant documents, with a score of 5 for example, then the documents with a score of 4, and finally the documents with a score of 0.

7.1.2 Ranking of Instances

In ranking of instances, the goal is to order the inputs, or observations by analogy with the classification framework, among themselves with respect to a fixed information need, q^*. In this case, the information need reflects a sought-after theme and may not be explicitly formulated in the form of a query. The goal is thus to order an input stream of documents so that those that are relevant to the information need are ordered above the irrelevant examples. The vector representation of the pairs (document, information need) is reduced in this case to the vector representation of the documents which is generally defined with the vector space model, proposed by Salton [146]. With this representation, we associate with each document x of a collection $\mathcal{C}$, a vector $\mathbf{x}$, whose dimension corresponds to the size of the vocabulary, which is obtained after a certain number of preprocessings aimed at keeping only the informative terms of the collection [5, Ch.1 & 2]. The vector space considered is therefore a vector space of terms in which each dimension is associated with a term of the collection.

$$\forall x \in \mathcal{C}, \mathbf{x} = (w_{ix})_{i \in \{1,\dots,d\}}. \tag{7.9}$$

In this case, w_{ix} is the weight that the term with index i of the vocabulary has in the document x. We will see later that there are several ways to calculate this weight. However, it is important to note here that all these methods assign a zero weight to a term that is absent from the document. With this representation, the order of appearance of the terms in the documents is not taken into account. For this reason, it is also known as the by bag of words representation. In information retrieval, the most common encoding of documents, known as the *tf-idf* encoding, is defined as:

$$\forall i \in \{1, \dots, d\};\ w_{ix} = \mathrm{tf}_{\theta_i, x} \times \ln \frac{m}{\mathrm{df}_{\theta_i}}, \tag{7.10}$$

where $\mathrm{tf}_{\theta_i, x}$ is the number of occurrences of the term θ_i of the vocabulary in the document x (or *term frequency*), m is the number of documents in the collection, and df_{θ_i} is the number of documents in which the term with index i of the vocabulary appears, called *document frequency*. Usually $\ln \frac{m}{\mathrm{df}_{\theta_i}}$ is denoted by (*Inverse Document Frequency*) idf_{θ_i}:

$$\mathrm{idf}_{\theta_i} = \ln \frac{m}{\mathrm{df}_{\theta_i}}. \tag{7.11}$$

With this vector representation of documents, the training sets, test sets, and examples have, in the case of ranking of instances, the same form as in classification or regression (according to the space $\mathcal{Y}$). The notable difference is the definition of the generalization error, which takes into account the relative scores between two observations, and no longer the agreement between the predicted value and the desired value.

For a set $S = \{(\mathbf{x}_1, y_1), \dots, (\mathbf{x}_m, y_m)\}$ of examples sampled according to a fixed but unknown probability distribution $\mathcal{D}$, and for a given score function $h : \mathcal{X} \to \mathbb{R}$, the empirical error is defined using a ranking error function:

$$\hat{R}_{oi}(h, S) = \ell_o(h(\Im, q^*), \mathbf{y}), \tag{7.12}$$

where $\Im = (\mathbf{x}_1, \dots, \mathbf{x}_m)$, $\mathbf{y} = (y_1, \dots, y_m)$ and $h(\Im, q^*) = (h(\mathbf{x}_1), \dots, h(\mathbf{x}_m))$. In the special case where the relevance judgments are reduced to binary judgments $\mathbf{y} \in \{-1, +1\}^m$, the framework is commonly called bipartite ranking. As mentioned above, the learning community has generally considered the error ℓ_{pc} (7.6) in theoretical analyses for ranking of instances, since in this case the generalization error associated with this error function is defined on a random crucial pair by:

$$R_{oi}(h) = \mathbb{E}\left[\mathbb{1}_{h(\mathbf{x}_i) \leqslant h(\mathbf{x}_j)} | y_i > y_j\right], \tag{7.13}$$

verifies $R_{oi}(h) = \mathbb{E}\left[\ell_{pc}(h(\Im, q^*), \mathbf{y})\right]$, which by analogy with the framework of binary classification (Chap. 1) allows to establish the consistency of the ERM principle according to this framework. The only notable exception is the work of Hill [78], which establishes probabilistic bounds on the difference in average precision

(ℓ_{pm}), between two sets of examples S sampled i.i.d. according to a probability distribution.

7.1.3 Ranking of Alternatives

The ranking of alternatives framework is the most widespread in Information Retrieval and it encompasses tasks such as document retrieval or automatic text summarization. In this case, we do not seek to order the system's inputs among themselves, but rather their associated *alternatives* in such a way that the predicted order on the latter reflects the relevance criterion for each of the inputs (Fig. 7.2). For example, in document retrieval, an input is a query and the goal is to order the documents (alternatives) of a given collection in such a way that the relevant documents for each of the inputs are ordered above the irrelevant documents. Formally, for each input q_i, let $\mathcal{A}_{q_i} = (x_1^{(i)}, \ldots, x_{m_i}^{(i)})$ be the set of its m_i candidate alternatives. In documentary research, this involves determining a subset of the initial collection of documents relating to a given query.

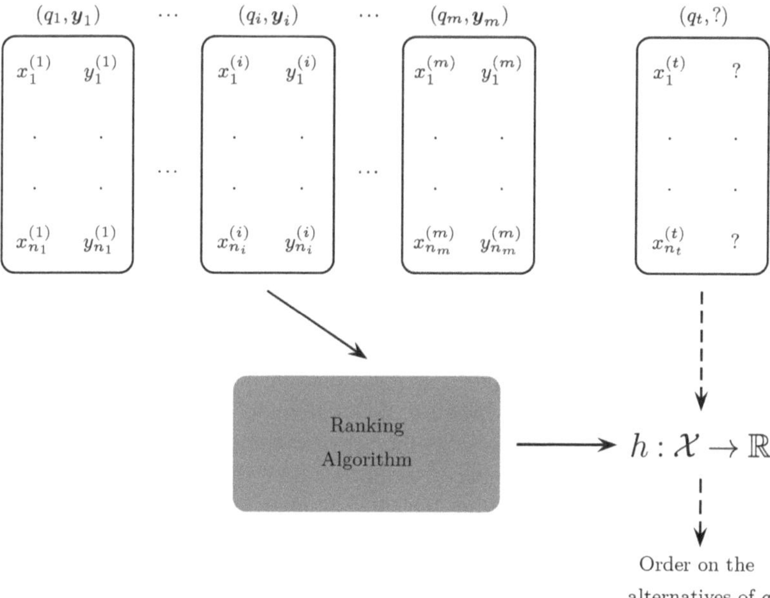

Fig. 7.2 Illustration of the ranking of alternatives framework. In the learning phase, a score function h is learned on a set of inputs $\{(q_1, \mathbf{y}_1), \ldots, (q_m, \mathbf{y}_m)\}$ where each input q_i is associated with a list of alternatives $(x_1^{(i)}, \ldots, x_{m_i}^{(i)})$ and corresponding relevance judgments $\mathbf{y}_i = (y_1^{(i)}, \ldots, y_{m_i}^{(i)})$. Once the function h is found, for a new input q_t, its list of alternatives is ordered according to the score values assigned by $h(.)$

In the supervised framework, we further assume that each observation is associated with a desired vector[2] $\boldsymbol{y}_i$. The output vector $\boldsymbol{y}_i = (y_1^{(i)}, \ldots, y_{m_i}^{(i)}) \subset \mathbb{R}^{m_i}$ thus defines the order that we seek to predict on the alternatives in $\mathcal{A}_{q_i}$. The score function f that is to predict this order takes as input an observation q_i and an alternative $x_j^{(i)}$ in $\mathcal{A}_{q_i}$ and returns a real score reflecting the *similarity* between the observation and its alternative, represented by the vector $\mathbf{x}_{j,q_i}^{(i)} \in \mathcal{X}$ i.e. $h : \mathcal{X} \to \mathbb{R}$.

We further assume that the pairs $(q, \boldsymbol{y})$ are generated independently according to a probability distribution $\mathcal{D}$.

For a given cost function $\ell_o : \mathbb{R}^{|\mathcal{A}_q|} \times \mathbb{R}^{|\mathcal{A}_q|} \to \mathbb{R}_+$, the empirical risk of the function f on a training set $S = \{(q_i, \boldsymbol{y}_i)\}_{i \in \{1,\ldots,m\}}$ of size m (assumed to be sampled i.i.d according to $\mathcal{D}$) is defined as the mean error in the sense of ℓ_o of f on S:

$$\hat{R}_{oa}(h, S) = \frac{1}{m} \sum_{i=1}^{m} \ell_o \left(h(\mathcal{A}_{q_i}, q_i), \boldsymbol{y}_i\right). \tag{7.14}$$

The generalization error is defined as the expectation along $\mathcal{D}$ of $\hat{R}_{oa}$:

$$R_{oa}(h) = \mathbb{E}_{(q,\boldsymbol{y}) \sim \mathcal{D}} \ell_o \left(h(\mathcal{A}_q, q), \boldsymbol{y}\right). \tag{7.15}$$

For example, if the chosen ranking error function is the ranking error of crucial pairs (7.6), the empirical risk of the function h would be written as:

$$\hat{R}_{oa}(h, S) = \frac{1}{m} \sum_{i=1}^{m} \frac{1}{\sum_{j,\ell} \mathbb{1}_{y_j^{(i)} > y_\ell^{(i)}}} \sum_{j,\ell : y_j^{(i)} > y_\ell^{(i)}} \mathbb{1}_{h(\mathbf{x}_{j,q_i}^{(i)}) \leqslant h(\mathbf{x}_{\ell,q_i}^{(i)})}. \tag{7.16}$$

In IR, the vector representing the pairs (alternative, input) or (document, query) is most often composed of simple features calculated from the terms of the vocabulary common to the document and the query.

In the two flagship projects on learning to rank functions driven by the research centers of *Microsoft*[3] and *Yahoo!*,[4] a few hundred of these features have been defined, the most common of which are shown in the following table.

In this case $\theta \in q \cap x$ represents the set of common terms in a query q and a document x, $\mathrm{tf}_{\theta,\mathcal{C}}$ is the number of occurrences of the term θ in the collection of documents $\mathcal{C}$, and $l_\mathcal{C}$ and l_x are the total number of occurrences of the terms in the collection $\mathcal{C}$ and the document $x \in \mathcal{C}$ respectively. The BM25(x, q) and SPL(x, q) features are scores assigned respectively by the OKAPI BM25 [141] and informational [33] models, which have become references in the development of

[2] For the sake of homogeneity, we have kept the same notation as in the previous chapters to designate the desired output vectors associated with the observations.

[3] http://research.microsoft.com/en-us/um/beijing/projects/letor/.

[4] http://jmlr.org/proceedings/papers/v14/.

	Some common features used in representation, $\mathbf{x}_q$, of (x, q)		
1.	$\sum_{\theta \in q \cap x} \ln(1 + \mathrm{tf}_{\theta,x})$	2.	$\sum_{\theta \in q \cap x} \ln\left(1 + \frac{l_C}{\mathrm{tf}_{\theta,C}}\right)$
3.	$\sum_{\theta \in q \cap x} \ln(\mathrm{idf}_\theta)$	4.	$\sum_{\theta \in q \cap x} \ln\left(1 + \frac{\mathrm{tf}_{\theta,x}}{l_x}\right)$
5.	$\sum_{\theta \in q \cap x} \ln\left(1 + \frac{\mathrm{tf}_{\theta,x}}{l_x}.\mathrm{idf}_\theta\right)$	6.	$\sum_{\theta \in q \cap x} \ln\left(1 + \frac{\mathrm{tf}_{\theta,x}}{l_x}.\frac{l_C}{\mathrm{tf}_{\theta,C}}\right)$
7.	$\sum_{\theta \in q \cap x} \mathrm{tf}_{\theta,x}$	8.	$\sum_{\theta \in q \cap x} \mathrm{tf}_{\theta,x}.\mathrm{idf}_\theta$
9.	$\mathrm{BM25}(x, q)$	10.	$\mathrm{SPL}(x, q)$

information retrieval systems with:

$$\mathrm{BM25}(x, q) = \sum_{\theta \in q \cap x} \frac{(k_3 + 1) \times \mathrm{tf}_{\theta,q}}{k_3 + \mathrm{tf}_{\theta,q}} \frac{(k_1 + 1) \times \mathrm{tf}_{t,x}}{k_1((1 - b) + b\frac{l_x}{r}) + \mathrm{tf}_{t,x}} \ln \frac{N - \mathrm{df}_\theta + 0,5}{\mathrm{df}_\theta + 0,5}, \tag{7.17}$$

where k_1, b and k_3 are constants set by default to $k_1 = 1.2$, $b = 0.75$ and $k_3 = 1000$. N is the number of documents in the collection and $r = \frac{1}{N} \sum_{x \in C} l_x$ represents the average of the sizes of the documents in the collection. The informational model SPL (for *Smoothed Power Law*) is defined by:

$$\mathrm{SPL}(d, q) = \sum_{\theta \in q \cap d} \mathrm{tf}_{\theta,q} \ln \frac{1 - \lambda_\theta}{\lambda_\theta^{\frac{\mathrm{tf}_{\theta,x}}{\mathrm{tf}_{\theta,x}+1}} - \lambda_\theta}, \tag{7.18}$$

where λ_θ is a parameter dependent on the collection, which can either be fixed, for example, as:

$$\lambda_\theta = \frac{\mathrm{df}_\theta}{N}, \tag{7.19}$$

or estimated from the collection.

7.2 Ranking Approaches

In this section, we will present the different approaches proposed for learning ranking functions, as well as the typical algorithms developed in each of these approaches.

7.2.1 Pointwise Ranking

Pointwise approaches assume that the vectors representing pairs of documents and queries, as well as the associated relevance judgments, are generated identically and independently according to a given probability distribution.

In practice, the algorithms resulting from this approach attempt to find the function that best associates the vectors of the inputs with their respective outputs, and not the one that respects the desired order that is induced a posteriori on the values of the outputs of the learned function. These algorithms are, for the most part, extensions of the techniques developed according to the frameworks of regression and classification and we also note that the i.i.d. assumption of pointwise approaches is only verified in the case of ranking of instance since, in the case of ranking of alternatives, the vector representations of the pairs constituted by the alternatives of an input and the latter all depend on the input.

In the remainder of this section, we will present the two most popular models developed following this approach, based on ordinal regression [112] and which take into account the numerical relationship between relevance judgments to learn the scoring function. In the case where there are k ordered categories, this function is found by setting suitable thresholds $b_1 \leqslant \ldots \leqslant b_{k-1} \leqslant b_k = \infty$ in order to distinguish the outputs of the function according to these categories.

7.2.1.1 PRrank

The PRrank (for Perceptron Ranking) model proposed by Crammer et al. [40] is an iterative online algorithm based on the Perceptron algorithm (Chap. 4, Sect. 4.1.2).

At a given iteration t and for a new example $(x^{(t)}, y^{(t)})$ chosen randomly, the model predicts the rank $\hat{y}^{(t)}$ of the latter, with its weight vector $\boldsymbol{w}^{(t)}$, and the set of current thresholds $\boldsymbol{b}^{(t)} = (b_1^{(t)}, \ldots, b_k^{(t)})$ by setting this rank to be the index of the smallest threshold $b_{\hat{y}^{(t)}}^{(t)}$ verifying $\left\langle \boldsymbol{w}^{(t)}, \mathbf{x}^{(t)} \right\rangle < b_{\hat{y}^{(t)}}^{(t)}$:[5]

$$\hat{y}^{(t)} = \min_{r \in \{1,\ldots,k\}} \left\{ r \middle| \left\langle \boldsymbol{w}^{(t)}, \mathbf{x}^{(t)} \right\rangle - b_r^{(t)} < 0 \right\}. \tag{7.20}$$

The PRank algorithm (Algorithm 19) then compares the predicted rank $\hat{y}^{(t)}$ to the true rank of the example, $y^{(t)}$, and updates its weights $\boldsymbol{w}^{(t)}$ and thresholds $\boldsymbol{b}^{(t)}$ so as to minimize the set of rank prediction errors made so far which, because the predicted and desired rank outputs are integers, is written as:

$$E_t = \sum_{t'=1}^{t} |\hat{y}^{(t')} - y^{(t')}|. \tag{7.21}$$

[5] We note that this minimum is well defined since $b_k = \infty$.

Input :

- A training set $S = ((\mathbf{x}_1, y_1), \ldots, (\mathbf{x}_m, y_m))$; Maximum number of iterations T;

Initialisation:

- $t \leftarrow 1$;
- Initialize the weights $\boldsymbol{w}^{(1)}$ // Generally $\boldsymbol{w}^{(1)} = \mathbf{0}$;
- Initialize the thresholds $b_1^{(1)} = 0, \ldots b_{k-1}^{(1)} = 0, b_k^{(1)} = \infty$;

while $t \leqslant T$ **do**
  Randomly draw an example $(\mathbf{x}^{(t)}, y^{(t)}) \in S$;
  Predict $\hat{y}^{(t)}$ the rank of the example $\mathbf{x}^{(t)}$ with $(\boldsymbol{w}^{(t)}, \boldsymbol{b}^{(t)})$ // $\triangleright$ (7.20);
  if $y^{(t)} \neq \hat{y}^{(t)}$ **then**
    for $r = 1, \ldots, k-1$ **do**
      if $y^{(t)} \leqslant r$ **then**
        $z_r^{(t)} \leftarrow -1$;
      else
        $z_r^{(t)} \leftarrow +1$;
    for $r = 1, \ldots, k-1$ **do**
      if $z_r^{(t)} \left(\langle \boldsymbol{w}^{(t)}, \mathbf{x}^{(t)} \rangle - b_r^{(t)} \right) \leqslant 0$ **then**
        $s_r^{(t)} \leftarrow z_r^{(t)}$;
      else
        $s_r^{(t)} \leftarrow 0$;
    $\bar{\boldsymbol{w}}^{(t+1)} \leftarrow \bar{\boldsymbol{w}}^{(t)} + \left(\sum_r s_r^{(t)} \right) \mathbf{x}^{(t)}$ // $\triangleright$ (7.23);
    for $r = 1, \ldots, k-1$ **do**
      $b_r^{(t+1)} \leftarrow b_r^{(t)} - s_r^{(t)}$;
  else
    $\boldsymbol{w}^{(t+1)} \leftarrow \boldsymbol{w}^{(t)}$;
    $\boldsymbol{b}^{(t+1)} \leftarrow \boldsymbol{b}^{(t)}$;
  $t \leftarrow t + 1$;

Output : Model parameters $\boldsymbol{w}^{(t)}$ and thresholds $\boldsymbol{b}^{(t)}$.

Algorithme 19: PRank

The algorithm update rule inspired by the perceptron algorithm only modifies the current weight in the case where the predicted rank $\hat{y}^{(t)}$ differs from the desired rank $y^{(t)}$ of the example $\mathbf{x}^{(t)}$. Thus, since the thresholds $\boldsymbol{b}^{(t)}$ are ordered, $b_1^{(t)} \leqslant b_2^{(t)} \leqslant \ldots \leqslant b_{k-1}^{(t)} \leqslant b_k^{(t)}$, the predicted rank is correct only if $\langle \boldsymbol{w}^{(t)}, \mathbf{x}^{(t)} \rangle > b_r^{(t)}$ for all rank values $r = 1, \ldots, y^{(t)} - 1$ and $\langle \boldsymbol{w}^{(t)}, \mathbf{x}^{(t)} \rangle < b_r^{(t)}$ for $r = y^{(t)}, \ldots, k-1$. These two inequalities can be expressed by a single inequality by introducing the binary variables $z_1^{(t)}, \ldots, z_{k-1}^{(t)}$ verifying:

$$\forall r, z_r^{(t)} = \begin{cases} +1, & \text{if } \langle \boldsymbol{w}^{(t)}, \mathbf{x}^{(t)} \rangle > b_r^{(t)}, \\ -1, & \text{if } \langle \boldsymbol{w}^{(t)}, \mathbf{x}^{(t)} \rangle < b_r^{(t)}. \end{cases} \tag{7.22}$$

Thus, the maximal rank for which $z_r^{(t)} = +1$ is $y^{(t)} - 1$ and the rank prediction of the example $\mathbf{x}^{(t)}$ is correct, if and only if $z_r^{(t)} \left(\langle \boldsymbol{w}^{(t)}, \mathbf{x}^{(t)} \rangle - b_r^{(t)} \right) > 0$ for all r. In case the algorithm makes a rank error for example $\mathbf{x}^{(t)}$, then there will be at least one threshold of index r for which the value of $\langle \boldsymbol{w}^{(t)}, \mathbf{x}^{(t)} \rangle$ is on the wrong side of $b_r^{(t)}$, i.e. $z_r^{(t)} \left(\langle \boldsymbol{w}^{(t)}, \mathbf{x}^{(t)} \rangle - b_r^{(t)} \right) \leqslant 0$. To correct the error, the algorithm then shifts the values of $\langle \boldsymbol{w}^{(t)}, \mathbf{x}^{(t)} \rangle$ and $b_r^{(t)}$ towards each other. Thus, the thresholds $b_r^{(t)}$ for which $z_r^{(t)} \left(\langle \boldsymbol{w}^{(t)}, \mathbf{x}^{(t)} \rangle - b_r^{(t)} \right) \leqslant 0$ are replaced by $b_r^{(t)} - z_r^{(t)}$ and, the weight vector $\boldsymbol{w}^{(t)}$ is modified with the following update rule:

$$\boldsymbol{w}^{(t+1)} = \boldsymbol{w}^{(t)} + \left(\sum_{r | z_r^{(t)} \left(\langle \boldsymbol{w}^{(t)}, \mathbf{x}^{(t)} \rangle - b_r^{(t)} \right) \leqslant 0} z_r^{(t)} \right) \mathbf{x}^{(t)}. \tag{7.23}$$

An interesting result of this algorithm is that if a solution $\boldsymbol{v}^* = (\boldsymbol{w}^*, \boldsymbol{b}^*)$ exists, then the ranking prediction error E_t, (7.21), is bounded. This result is stated as follows:

Theorem 7.1 (Bounded Ranking Error [40]) *Let* $\left((\mathbf{x}^{(t)}, y^{(t)}) \right)_{t \in [\![1,T]\!]}$ *be an input sequence of the PRank algorithm (19) such that the examples are contained in a hypersphere of radius* R*;* $\forall t, ||\mathbf{x}^{(t)}|| \leqslant R$*. Suppose there exists an ranking rule* $\boldsymbol{v}^* = (\boldsymbol{w}^*, \boldsymbol{b}^*)$ *such that* $b_1^* \leqslant \ldots \leqslant b_{k-1}^*$ *and* $||\boldsymbol{w}^*|| = 1$ *which correctly orders the examples with a certain margin* $\rho = \min_{r,t} \left\{ z_r^{(t)} \left(\langle \boldsymbol{w}^*, \mathbf{x}^{(t)} \rangle - b_r^* \right) \right\} > 0$*. We further assume that* $\boldsymbol{w}^{(1)} = \mathbf{0}$ *and* $b_1^{(1)} = 0, \ldots, b_{k-1}^{(1)} = 0, b_k^{(1)} = \infty$*. Then, in this case, the rank prediction error of the algorithm is bounded by:*

$$\sum_{t=1}^{T} |\hat{y}^{(t)} - y^{(t)}| \leqslant \frac{(k-1)(R^2+1)}{\rho^2}$$

Proof The proof of this theorem is quite similar in principle to that of Novikoff's theorem (Chap. 4, Theorem 4.1) and it consists in maximizing and minimizing the norm of the solution vector $\boldsymbol{v}^{(T+1)} = (\boldsymbol{w}^{(T+1)}, \boldsymbol{b}^{(T+1)})$ found by the Algorithm 19 after T iterations.

Let $(\mathbf{x}^{(t)}, y^{(t)})$ be the example randomly chosen by the algorithm at iteration $t \in \{1, \dots, T\}$. The update rule then produces the ranking rule, $\boldsymbol{v}^{(t+1)} = (\boldsymbol{w}^{(t+1)}, \boldsymbol{b}^{(t+1)})$, verifying the following relation:

$$\boldsymbol{w}^{(t+1)} = \boldsymbol{w}^{(t)} + \left(\sum_r s_r^{(t)}\right)\mathbf{x}^{(t)}, \tag{7.24}$$

$$\forall r \in \{1, \dots, k-1\}, b_r^{(t+1)} = b_r^{(t)} - s_r^{(t)}. \tag{7.25}$$

Let $n^{(t)} = |\hat{y}^{(t)} - y^{(t)}|$ be the absolute error between the predicted and desired ranks for the chosen example $\mathbf{x}^{(t)}$. Let $\boldsymbol{v}^* = (\boldsymbol{w}^*, \boldsymbol{b}^*)$ be such that $b_1^* \leqslant \dots \leqslant b_{k-1}^*$ and $||\boldsymbol{w}^*|| = 1$, the ranking rule that correctly orders all examples with a certain margin $\rho = \min_{r,t}\left\{z_r^{(t)}\left(\langle \boldsymbol{w}^*, \mathbf{x}^{(t)}\rangle - b_r^*\right)\right\} > 0$. From the bilinearity of the scalar product, we obtain:

$$\left\langle \boldsymbol{v}^*, \boldsymbol{v}^{(t+1)}\right\rangle = \left\langle \boldsymbol{v}^*, \boldsymbol{v}^{(t)}\right\rangle + \sum_{r=1}^{k-1} s_r^{(t)}\left(\left\langle \boldsymbol{w}^*, \mathbf{x}^{(t)}\right\rangle - b_r^*\right). \tag{7.26}$$

Moreover, for the rank prediction $\hat{y}^{(t)}$ of the example $\mathbf{x}^{(t)}$, there are two cases to consider:

- if this prediction is correct, $\hat{y}^{(t)} = y^{(t)}$, we have then $s_r^{(t)} = 0$ and so $s_r^{(t)}\left(\langle \boldsymbol{w}^*, \mathbf{x}^{(t)}\rangle - b_r^*\right) = 0$,
- otherwise, $\hat{y}^{(t)} \neq y^{(t)}$, we have $z_r^{(t)} = s_r^{(t)}$ and therefore, according to the hypothesis that the rule $\boldsymbol{v}^* = (\boldsymbol{w}^*, \boldsymbol{b}^*)$ order the examples well, we obtain:

$$s_r^{(t)}\left(\left\langle \boldsymbol{w}^*, \mathbf{x}^{(t)}\right\rangle - b_r^*\right) = z_r^{(t)}\left(\left\langle \boldsymbol{w}^*, \mathbf{x}^{(t)}\right\rangle - b_r^*\right) \geqslant \rho.$$

By summing the terms $s_r^{(t)}\left(\langle \boldsymbol{w}^*, \mathbf{x}^{(t)}\rangle - b_r^*\right)$ over all ranks $r \in \{1, \dots, k-1\}$, we then have:

$$\sum_{r=1}^{k-1} s_r^{(t)}\left(\left\langle \boldsymbol{w}^*, \mathbf{x}^{(t)}\right\rangle - b_r^*\right) \geqslant n^{(t)}\rho. \tag{7.27}$$

According to the two inequalities (7.26) and (7.27), we have $\left\langle \boldsymbol{v}^*, \boldsymbol{v}^{(t+1)}\right\rangle \geqslant \left\langle \boldsymbol{v}^*, \boldsymbol{v}^{(t)}\right\rangle + n^{(t)}\rho$. Since the ranking rule is initialized to the zero vector, after T iterations the emitted rule then verifies $\left\langle \boldsymbol{v}^*, \boldsymbol{v}^{(T+1)}\right\rangle \geqslant \left(\sum_t n^{(t)}\right)\rho$. That is, according to the Cauchy-Schwarz inequality and the fact that the ranking rule $\boldsymbol{v}^*$ is a unit vector, the norm of the vector $\boldsymbol{v}^{(T+1)}$ is then lower bounded by:

$$\|\boldsymbol{v}^{(T+1)}\| \geqslant \left(\sum_t n^{(t)}\right)\rho.$$

We will now increase the norm of $\boldsymbol{v}^{(T+1)}$ by replacing the values of the vectors $\boldsymbol{w}^{(t+1)}$ and $\boldsymbol{b}^{(t+1)}$ by what is given by the update rules (7.24) and (7.25).

$$\|\boldsymbol{v}^{(t+1)}\|^2 = \|\boldsymbol{w}^{(t)}\|^2 + \|\boldsymbol{b}^{(t)}\|^2 + \left(\sum_r s_r^{(t)}\right)^2 \|\mathbf{x}^{(t)}\|^2 + \sum_r \left(s_r^{(t)}\right)^2 + 2\sum_r s_r^{(t)}\left(\left\langle \boldsymbol{w}^{(t)}, \mathbf{x}^{(t)}\right\rangle - b_r^{(t)}\right).$$

Using the fact that the examples are all contained in a hypersphere of radius R and since $\forall r,\ s_r^{(t)} \in \{-1, 0, +1\}$, we have $\sum_r \left(s_r^{(t)}\right)^2 = n^{(t)}$ and $\left(\sum_r s_r^{(t)}\right)^2 \leqslant (n^{(t)})^2$ and $\sum_r s_r^{(t)}(\langle \boldsymbol{w}^{(t)}, \mathbf{x}^{(t)}\rangle - b_r^{(t)}) = \sum_r \mathbb{1}_{\hat{y}^{(t)} \neq y^{(t)}} z_r^{(t)}(\langle \boldsymbol{w}^{(t)}, \mathbf{x}^{(t)}\rangle - b_r^{(t)}) \leqslant 0$. We then have $\|\boldsymbol{v}^{(t+1)}\|^2 \leqslant \|\boldsymbol{v}^{(t)}\|^2 + (n^{(t)})^2 R^2 + n^{(t)}$, and therefore:

$$\|\boldsymbol{v}^{(T+1)}\|^2 \leqslant R^2 \sum_t (n^{(t)})^2 + \sum_t n^{(t)}.$$

The lower and upper bounding the norm of the vector corresponding to the ranking function after T iterations of the algorithm thus results in:

$$\rho^2 \left(\sum_t n^{(t)}\right)^2 \leqslant \|\boldsymbol{v}^{(T+1)}\|^2 \leqslant R^2 \sum_t \left(n^{(t)}\right)^2 + \sum_t n^{(t)}. \tag{7.28}$$

Dividing the inequalities by $\rho^2 \sum_t n^{(t)}$ gives $\sum_t n^{(t)} \leqslant \frac{R^2 \sum_t (n^{(t)})^2 / \sum_t n^{(t)} + 1}{\rho^2}$. The result follows by noting that $n^{(t)} \leqslant (k-1)$, or:

$$\sum_t n^{(t)} \leqslant \frac{R^2(k-1)+1}{\rho^2},$$

and $(k-1)R^2 + 1 \leqslant (k-1)(R^2+1)$. □

7.2.1.2 Ranking Based on Margin Maximization

According to Theorem 7.1, we notice that the margin still plays a preponderant role in the learning of the ranking rule, as it was defined in the previous section. Thus, for two training sets included in two hyperspheres of the same radius, if there are two ranking rules allowing to sort the examples perfectly in the two bases, the cumulative prediction errors of the examples at any iteration of the Algorithm 19 will be lower for the set where the margin corresponding to the associated rule is the largest.

This observation motivated other works following the same framework as previously for ranking of instances, notably that of Shashua et al. [154], who proposed an adaptation of SVMs (Chap. 5) for ordinal regression. The starting point of this work is based on the observation that the ranking rule defined with a single weight vector $\boldsymbol{w}$ and thresholds $\boldsymbol{b} = \{b_1, \dots, b_{k-1}, b_k\}$ and which consists in predicting the rank r of an example $\mathbf{x}$ as the index of the smallest threshold b_r for which $\langle \boldsymbol{w}, \mathbf{x} \rangle < b_r$, divides the input space into regions of equal ranks: all examples verifying the inequalities:

$$b_{r-1} < \langle \boldsymbol{w}, \mathbf{x} \rangle < b_r, \tag{7.29}$$

have the same predicted rank $\hat{y} = r$ and, as in the case of classification, the margin between two ranks r and $r+1$ is $2/|\boldsymbol{w}|$. Maximizing the margin subject to rank separability constraints that respect inequalities (7.29) for all examples $\mathbf{x}$ of rank r and on a training base $S = \{(\mathbf{x}_i, y_i); i \in \{1, \dots, m\}\}$, results in the following optimization problem:

$$\min_{\boldsymbol{w}, \boldsymbol{b}, \xi_{i,r}, \xi^*_{i,r}} \frac{1}{2}||\bar{\boldsymbol{w}}||^2 + C \sum_{r=1}^{k-1} \sum_{\mathbf{x}_i | y_i = r} (\xi_{i,r} + \xi^*_{i,r})$$

$$\begin{aligned}
\text{s.t. } \forall r, \forall i, \text{ if } y_i = r; &\langle \boldsymbol{w}, \mathbf{x}_i \rangle - b_r \leqslant -1 + \xi_{i,r} \\
&\langle \boldsymbol{w}, \mathbf{x}_i \rangle - b_{r-1} \geqslant 1 - \xi^*_{i,r} \\
&\xi_{i,r} \geqslant 0, \xi^*_{i,r} \geqslant 0.
\end{aligned}$$

Other SVM-based techniques for ordinal regression have proposed to add implicit and explicit constraints to the thresholds $\boldsymbol{b}$ in the optimization problem [32], with explicit constraints taking the form $b_{r-1} \leqslant b_r$, while implicit constraints use redundancy in the training examples to ensure the ordinal relationship between the thresholds.

7.2.2 *Pairwise Ranking*

Pairwise approaches consider the relative order between two examples in the sorted list to learn a score function. They are therefore similar to algorithms based on the reduction of some ranking problems to the classification of crucial pairs, presented in the previous section. The major difference is however that, without going through the reduction to the classification of pairs, it is possible to easily extend the algorithms initially developed for the ranking of instances to the case of ranking of alternatives. It should also be noted that, other approaches, called listwise approaches, consider the absolute position of the examples in the outputs to be ordered to find a score function. The first works following this approach attempted to directly optimize the ranking error measures such as the MAP (7.5) or the NDCG (7.8) presented in Sect. 7.1.1, by working on continuous and derivable bounds of these measures [138, 165, 188–190]. The confusion of his works is that they proceeded by analogy with the framework of classification and the search for a classifier by optimizing a convex, continuous and derivable bound of the classification error. However, in the case of ranking [26] demonstrated that none of the MAP or NDCG measures admits a convex, derivable and continuous bound that would have the same minimizer as the considered measure.

In the remainder of this section, we will first present in detail an algorithm representative of pairwise approaches for ranking of instances, as well as its extension to the case of ranking of alternatives. We then present the idea supporting some other algorithms developed following this approach and which have considered the problem as a crucial pair classification problem.

7.2.2.1 RankBoost

RankBoost [60] is one of the first algorithms proposed following the pairwise approach for ranking of instances. Due to its simplicity of implementation and its linear complexity in number of examples in the bipartite case, this algorithm has become a reference in the development of ranking models.

RankBoost is based on the algorithm discrete binary AdaBoost (Chap. 6, Sect. 6.1), and like it, it iteratively constructs a linear combination of basis functions, adapting at each iteration a probability distribution on the set of crucial pairs, so that the more the current combination reverses the order of a pair, the higher the weight assigned to the pair will be. Thus, the algorithm iteratively determines the weights $\{a_t\}_{t\in\{1,\dots,T\}}$ and the basis functions $\{f_t\}_{t\in\{1,\dots,T\}}$ (with values in $\{0, 1\}$), so that for each crucial pair $(\mathbf{x}_i, y_i), (\mathbf{x}_{i'}, y_{i'})$ of a training base $S = \{(\mathbf{x}_1, y_1), \dots, (\mathbf{x}_m, y_m)\}$, with $(y_i, y_{i'}) \in \mathbb{R}^2$ and $y_i > y_{i'}$, we have:

$$\sum_{t=1}^{T} a_t f_t(\mathbf{x}_i) - \sum_{t=1}^{T} a_t f_t(\mathbf{x}_{i'}) > 0 \Leftrightarrow \sum_{t=1}^{T} a_t (f_t(\mathbf{x}_i) - f_t(\mathbf{x}_{i'})) > 0. \tag{7.30}$$

Thus, at each iteration t, a distribution $D^{(t)}$ is maintained over the set of crucial pairs $(\mathbf{x}_i, y_i), (\mathbf{x}_{i'}, y_{i'}) \in S^2$, as follows:

$$\begin{aligned}\forall (i,i') \in \{1,\ldots,m\}^2;\ y_i > y_{i'},\ & D^{(t+1)}(i,i') \\ &= \frac{D^{(t)}(i,i')\exp(a_t(f_t(\mathbf{x}_{i'}) - f_t(\mathbf{x}_i)))}{Z^{(t)}},\end{aligned} \tag{7.31}$$

where $Z^{(t)} = \sum_{i,i'} D^{(t)}(i,i')e^{a_t(f_t(\mathbf{x}_{i'})-f_t(\mathbf{x}_i))}$ is the normalization coefficient and f_t is a binary-valued function, chosen by another basic algorithm (assumed to be determined at this stage) and minimizing the ranking error over the set of pairs weighted by $D^{(t)}$. Applying this updating rule from iteration to iteration, from the final step T to the initial step, 1 we get:

$$\forall (i,i') \in \{1,\ldots,m\}^2;\ y_i > y_{i'},\ D^{(T+1)}(i,i') = \frac{D^{(1)}(i,i')\exp(h(\mathbf{x}_{i'}) - h(\mathbf{x}_i))}{\prod_{t=1}^{T} Z^{(t)}}, \tag{7.32}$$

where $h = \sum_{t=1}^{T} a_t f_t$ is the final ranking function found by the algorithm.

Thus, the empirical ranking error of instances of the function h on the training base S, $\hat{R}_{oi}(h,S) = \sum_{i,i':y_i>y_{i'}} D^{(1)}(i,i')\mathbb{1}_{h(\mathbf{x}_i)\leqslant h(\mathbf{x}_{i'})}$ verifies:

$$\hat{R}_{oi}(h,S) = \sum_{i,i':y_i>y_{i'}} D^{(1)}(i,i')\mathbb{1}_{h(\mathbf{x}_i)-h(\mathbf{x}_{i'})\leqslant 0} \leqslant \sum_{i,i':y_i>y_{i'}} D^{(1)}(i,i')e^{h(\mathbf{x}_{i'})-h(\mathbf{x}_i)},$$

where $\forall z \in \mathbb{R}$, $\mathbb{1}_{z\leqslant 0} \leqslant e^{-z}$. According to the Eq. (7.32), this gives:

$$\hat{R}_{oi}(h,S) \leqslant \underbrace{\left(\sum_{i,i':y_i>y_{i'}} D^{(T+1)}(i,i')\right)}_{=1} \prod_{t=1}^{T} Z^{(t)} \tag{7.33}$$

$$\leqslant \prod_{t=1}^{T} Z^{(t)}. \tag{7.34}$$

To minimize this empirical error, the choice of parameters a_t is then made by minimizing the normalization factor $Z^{(t)}$, at each iteration t. Using the following inequality, which is obtained thanks to the convexity of $x \mapsto e^{ax}$ and Jensen's inequality:

$$\forall (z,a) \in \mathbb{R}^2,\ e^{az} \leqslant \left(\frac{1+z}{2}\right)e^{a} + \left(\frac{1-z}{2}\right)e^{-a},$$

we can upper-bound $Z^{(t)}$ by:

$$\forall t, Z^{(t)} \leqslant \sum_{i,i'} D^{(t)}(i,i')\left[\left(\frac{1+f_t(\mathbf{x}_{i'})-f_t(\mathbf{x}_i)}{2}\right)e^{a_t} + \left(\frac{1-f_t(\mathbf{x}_{i'})+f_t(\mathbf{x}_i)}{2}\right)e^{-a_t}\right]$$

$$= \left(\frac{1-r_t}{2}\right)e^{a_t} + \left(\frac{1+r_t}{2}\right)e^{-a_t}, \tag{7.35}$$

where

$$r_t = \sum_{i,i'} D^{(t)}(i,i')(f_t(\mathbf{x}_i) - f_t(\mathbf{x}_{i'})). \tag{7.36}$$

The minimum of the second term in the inequality (7.35) is then obtained for:

$$a_t = \frac{1}{2}\ln\left(\frac{1+r_t}{1-r_t}\right). \tag{7.37}$$

Moreover, with this value of a_t, the inequality (7.35) is written:

$$\forall t, Z^{(t)} \leqslant \sqrt{1-r_t^2}. \tag{7.38}$$

To minimize the bound $Z^{(t)}$, at each iteration, we must maximize the absolute value of r_t as defined in the Eq. (7.36) and also fix the value of the weight a_t as in (7.37). In the general case, the computation of r_t requires performing a sum over all crucial pairs; the complexity of the algorithm is therefore of the order of the number of crucial pairs.

The major innovation described in [60] is to use the structure of crucial pairs in the most important ranking case, which is the bipartite case. The complexity then becomes linear as a function of the number of examples, and not as a function of the number of crucial pairs.

7.2.2.2 Bipartite Case

We recall that in the bipartite case, the relevance judgments are binary, taking values in the set $\{-1,+1\}$. In this case, the central idea of using the crucial pairs structure is to maintain a distribution $\nu^{(t)}$ over the examples (and not over the pairs of examples) and to rewrite the distribution $D^{(t)}$ as:

$$\forall (i,i') \in \{1,\dots,m\}^2 \text{ such that } y_i = 1 \text{ and } y_{i'} = -1, D^{(t)}(i,i') = \nu^{(t)}(i)\nu^{(t)}(i'), \tag{7.39}$$

with the initial distribution of examples defined by:

$$\forall i \in \{1, \dots, m\}, \nu^{(1)}(i) = \begin{cases} \frac{1}{n_+} \text{ if } y_i = 1, \\ \frac{1}{n_-} \text{ if } y_i = -1, \end{cases} \tag{7.40}$$

where n_- (respectively n_+) is the number of irrelevant (respectively relevant) examples of S.

The decomposition (7.39) is motivated by the property of the exponential function that transforms sums into products: $\forall(a, b) \in \mathbb{R}^2, e^{a+b} = e^a e^b$, $Z^{(t)}$ is written in this case:

$$\begin{aligned} Z^{(t)} &= \sum_{i,i'} D^{(t)}(i, i') e^{a_t(f_t(\mathbf{x}_{i'}) - f_t(\mathbf{x}_i))} = \sum_{i,i'} \nu^{(t)}(i') e^{a_t f_t(\mathbf{x}_{i'})} \nu^{(t)}(i) e^{-a_t f_t(\mathbf{x}_i)} \\ &= \underbrace{\sum_{i:y_i=1} \nu^{(t)}(i) e^{-a_t f_t(\mathbf{x}_i)}}_{Z_1^{(t)}} \underbrace{\sum_{i':y_{i'}=-1} \nu^{(t)}(i') e^{a_t f_t(\mathbf{x}_{i'})}}_{Z_{-1}^{(t)}}. \end{aligned} \tag{7.41}$$

This decomposition and the property of the exponential function thus ensure that the Eq. (7.39) remains true at iteration $t + 1$.

Algorithm 20 exhibits the different steps of RankBoost in the bipartite case. Thus, the initial decomposition of $D^{(t)}$ (7.39) in the form of products allows to calculate the function r_t (7.36) with a complexity linear in number of examples. Moreover, since the update rule of the $\nu^{(t)}$ and the calculation of the optimal a_t weights (7.37) are also done with this complexity, the complexity of the RankBoost algorithm in the bipartite case is $O(T \times m \times \mathfrak{C})$, where T is the number of iterations and $\mathfrak{C}$ the complexity of the selection algorithm of the basis functions f_t.

The second innovation of RankBoost is to propose an efficient basis function selection algorithm when the number of basis functions is finite.

The algorithm described below allows to find, with a linear complexity in number of examples, a function f_t that minimizes r_t (7.36), in a particular case where the functions f_t are Boolean and created by thresholding real characteristics associated with the examples.

Basis Functions

Suppose that the feature set of each observation $\mathbf{x}$ is provided by real functions $\forall j \in \{1, \dots, d\}; \varphi_j : \mathcal{X} \to \mathbb{R}$. In the case where, at the time of learning, the value $\varphi_j(\mathbf{x})$ is unknown to the system, we denote $\varphi_j(\mathbf{x}) = \perp$ and [60] proposed the following binary basis functions, also called *decision stumps*:

$$\forall \mathbf{x} \in \mathcal{X};\ f_{j,\theta,nd}(\mathbf{x}) = \begin{cases} 1, \text{ if } \varphi_j(\mathbf{x}) > \theta \\ 0, \text{ if } \varphi_j(\mathbf{x}) \leqslant \theta \\ nd, \text{ if } \varphi_j(\mathbf{x}) = \perp. \end{cases} \tag{7.42}$$

where $\theta \in \mathbb{R}$ and $nd \in \{0, 1\}$. Thus, it is a question of using a thresholding of the characteristic φ_j to create binary values. These functions also allocate the same value ($nd \in \{0, 1\}$) to all observations for which φ_j is unknown.

The set of basic functions to be combined is then created by defining a priori a set of thresholds $\{\theta_\ell\}_{\ell=1}^{p}$ with $\theta_1 > \ldots > \theta_p$. Generally speaking, these thresholds depend on the characteristic φ_j considered, and the objective is to determine the values of j, ℓ and nd such that $f_{j,\theta_\ell,nd}$ is the basis function maximizing r_t (7.36). In all the following calculations, we assume that j, ℓ and nd are fixed and we simply denote $\theta = \theta_\ell$ and $f = f_{j,\theta_\ell,nd}$ and $r = r_t$. For the function f considered (7.42) and the decomposition of the distribution D (7.39), we then have:

$$
\begin{aligned}
r &= \sum_{\mathbf{x}_i:y_i=1} \sum_{\mathbf{x}_{i'}:y_{i'}=-1} D(i,i')(f(\mathbf{x}_i) - f(\mathbf{x}_{i'})) \\
&= \sum_{\mathbf{x}_i:y_i=1} \sum_{\mathbf{x}_{i'}:y_{i'}=-1} \nu(i)\nu(i')(y_i f(\mathbf{x}_i) + y_{i'} f(\mathbf{x}_{i'})) \\
&= \sum_{\mathbf{x}_i:y_i=1} \left(\nu(i) \underbrace{\sum_{\mathbf{x}_{i'}:y_{i'}=-1} \nu(i')}_{=1} \right) y_i f(\mathbf{x}_i) + \sum_{\mathbf{x}_{i'}:y_{i'}=-1} \left(\nu(i') \underbrace{\sum_{\mathbf{x}_i:y_i=1} \nu(i)}_{=1} \right) y_{i'} f(\mathbf{x}_{i'}) \\
&= \sum_{i=1}^{m} \pi(\mathbf{x}_i) f(\mathbf{x}_i),
\end{aligned}
$$

where,

$$
\pi(\mathbf{x}_i) = y_i \nu(i).
$$

Using the definition of f (7.42), we obtain:

$$
r = \sum_{\mathbf{x}_i:\varphi_j(\mathbf{x}_i)>\theta} \pi(\mathbf{x}_i) + nd \sum_{\mathbf{x}_i:\varphi_j(\mathbf{x}_i)=\perp} \pi(\mathbf{x}_i) = \sum_{\mathbf{x}_i:\varphi_j(\mathbf{x}_i)>\theta} \pi(\mathbf{x}_i) - nd \sum_{\mathbf{x}_i:\varphi_j(\mathbf{x}_i)\neq\perp} \pi(\mathbf{x}_i).
$$

The last equality is obtained by noting that:

$$
\sum_{\mathbf{x}_i} \pi(\mathbf{x}_i) = \sum_{\mathbf{x}_i:\varphi_j(\mathbf{x}_i)=\perp} \pi(\mathbf{x}_i) + \sum_{\mathbf{x}_i:\varphi_j(\mathbf{x}_i)\neq\perp} \pi(\mathbf{x}_i) = 0.
$$

Input : A training set $S = \{(\mathbf{x}_i, y_i); i \in \{1, \ldots, m\}\}$ where there are n_+ relevant examples and n_- irrelevant examples$(m = n_+ + n_-)$.

Initialization:

$$\forall i \in \{1, \ldots, m\}, \nu^{(1)}(i) = \begin{cases} \frac{1}{n_+} \text{ if } y_i = 1, \\ \frac{1}{n_-} \text{ if } y_i = -1. \end{cases}$$

for $t = 1, \ldots, T$ **do**

- Select a rule f_t from $D^{(t)}$; // $\triangleright$ For example with the algorithm algo_WeakLearner
- Calculate a_t which minimizes an upper bound on $Z^{(t)}$; // $\triangleright$ (7.37)
- Update // $\triangleright$ (7.31), (7.39) and (7.41)

$$\nu^{(t+1)}(i) = \begin{cases} \frac{\nu^{(t)}(i)e^{-a_t f_t(\mathbf{x}_i)}}{Z_1^{(t)}} \text{ if } y_i = 1, \\ \frac{\nu^{(t)}(i)e^{a_t f_t(\mathbf{x}_i)}}{Z_{-1}^{(t)}} \text{ if } y_i = -1. \end{cases}$$

with $Z_1^{(t)}$ and $Z_{-1}^{(t)}$ defined by:

$$Z_1^{(t)} = \sum_{i:y_i=1} \nu^{(t)}(i)e^{-a_t f_t(\mathbf{x}_i)},$$

$$Z_{-1}^{(t)} = \sum_{i':y_{i'}=-1} \nu^{(t)}(i')e^{a_t f_t(\mathbf{x}_{i'})}.$$

- Set
  $\forall(i, i')$ such that $y_i = 1$ and $y_{i'} = -1$ $D^{(t+1)}(i, i') = \nu^{(t+1)}(i)\nu^{(t+1)}(i')$;

Output : The score function $h(\mathbf{x}) = \sum_{t=1}^{T} a_t f_t(\mathbf{x})$.

Algorithme 20: Bipartite RankBoost

Which gives:

$$r = \underbrace{\sum_{\mathbf{x}_i:\varphi_j(\mathbf{x}_i)>\theta} \pi(\mathbf{x}_i)}_{L} - nd \underbrace{\sum_{\mathbf{x}_i:\varphi_j(\mathbf{x}_i)\neq\perp} \pi(\mathbf{x}_i)}_{R}. \tag{7.43}$$

The objective is to choose the parameters j, θ and nd such that the absolute value of this expression is maximal (7.38). The Algorithm 21 describes the method proposed by [60]: the features φ_j are processed one after the other. For each j, the right-hand term R does not depend on either θ_ℓ or nd; it is therefore calculated at the beginning. Then, the left-hand sum L is calculated incrementally for the different

---

Input :

- A distribution $D(i, i') = \nu(i)\nu(i')$ over crucial pairs $(\mathbf{x}_i, \mathbf{x}_{i'}) \in S \times S$;
- A set of features $\{\varphi_j\}_{j=1}^d$;
- For each φ_j a set of thresholds $\{\theta_\ell\}_{\ell=1}^p$ such that $\theta_1 \geqslant \ldots . \geqslant \theta_p$;

Initialization:

- $\forall \mathbf{x}_i, \pi(\mathbf{x}_i) \leftarrow y_i \nu(i)$;
- $r^* \leftarrow 0$;

```
for j = 1, ..., d do
    L ← 0;
    R ← Σ_{x_i : φ_j(x_i) ≠ ⊥} π(x_i);
    θ_0 ← ∞;
    for ℓ = 1, ..., p do
        L ← L + Σ_{x_i : θ_{ℓ-1} ⩾ φ_j(x_i) > θ_ℓ} π(x_i);
        if |L| > |L − R| then
            nd ← 0;
        else
            nd ← 1;
        if |L − nd × R| > |r*| then
            r* ← L − nd × R;
            j* ← j ;
            θ* ← θ_ℓ;
            nd* ← nd;
```

Output : $(\varphi_{j^*}, \theta^*, nd^*)$.

Algorithme 21: Search for the basis functions

possible threshold values $\{\theta_\ell\}_{\ell=1}^p$, starting with the largest thresholds.[6] For each of these thresholds, the algorithm determines the best value $nd \in \{0, 1\}$, then keeps in memory the triplet (j^*, θ^*, nd^*) that maximizes r, among the triplets (j, θ_ℓ, nd) already calculated.

The estimation of the right-hand term R, in the equation is therefore done with a linear complexity relative to the number of observations for which the real value of φ_j is known. At each iteration, the complexity $\mathfrak{C}$ of this search algorithm, whose pseudo-code is described Algorithm 21, is therefore $O(m \times p \times d)$.

In the following section, we will present the extension of the RankBoost algorithm to the case of ranking of alternatives.

---

[6] This calculation is performed efficiently by sorting, once and for all, the $\varphi_j(\mathbf{x}_i)_{i=1}^m$ in descending order.

7.2.2.3 Extension to Ranking of Alternatives

Usunier [173] extended the RankBoost algorithm to the case of ranking of alternatives by maintaining a distribution for each entry q of the training base, as well as, for each of these examples, a distribution on its corresponding crucial pairs. To simplify the presentation, we consider in the following the case where the relevance judgments associated with the alternatives are binary. Thus, for a training base $S = \{(q_1, \mathbf{y}_1), \ldots, (q_m, \mathbf{y}_m)\}$, each entry q_i is associated with a list of alternatives $(x_1^{(i)}, \ldots, x_{m_i}^{(i)})$ and corresponding binary relevance judgments $\mathbf{y}_i = (y_1^{(i)}, \ldots, y_{m_i}^{(i)}) \in \{-1, +1\}^{m_i}$. We recall that the score function h takes as input the representation $\mathbf{x}_{j,q_i}^{(i)}$ associated with the pair $(x_j^{(i)}, q_i)$, where q_i is an input and $x_j^{(i)}$ one of its alternatives (Sect. 7.1.3), and that the ranking error (7.16) in the considered case is written:

$$\hat{R}_{oa}(h, S) = \frac{1}{m}\sum_{i=1}^{m} \frac{1}{n_-^{(i)} n_+^{(i)}} \sum_{j:y_j^{(i)}=1} \sum_{\ell:y_\ell^{(i)}=-1} \mathbb{1}_{h(\mathbf{x}_{j,q_i}^{(i)}) \leqslant h(\mathbf{x}_{\ell,q_i}^{(i)})}, \tag{7.44}$$

where, $n_+^{(i)}$ (respectively $n_-^{(i)}$) denotes the number of relevant (non-relevant) alternatives for example q_i.

The extension of RankBoost is given in Algorithm (22). At each iteration t, the algorithm maintains a distribution $\lambda^{(t)}$ over the examples in the training set, a distribution $\nu_i^{(t)}$ over the alternatives associated with example q_i, and a distribution $D_i^{(t)}$ over the crucial pairs of alternatives, represented by a distribution over the pairs (j, ℓ) such that $y_j^{(i)} = 1$ and $y_\ell^{(i)} = -1$. For each example q_i, this latter distribution is defined based on the other two; $\forall i \in \{1, \ldots, m\}, \forall (j, \ell) \in \{1, \ldots, m_i\}^2$, such that $y_j^{(i)} = 1, y_\ell^{(i)} = -1$:

$$D_i^{(t)}(j, \ell) = \lambda_i^{(t)} \nu_i^{(t)}(j) \nu_i^{(t)}(\ell).$$

A high weight assigned to a pair means that the score function f_t learned so far orders, for an example $q_i \in S$, its irrelevant alternative above its relevant alternative. As in the case of ranking of instances or classification, these distributions are initialized in a uniform way:

$$\forall i \in \{1, \ldots, m\}, \lambda_i^{(1)} = \frac{1}{m},$$

$$\nu_i^{(1)}(j) = \begin{cases} \frac{1}{n_+^{(i)}} \text{ if } y_j^{(i)} = 1, \\ \frac{1}{n_-^{(i)}} \text{ if } y_j^{(i)} = -1. \end{cases}$$

Input : A training set $S = \{(q_i, \mathbf{y}_i); i \in \{1, \dots, m\}\}$, where for each example q_i, there are m_i alternatives $(x_1^{(i)}, \dots, x_{m_i}^{(i)})$.

Initialisation:

$$\forall i \in \{1, \dots, m\}, \lambda_i^{(1)} = \frac{1}{m}$$

$$\nu_i^{(1)}(k) = \begin{cases} 1/p_i \text{ if } y_i^k = 1, \\ 1/n_i \text{ if } y_i^k = -1. \end{cases}$$

for $t = 1, \dots, T$ **do**

- Select the rule f_t from the distribution $D^{(t)}$.
- Estimate a_t which maximizes the upper bound on $Z^{(t)}$.
- $\forall i \in \{1, \dots, m\}, \forall (k, l) \in \{1, \dots, m_i\}^2$ such that $y_j^{(i)} = 1$ and $y_\ell^{(i)} = -1$: Update $D_i^{(t+1)}$: $D_i^{(t+1)}(j, \ell) = \lambda_i^{(t+1)} \nu_i^{(t+1)}(j) \nu_i^{(t+1)}(\ell)$.

$$\forall i \in \{1, \dots, m\}, \lambda_i^{(t+1)} = \frac{\lambda_i^{(t)} Z_{-1i}^{(t)} Z_{1i}^{(t)}}{Z^{(t)}}.$$

$$\nu_i^{(t+1)}(j) = \begin{cases} \frac{\nu_i^{(t)}(j) \exp(-a_t f_t(\mathbf{x}_{j,q_i}^{(i)}))}{Z_{1i}^{(t)}} \text{ if } y_j^{(i)} = 1, \\ \frac{\nu_i^{(t)}(j) \exp(a_t f_t(\mathbf{x}_{j,q_i}^{(i)}))}{Z_{-1i}^{(t)}} \text{ if } y_j^{(i)} = -1. \end{cases}$$

where $Z_{-1i}^{(t)}, Z_{1i}^{(t)}$ and $Z^{(t)}$ are defined by:

$$Z_{1i}^{(t)} = \sum_{j: y_j^{(i)} = 1} \nu_i^{(t)}(j) e^{-a_t f_t(\mathbf{x}_{j,q_i}^{(i)})},$$

$$Z_{-1i}^{(t)} = \sum_{\ell: y_\ell^{(i)} = -1} \nu_i^{(t)}(\ell) e^{a_t f_t(\mathbf{x}_{\ell,q_i}^{(i)})},$$

$$Z^{(t)} = \sum_{i=1}^{m} \lambda_i^{(t)} Z_{-1i}^{(t)} Z_{1i}^{(t)}.$$

Output : The score function $h = \sum_{t=1}^{T} a_t f_t$.

Algorithme 22: Adaptation of RankBoost to the ranking of alternatives

These distributions are updated using f_t and its weight a_t for each example q_i of the training base and each alternative associated with q_i:

$$\forall i \in \{1, \dots, m\}, \lambda_i^{(t+1)} = \frac{\lambda_i^{(t)} Z_{-1i}^{(t)} Z_{1i}^{(t)}}{Z^{(t)}}, \tag{7.45}$$

$$
\nu_i^{(t+1)}(j) = \begin{cases} \frac{\nu_i^{(t)}(j)\exp(-a_t f_t(\mathbf{x}_{j,q_i}^{(i)}))}{Z_{1i}^{(t)}} \text{ if } y_j^{(i)} = 1, \\ \frac{\nu_i^{(t)}(k)\exp(a_t f_t(\mathbf{x}_{j,q_i}^{(i)}))}{Z_{-1i}^{(t)}} \text{ if } y_j^{(i)} = -1, \end{cases}
$$

where $Z_{-1i}^{(t)}$, $Z_{1i}^{(t)}$ and $Z^{(t)}$ are defined as:

$$
Z_{1i}^{(t)} = \sum_{j:y_j^{(i)}=1} \nu_i^{(t)}(j)\exp(-a_t f_t(\mathbf{x}_{j,q_i}^{(i)})),
$$

$$
Z_{-1i}^{(t)} = \sum_{l:y_\ell^{(i)}=-1} \nu_i^{(t)}(\ell)\exp(a_t f_t(\mathbf{x}_{\ell,q_i}^{(i)})),
$$

$$
Z^{(t)} = \sum_{i=1}^{m} \lambda_i^{(t)} Z_{-1i}^{(t)} Z_{1i}^{(t)}.
$$

Recall that $\forall q \in S, \forall x \in \mathcal{A}_q$, $\mathbf{x}_q$ denotes the vector representation of (x, q). The learning criterion defined for the final score function h is:

$$
\begin{aligned}
\hat{R}_{oa}(h, S) &= \frac{1}{m}\sum_{i=1}^{m} \frac{1}{n_-^{(i)} n_+^{(i)}} \sum_{j:y_j^{(i)}=1} \sum_{\ell:y_\ell^{(i)}=-1} \mathbb{1}_{h(\mathbf{x}_{j,q_i}^{(i)}) \leqslant h(\mathbf{x}_{\ell,q_i}^{(i)})} \\
&= \sum_{i=1}^{m} \sum_{j,\ell:y_j^{(i)} > y_\ell^{(i)}} D_i^{(1)}(j,\ell) \mathbb{1}_{h(\mathbf{x}_{j,q_i}^{(i)}) \leqslant h(\mathbf{x}_{\ell,q_i}^{(i)})}.
\end{aligned}
$$

As in the case of ranking of instances, given the upper bound $\mathbb{1}_{x\leqslant 0} \leqslant e^{-x}$, the definitions $D_i^{(1)}(k,l) = \lambda_i^{(1)} \nu_i^{(1)}(k) \nu_i^{(1)}(\ell)$ and $h(x,k) = \sum_t a_t f_t(x,k)$, along with the update rule $\lambda^{(t+1)}$ (7.45), it is immediate that the function $\hat{R}_{oa}(f, S)$ is upper bounded by:

$$
\hat{R}_{oa}(h, S) \leqslant \prod_t Z^{(t)}.
$$

As for the RankBoost algorithm in the bipartite case, a_t is chosen to minimize $Z^{(t)}$ at each iteration t. The detailed algorithm therefore iteratively decreases the upper bound on the cost function $\hat{R}_{oa}(f, S)$. With the decomposition of the distribution $D_i^{(t)}$ in the form of products, the complexity of the RankBoost adaptation is, as in the case of the bipartite ranking, linear in the number of alternatives for each example of the training base. Thus, by denoting T the number of iterations, the complexity of the algorithm is $O(\mathfrak{C} \times T \times \sum_{i=1}^{m}(n_i + p_i))$, where $\mathfrak{C}$ is the complexity of the algorithm for selecting the basis functions f_t.

The RankBoost selection algorithm (Algorithm 21) in the case of bipartite ranking of instances can easily be adapted to this case. We will not detail it here for the sake of remaining concise in our presentation. Generally speaking, the algorithm obtained therefore has a linear complexity with respect to the number of total alternatives to be ranked.

7.2.2.4 Reduction to Binary Classification of Pairs

Other studies adopting the pairwise approach have observed that, in the context of ranking instances and certain tasks related to ranking alternatives, the minimization of empirical risk based on crucial pairs (7.6) can be achieved using a binary classification algorithm applied to a transformed version of the original dataset [76, 139, 186]. Indeed, for a training set S of size m, if we consider the following transformed set:

$$\begin{aligned}\mathfrak{T}(S) &= \left\{((\mathbf{x}_i, \mathbf{x}_j), 1) \middle| (i, j) \in \{1, \ldots, m\}^2 \text{ and } y_i > y_j\right\} \\ &= \left((\boldsymbol{\xi}_1, d_1), \ldots, (\boldsymbol{\xi}_M, d_M)\right),\end{aligned}$$

where $M = \sum_{i,j} \mathbb{1}_{y_i > y_j}$ is the total number of crucial pairs that can be created on S and the desired output d_ℓ associated with the pair $\boldsymbol{\xi}_\ell$, is equal to $+1$ if the first example of the crucial pair is relevant to the topic considered and the second example is not relevant and -1 otherwise. We can then associate a classifier $c_h : \mathcal{X} \times \mathcal{X} \to \{-1, +1\}$ on the pairs of examples, defined from a score function $h : \mathcal{X} \to \mathbb{R}$ as follows:

$$c_h(\mathbf{x}, \mathbf{x}') = c_h(\boldsymbol{\xi}) = \operatorname{sign}(h(\mathbf{x}) - h(\mathbf{x}')). \tag{7.46}$$

In this case, we can see that the empirical ranking error of h on $S = \{(\mathbf{x}_i, y_i) \mid i \in \{1, \ldots, m\}\}$, $\hat{R}_{oi}(h, S)$, is equal to the error of the associated classifier c_h on $\mathfrak{T}(S)$, denoted by $\hat{\mathcal{L}}_m m^{\mathfrak{T}}(c_h, \mathfrak{T}(S))$:

$$\hat{R}_{oi}(h(\mathfrak{J}, q^*), \mathbf{y}) = \frac{1}{\sum_{i,j} \mathbb{1}_{y_i > y_j}} \sum_{i,j: y_i > y_j} \mathbb{1}_{h(\mathbf{x}_i) \leqslant h(\mathbf{x}_j)} = \underbrace{\frac{1}{M} \sum_{\ell=1}^{M} \mathbb{1}_{d_\ell c_h(\boldsymbol{\xi}_\ell) \leqslant 0}}_{\hat{\mathcal{L}}_m^{\mathfrak{T}}(c_h, \mathfrak{T}(S))}, \tag{7.47}$$

where $\mathfrak{J} = (\mathbf{x}_1, \ldots, \mathbf{x}_m)$, q^* is the topic associated with the ranking of instances task and $\mathbf{y} = (y_1, \ldots, y_m)$. The major problem with this reduction is that the pairs $\xi_\ell \in \mathfrak{T}(S)$ containing the same examples of S are dependent on each other and the empirical risk $\hat{\mathcal{L}}_m^{\mathfrak{T}}(c_h, \mathfrak{T}(S))$ is thus made up of a sum of interdependent random variables, which implies that the classical results in statistics, based on the assumption of independence of random variables, are no longer applicable in this case. We will study this point in the next section.

The major difference between point-wise approaches and the approach based on the reduction of the ranking of instances problem to a binary classification problem is that, for the latter, the criterion considered is indeed a ranking criterion (7.6) but the reduction introduces an interdependence between the training examples, whereas for algorithms based on the point-wise approach, the examples are i.i.d. but the ranking is deduced on the output of a prediction function learned with a classification or regression criterion.

7.3 Learning with Interdependent Data

The consideration of interdependence between variables in the learning of classification functions, in the case of this reduction and where the transformation T is explicit, has been studied in [174]. This study is based on the use of Janson's theorem, [88], which extends the theorem of Hoeffding [81] by revealing a measure of interdependence between variables defined in relation to the number of subsets of interdependent variables constructed on this basis. This degree is all the more important as there are subsets of independent variables.

Theorem 7.2 (Deviation Bound for the Sum of Bounded Interdependent Random Variables [88]) *Let $Y_1, \dots, Y_M$, M be random variables such that $\forall i \in \mathcal{M} = \{1, \dots, M\}, \exists (a_i, b_i) \in \mathbb{R}^2, Y_i \in [a_i, b_i]$. Let $\chi^*(\mathcal{M})$ be the pure chromatic number associated with the set $\mathcal{M}$. We then have:*

$$\forall \epsilon > 0,\ \mathbb{P}\left(\mathbb{E}\left(\sum_{i=1}^{M} Y_i\right) - \sum_{i=1}^{M} Y_i > \epsilon\right) \leqslant \exp\left(-\frac{2\epsilon^2}{\chi^*(\mathcal{M}) \sum_{i=1}^{M} (b_i - a_i)^2}\right).$$

In order to better understand the result of Janson's theorem [88], we illustrate on a simple example of bipartite ranking (Fig. 7.3) the definitions given in this publication. To simplify the presentation, we will assimilate in this example the transformed set $\mathfrak{T}(S)$ to the set of indices $\mathcal{M}$ of the interdependent examples. In this example, the set S contains $n_+ = 2$ relevant examples and $n_- = 3$ irrelevant examples, the set $\mathfrak{T}(S)$ contains all the crucial pairs ($n_- \times n_+ = 6$ in total), each formed by a relevant example and an irrelevant example. Figure 7.3a shows a covering of $\mathfrak{T}(S)$ by a family of three subsets $\mathfrak{M}_1$, $\mathfrak{M}_2$ and $\mathfrak{M}_3$, i.e. $\cup_j \mathfrak{M}_j = \mathfrak{T}(S)$. Figure 7.3b shows a family $\{(\mathfrak{M}_j, \omega_j)\}$ with $\mathfrak{M}_j \subset \mathfrak{T}(S)$ and $\omega_j \in [0, 1]$ that forms a fractional covering of $\mathfrak{T}(S)$ since $\forall i \in \mathfrak{T}(S), \sum_j \omega_j \mathbb{1}_{i \in \mathfrak{M}_j} \geqslant 1$.

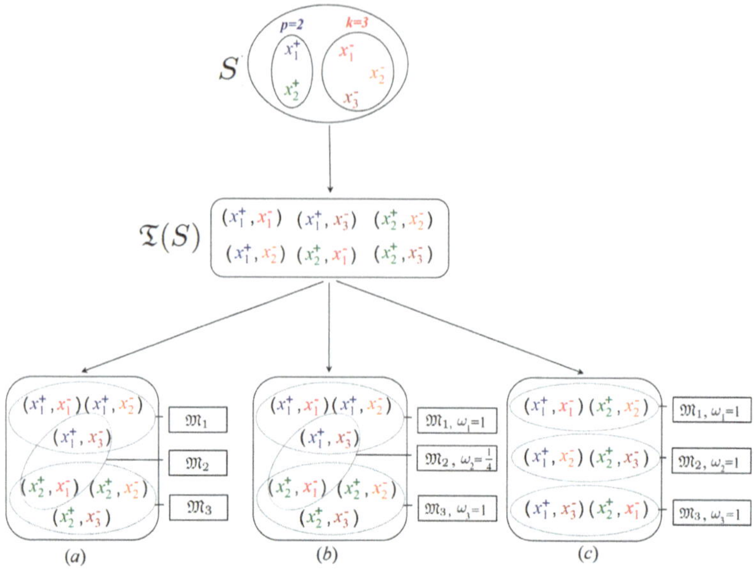

Fig. 7.3 Example of three coverings of a transformed set $\mathfrak{T}(S)$ of interdependent examples corresponding to a bipartite ranking problem. In (**a**) the three subsets $\mathfrak{M}_1$, $\mathfrak{M}_2$ and $\mathfrak{M}_3$ form a covering of $\mathfrak{T}(S)$. In (**b**) the sets $\{\mathfrak{M}_j, \omega_j\}_{j\in\{1,2,3\}}$ form a fractional covering of $\mathfrak{T}(S)$ and in (**c**) the sets $\{\mathfrak{M}_j, \omega_j\}_{j\in\{1,2,3\}}$ form an exact proper covering of $\mathfrak{T}(S)$

Figure 7.3c shows an exact proper fractional covering of $\mathfrak{T}(S)$ consisting of subsets $\mathfrak{M}_j$ containing independent variables and $\forall i \in \mathfrak{T}(S), \sum_j \omega_j \mathbb{1}_{i\in\mathfrak{M}_j} = 1$.

The chromatic number $\chi(\mathcal{M})$ is the smallest integer such that there exists a proper covering $\{\mathfrak{M}_j\}_j$ of $\mathcal{M}$ and the pure chromatic number, $\chi^*(\mathcal{M})$, is the minimum of $\sum_j \omega_j$ over the set of proper fractional coverings of $\mathcal{M}$.

Since a proper cover is also a fractional cover such that $\forall j, \omega_j = 1$, this implies that $\chi^*(\mathcal{M}) \leqslant \chi(\mathcal{M})$. In our example, we have $\chi^*(\mathcal{M}) = \chi(\mathcal{M}) = 3$.

We can see that $\chi^*(\mathcal{M}) = \chi(\mathcal{M}) = 1$ means that there exists a unique covering of $\mathcal{M}$ composed of independent random variables. This implies that Janson's theorem [88] is a direct extension of Hoeffding's theorem [81] since for a particular value of $\chi^*(\mathcal{M})$, involving the assumptions of Hoeffdings theorem, we obtain the latter theorem.

7.3.1 Test Error Bound

The generic test bound that we can obtain in the context of studying the classification of interdependent data is a direct application of Theorem 7.2 on the sum of random variables $Y_i = \frac{1}{M}\mathbb{1}_{d_i c_h(\mathfrak{x}_i)\leqslant 0}$ of (7.47). In this case, we have $\forall i,\ Y_i \in [0, \frac{1}{M}]$ and the test bound on a basis T is written:

$$\forall \delta > 0, \mathbb{P}\left(\mathcal{L}^{\mathfrak{T}}(c_h) \leqslant \hat{\mathcal{L}}_m^{\mathfrak{T}}(c_h, \mathfrak{T}(T)) + \sqrt{\frac{\chi^*(\mathfrak{T})\ln(\frac{1}{\delta})}{2M}}\right) \geqslant 1-\delta. \tag{7.48}$$

where $\mathcal{L}^{\mathfrak{T}}(c_h) = \mathbb{E}_{\mathfrak{T}(S)}\hat{\mathcal{L}}_m^{\mathfrak{T}}(c_h, \mathfrak{T}(S))$ is the generalization error of c_h defined as its error probability on a crucial pair. In the reduction of the ranking of instances problem to that of pair classification, the set of interdependent data is constructed on the basis of a transformation function $\mathfrak{T}$. It is then possible to link the degree of interdependence of the data $\chi^*(\mathfrak{T})$ to the latter.

7.3.2 Generalization Error Bound

As in the classification case, the test bound (7.48) has restrictions (Chap. 1, Sect. 1.3.1).

Its interpretation is that, for a given classifier c_h, there exists a transformed set $\mathfrak{T}(S)$ on which the inequality $\mathcal{L}^{\mathfrak{T}}(c_h) - \hat{\mathcal{L}}_m^{\mathfrak{T}}(c_h, \mathfrak{T}(S)) \leqslant \sqrt{\frac{\chi^*(\mathfrak{T})\ln(\frac{1}{\delta})}{2M}}$ holds with a probability of at least $1-\delta$. These sets $\mathfrak{T}(S)$ can be different for different classifiers; in other words, there are only a certain number of classifiers satisfying this inequality.

The generalization bound that we seek to have must be true for any classifier of a given set of functions. The idea is to consider the uniform deviations between the generalization error and the empirical error of any classifier on any training set.

To derive this generalization bound, Usunier et al. [174] proposed to extend Rademacher's theory studied in classification of independent data (Chap. 2) to the case of classification with interdependent data.

Let $\mathcal{F}$ be a class of functions with values in $\mathcal{Y} = \{-1, 1\}$ trained on i.i.d. sampled data following a distribution $\mathcal{D}$.[7]

For a given cost function $\ell : \mathcal{Y} \times \mathcal{Y} \to \mathbb{R}_+$, and a classifier $c_h : \mathcal{X} \to \mathcal{Y}$, let $\hat{\mathcal{L}}_m(c_h, S) = \frac{1}{m}\sum_{i=1}^m \ell(c_h(\mathbf{x}_i), y_i)$, the empirical risk of the classifier c_h on a training set S of size m and $\mathcal{L}(c_h) = \mathbb{E}_{(\mathbf{x},y)\sim\mathcal{D}}\ell(c_h(\mathbf{x}), y)$, the generalization error of c_h. For a classifier $c_h \in \mathcal{F}$ trained on S, we have seen that the statistical tool allowing to relate the supremum on $\mathcal{F}$ of $\mathcal{L}(c_h) - \hat{\mathcal{L}}_m(c_h, S)$ to its expectation (Chap. 1, Sect. 2.2.1), and thus to derive a bound on its generalization error, is the theorem of Mcdiarmid [114]. However, this theorem is not applicable in the state to the binary classification of pairs of examples resulting from the reduction of a

[7] This is not a restriction: if the set $\mathcal{F}$ considered is a set of real-valued functions, then we will take the set of associated classifiers $\{x \mapsto \text{sign}(h(\mathbf{x})) \mid h \in \mathcal{H}\}$.

ranking problem since it assumes that the input variables are independent; it cannot therefore be used on the function:

$$\mathfrak{T}(S) \mapsto \sup_{c_h \in \mathcal{F}} [\mathcal{L}^{\mathfrak{T}}(c_h) - \hat{\mathcal{L}}_m^T(c_h, \mathfrak{T}(S))], \tag{7.49}$$

because it is not decomposable into a sum of functions taking into account independent transformed examples.

To make this connection, it would be necessary to extend the theorem of Mcdiarmid [114] using the decomposition approach of Janson [88], so as to be able to take into account exact proper coverings of a transformed set that contain independent random variables.

7.3.2.1 Extension of Mcdiarmid's Theorem to Interdependent Random Variables

We will start the presentation by introducing a function Φ that upper-bounds $\mathfrak{T}(S) \mapsto \sup_{c_h \in \mathcal{F}} [\mathcal{L}^{\mathfrak{T}}(c_h) - \hat{\mathcal{L}}_m^{\mathfrak{T}}(c_h, \mathfrak{T}(S))]$ and then we will present the extension of the theorem of [114] that applies to the function Φ.

Let S be a training set consisting of independent random examples and $\{\mathfrak{M}_j, \omega_j\}_{j \in \{1,\dots,p\}}$ be an exact proper covering (the sets $\mathfrak{M}_j$ are independent and $\sum_{j=1}^p \omega_j \mathbb{1}_{i \in \mathfrak{M}_j} = 1$) of the transformed set $\mathfrak{T}(S)$. In this case, any sum of the form $\sum_{i=1}^M t_i$ can be written:

$$\sum_{i=1}^M t_i = \sum_{i=1}^M \sum_{j=1}^p \omega_j \mathbb{1}_{i \in \mathfrak{M}_j} t_i = \sum_{j=1}^p \omega_j \sum_{i \in \mathfrak{M}_j} t_i. \tag{7.50}$$

Now rewriting $\mathcal{L}^{\mathfrak{T}}(c_h) - \hat{\mathcal{L}}_m^T(c_h, \mathfrak{T}(S))$ as:

$$\mathcal{L}^{\mathfrak{T}}(c_h) - \hat{\mathcal{L}}_m^T(c_h, \mathfrak{T}(S)) = \frac{1}{M} \sum_{i=1}^M \left(\mathbb{E}_{\mathfrak{T}(\tilde{S})} \left[\ell(c_h(\tilde{\boldsymbol{\xi}}_i), \tilde{d}_i) \right] - \ell(c_h(\boldsymbol{\xi}_i), d_i) \right),$$

we can apply the previous result (7.50) to obtain:

$$\begin{aligned} &\mathcal{L}^{\mathfrak{T}}(c_h) - \hat{\mathcal{L}}_m^T(c_h, \mathfrak{T}(S)) \\ &= \frac{1}{M} \sum_{j=1}^p \omega_j \left(\mathbb{E}_{\mathfrak{T}(\tilde{S})} \left[\sum_{i \in \mathfrak{M}_j} \ell(c_h(\tilde{\boldsymbol{\xi}}_i), \tilde{d}_i) \right] - \sum_{i \in \mathfrak{M}_j} \ell(c_h(\boldsymbol{\xi}_i), d_i) \right). \end{aligned}$$

Taking the supremum over the set of functions and noting that the supremum of a sum is less than the sum of the supremums, we obtain:

$$\sup_{c_h \in \mathcal{F}} [\mathcal{L}^{\mathfrak{T}}(c_h) - \hat{\mathcal{L}}_m^T(c_h, \mathfrak{T}(S))] \leqslant$$

$$\sum_{j=1}^{p} \frac{\omega_j}{M} \sup_{c_h \in \mathcal{F}} \left(\mathbb{E}_{\mathfrak{T}(\tilde{S})} \left[\sum_{i \in \mathfrak{M}_j} \ell(c_h(\tilde{\boldsymbol{\xi}}_i), \tilde{d}_i) \right] - \sum_{i \in \mathfrak{M}_j} \ell(c_h(\boldsymbol{\xi}_i), d_i) \right).$$

This upper bound is the first step in obtaining the generalization bound. We note that the second term of the previous inequality is written as a weighted sum of independent variables of the form $\sum_{j=1}^{p} \omega_j \phi_j((\boldsymbol{\xi}_1, d_1), \ldots, (\boldsymbol{\xi}_{|\mathfrak{M}_j|}, d_{|\mathfrak{M}_j|}))$ with:

$$\phi_j : (\boldsymbol{\xi}_i, d_i)_{i=1}^{|\mathfrak{M}_j|} \mapsto \frac{1}{M} \sup_{c_h \in \mathcal{F}} \mathbb{E}_{\mathfrak{T}(\tilde{S})} \left[\sum_{i \in \mathfrak{M}_j} \ell(c_h(\tilde{\boldsymbol{\xi}}_i), \tilde{d}_i) \right] - \sum_{i \in \mathfrak{M}_j} \ell(c_h(\boldsymbol{\xi}_i), d_i).$$

Indeed, each of the ϕ_j is only a function of independent variables: the only indices i of the variables $\boldsymbol{\xi}_i$ on which ϕ_j will depend are those belonging to $\mathfrak{M}_j$.

7.3.2.2 Extension of Mcdiarmid's Theorem

Usunier [173] extended the theorem of Mcdiarmid [114] by considering functions Φ starting from a set of interdependent examples and which can be written as a combination of functions taking independent data as input.

Theorem 7.3 (Extension of the Bounded Difference Bound for Interdependent Random Variables [173]) *Let $X_1, \ldots, X_m$, m be independent random variables with values in $\mathcal{X}$. Let $\mathfrak{T} : \mathcal{X}^m \to \mathfrak{X}^M$, $\{\mathfrak{M}_j, \omega_j\}_{j=1}^p$ and $\Phi : \mathfrak{X}^M \to \mathbb{R}$ be such that:*

1. *$\{\mathfrak{M}_j, \omega_j\}_{j=1}^p$ is an exact proper cover of $\{1, \ldots, M\}$ for random variables $\{z_i\}_{i=1}^M$ with value in $\mathfrak{X}$ defined by $\mathfrak{T}(X_1, \ldots, X_m) = (z_1, \ldots, z_M)$. We then note:*

$$M_j = |\mathfrak{M}_j| \text{ and } \mathfrak{M}_j = \{\mu_{j,1}, \ldots, \mu_{j,|\mathfrak{M}_j|}\}.$$

2. *$\sum_{j=1}^p \omega_j = \chi^*(\mathfrak{T})$,*
3. *There exist p functions $\phi_1, \ldots, \phi_p$ such that:*

(continued)

Theorem 7.3 (continued)

- $\forall j \in \{1, \dots, p\}, \phi_j : \mathfrak{X}^{M_j} \to \mathbb{R}$,
- $\forall z = (z_1, \dots, z_M) \in \mathfrak{X}^M, \Phi(z) = \sum_{j=1}^{p} \omega_j \phi_j(z_{\mu_{j,1}}, \dots, z_{\mu_{j,M_j}})$,
- $\exists (a_1, \dots, a_M) \in \mathbb{R}^M$ *such that:*

$$\forall j \in \{1, \dots, p\}, \forall i \in \{1, \dots, M_j\}, \forall z = (z_1, \dots, z_{M_j}) \in \mathfrak{X}^{M_j}, \forall z' \in \mathfrak{X}$$
$$|\phi_j(z_1, .., z_{M_j}) - \phi_j(z_1, .., z_{i-1}, z', z_{i+1}, .., z_{M_j})| \leqslant a_{\mu_{j,i}}.$$

Hence:

$$\forall t > 0, \mathbb{P}\left(\mathbb{E}_S(\phi \circ \mathfrak{T}) - (\phi \circ \mathfrak{T})(S) > t\right) \leqslant \exp\left(\frac{-2t^2}{\chi^*(\mathfrak{T}) \sum_{i=1}^{M} a_i^2}\right).$$

Proof By applying the Chernoff's inequality [31] to the transformation $\phi \circ \mathfrak{T}(S)$ we get:

$$\forall t > 0, s > 0, \mathbb{P}_S\left(\phi \circ \mathfrak{T}(S) - \mathbb{E}\left[\phi \circ \mathfrak{T}\right] > t\right) \leqslant e^{-st} \mathbb{E}_S\left[e^{s((\phi \circ \mathfrak{T})(S) - \mathbb{E}\phi \circ \mathfrak{T})}\right]. \tag{7.51}$$

The additivity of ϕ in the right term of the inequality of [31], combined with the decomposition from [88] for fixed $t > 0$ and $s > 0$, yields:

$$\mathbb{E}_S\left[e^{s(\phi \circ \mathfrak{T}(S) - \mathbb{E}[\phi \circ \mathfrak{T}])}\right] = \mathbb{E}\left[e^{s \sum_{j=1}^{p} \omega_j \left(\phi_j(\tau_{\mathfrak{M}_j}) - \mathbb{E}[\phi_j]\right)}\right],$$

where $\forall j = 1, \dots, p;\ \tau_{\mathfrak{M}_j} = (z_{\mu_{j,1}}, \dots, z_{\mu_{j,M_j}})$. Let $b_1, \dots, b_m$, be m arbitrary strictly positive real numbers such that $\sum_{j=1}^{m} b_j = 1$. By Jensen's inequality, we have:

$$\mathbb{E}_S e^{s(\phi \circ \mathfrak{T}(S) - \mathbb{E}[\phi \circ \mathfrak{T}])} \leqslant \sum_{j=1}^{m} b_j \mathbb{E}_{\tau_{\mathfrak{M}_j}}\left[e^{\frac{s\omega_j}{b_j}\left(\phi_j(\tau_{\mathfrak{M}_j}) - \mathbb{E}[\phi_j]\right)}\right].$$

Applying the following lemma, related to the proof of bounded difference theorem [114] also proven in [164, appendix A.1] to each ϕ_j with the independent random variables of $\tau_{\mathfrak{M}_j}$ and the corresponding coefficients $a_{\mu_{j,i}}$:

Lemma 7.4 (Exponential Bound for Functions Under McDiarmid's Conditions) *Let $X_1, \ldots, X_m$ be m random variables and ϕ be a function verifying the hypotheses of bounded difference theorem [114]. Considering the coefficients of Theorem 2.4, we then have for all $s > 0$:*

$$\mathbb{E}_{X_1,\ldots,X_m}\left[e^{s(\phi(X_1,\ldots,X_m)-\mathbb{E}[\phi])}\right] \leqslant e^{\frac{s^2}{8}\sum_{i=1}^m a_i^2}.$$

We have:

$$\mathbb{E}_S e^{s(\phi\circ\mathfrak{T}(S)-\mathbb{E}\phi\circ\mathfrak{T})} \leqslant \sum_{j=1}^{p} b_j e^{\frac{s^2\omega_j^2}{8p_j^2}\sum_{i\in\mathfrak{M}_j} a_i^2}.$$

By estimating the b_j and s for which the previous bound is optimal. This step is identical to the last step of the proof of Janson's theorem [88]. Let us take the Eq. (7.51), let us denote $C_j = \sum_{i\in\mathfrak{M}_j} c_i^2$, $C = \sum_{j=1}^n \omega_j\sqrt{C_j}$ and let us choose $p_j = \omega_j\sqrt{C_j}/C$. We have:

$$\forall t > 0, s > 0, \mathbb{P}_S\left(\phi\circ\mathfrak{T}(S) - \mathbb{E}\phi\circ\mathfrak{T} > t\right) \leqslant e^{-st}\sum_{j=1}^{n} p_j e^{\frac{1}{8}s^2C^2} = e^{-st+\frac{1}{8}s^2C^2}.$$

and the optimal choice for s is $4t/C^2$, which gives:

$$\forall t > 0, \mathbb{P}_S\left(\phi\circ\mathfrak{T}(S) - \mathbb{E}\phi\circ\mathfrak{T} > t\right) \leqslant e^{-2t^2/C^2}.$$

The following calculation, using the Cauchy-Schwarz inequality, completes the proof:

$$C^2 = \left(\sum_{j=1}^{p}\omega_j\sqrt{C_j}\right)^2 \leqslant \left(\sum_{j=1}^{p}\omega_j\right)\left(\sum_{j=1}^{p}\omega_j C_j\right) = \chi^*(\mathfrak{T})\left(\sum_{i=1}^{M} a_i^2\right),$$

where $\sum_{j=1}^p \omega_j = \chi^*(\mathfrak{T})$. □

Thus, the function $\mathfrak{T}$ creates a set of interdependent random variables z_i from a training base S formed of independent variables. With the decomposition introduced in [88], we decompose the transformed training set $\mathfrak{T}(S)$ into subsets $j \in \{1, \ldots, p\}, \mathfrak{M}_j = \{\mu_{j,1}, \ldots, \mu_{j,M_j}\}$ of independent variables. If now we can find a function Φ on the transformed set that can be written as a weighted sum of functions ϕ_j taking their value on each independent subset $\mathfrak{M}_j$ and such that each ϕ_j admits bounded differences on each of its input variables, then we can bound the function $S \mapsto (\phi \circ \mathfrak{T})(S) - \mathbb{E}_S(\phi \circ \mathfrak{T})$. This is indeed the case of the upper bound of $\sup_{c_h \in \mathcal{F}}[\mathcal{L}^{\mathfrak{T}}(c_h) - \hat{\mathcal{L}}_m^T(c_h, \mathfrak{T}(S))]$ which is written as $\sum_{j=1}^p \omega_j \phi_j((\boldsymbol{\xi}_1, d_1), \ldots, (\boldsymbol{\xi}_{M_j}, d_{M_j}))$. The coefficients of the theorem are then equal to $a_i = \frac{1}{M}$. In this case, for all $\delta \in]0, 1]$ and with a probability at least $1 - \delta$:

$$\sup_{c_h \in \mathcal{F}} \left(R^{\mathfrak{T}}(c_h) - R_M^T(c_h, \mathfrak{T}(S))\right) \leqslant \sqrt{\frac{\chi^*(\mathfrak{T}) \ln(1/\delta)}{2M}} + \mathbb{E}_{\mathfrak{T}(S)} \left(\sum_{j=1}^p \frac{\omega_j}{M} \sup_{c_h \in \mathcal{F}} \left[\mathbb{E}_{\mathfrak{T}(\tilde{S})} \sum_{i \in \mathfrak{M}_j} \ell(c_h(\tilde{\boldsymbol{\xi}}_i), \tilde{d}_i) - \sum_{i \in \mathfrak{M}_j} \ell(c_h(\boldsymbol{\xi}_i), d_i)\right]\right). \tag{7.52}$$

7.3.2.3 Fractional Rademacher Complexity

Consider the expression that occurs in the second term of the previous inequality:

$$\mathbb{E}_{\mathfrak{T}(S)} \left(\sum_{j=1}^p \frac{\omega_j}{M} \sup_{c_h \in \mathcal{F}} \left[\mathbb{E}_{\mathfrak{T}(\tilde{S})} \sum_{i \in \mathfrak{M}_j} \ell(c_h(\tilde{\boldsymbol{\xi}}_i), \tilde{d}_i) - \sum_{i \in \mathfrak{M}_j} \ell(c_h(\boldsymbol{\xi}_i), d_i)\right]\right). \tag{7.53}$$

Taking out the supreme from the expectation, we obtain:

$$\mathbb{E}_{\mathfrak{T}(S)} \left(\sum_{j=1}^p \frac{\omega_j}{M} \sup_{c_h \in \mathcal{F}} \left[\mathbb{E}_{\mathfrak{T}(\tilde{S})} \sum_{i \in \mathfrak{M}_j} \ell(c_h(\tilde{\boldsymbol{\xi}}_i), \tilde{d}_i) - \sum_{i \in \mathfrak{M}_j} \ell(c_h(\boldsymbol{\xi}_i), d_i)\right]\right)$$
$$\leqslant \mathbb{E}_{\mathfrak{T}(S), \mathfrak{T}(\tilde{S})} \sum_{j=1}^p \frac{\omega_j}{M} \sup_{c_h \in \mathcal{F}} \left[\sum_{i \in \mathfrak{M}_j} \left(\ell(c_h(\tilde{\boldsymbol{\xi}}_i), \tilde{d}_i) - \ell(c_h(\boldsymbol{\xi}_i), d_i)\right)\right].$$

Considering $\boldsymbol{\sigma} = (\sigma_1, \ldots, \sigma_M)$, a realization of M independent Rademacher variables, we have for each sum over $\mathfrak{M}_j$:

$$\mathbb{E}_{\mathfrak{T}(S), \mathfrak{T}(\tilde{S})} \sup_{c_h \in \mathcal{F}} \left[\sum_{i \in \mathfrak{M}_j} \sigma_i \left(\ell(c_h(\tilde{\boldsymbol{\xi}}_i), \tilde{d}_i) - \ell(c_h(\boldsymbol{\xi}_i), d_i)\right)\right]$$
$$= \mathbb{E}_{\mathfrak{T}(S), \mathfrak{T}(\tilde{S})} \sup_{c_h \in \mathcal{F}} \left[\sum_{i \in \mathfrak{M}_j} \left(\ell(c_h(\tilde{\boldsymbol{\xi}}_i), \tilde{d}_i) - \ell(c_h(\boldsymbol{\xi}_i), d_i)\right)\right].$$

The introduction of the variables σ_i for each of these sums does not change anything and the term $-\sigma_i$ corresponds to an exchange of the examples $(\tilde{\boldsymbol{\xi}}_i, \tilde{d}_i)$ and $(\boldsymbol{\xi}_i, d_i)$. In each of the sums, $(\boldsymbol{\xi}_i, d_i)$ are independent. Therefore this exchange, if applicable, will have no effect on the other terms of the sum. Furthermore, when we take the expectation on $\mathfrak{T}(S)$ and $\mathfrak{T}(\tilde{S})$, the value of σ_i becomes without effect on the term considered because $(\tilde{\boldsymbol{\xi}}_i, \tilde{d}_i)$ and $(\boldsymbol{\xi}_i, d_i)$ have the same distribution. Since the σ_i have the same probability distribution, we have as in the case of binary classification:

$$\begin{aligned}
&\mathbb{E}_{\mathfrak{T}(S),\mathfrak{T}(\tilde{S})} \sup_{c_h\in\mathcal{F}} \left[\sum_{i\in\mathfrak{M}_j} \sigma_i \left(\ell(c_h(\tilde{\boldsymbol{\xi}}_i), \tilde{d}_i) - \ell(c_h(\boldsymbol{\xi}_i), d_i)\right)\right] \\
&\quad = \mathbb{E}_{\mathfrak{T}(S),\mathfrak{T}(\tilde{S})}\mathbb{E}_{\sigma} \sup_{c_h\in\mathcal{F}} \left[\sum_{i\in\mathfrak{M}_j} \sigma_i \left(\ell(c_h(\tilde{\boldsymbol{\xi}}_i), \tilde{d}_i) - \ell(c_h(\boldsymbol{\xi}_i), d_i)\right)\right].
\end{aligned}$$

Using the triangle inequality on the supremum in the second inequality, we can upper-bound (7.53):

$$\begin{aligned}
&\mathbb{E}_{\mathfrak{T}(S)}\left(\sum_{j=1}^{p} \frac{\omega_j}{M} \sup_{c_h\in\mathcal{F}} \left[\mathbb{E}_{\mathfrak{T}(\tilde{S})} \sum_{i\in\mathfrak{M}_j} \ell(c_h(\tilde{\boldsymbol{\xi}}_i), \tilde{d}_i) - \sum_{i\in\mathfrak{M}_j} \ell(c_h(\boldsymbol{\xi}_i), d_i)\right]\right) \qquad (7.54) \\
&\qquad \leqslant \mathbb{E}_{\mathfrak{T}(S)} \frac{2}{M} \mathbb{E}_{\sigma} \sum_{j=1}^{p} \omega_j \sup_{c_h\in\mathcal{F}} \sum_{i\in\mathfrak{M}_j} \sigma_i \ell(c_h(\boldsymbol{\xi}_i), d_i)
\end{aligned}$$

Definition 7.5 The empirical fractional Rademacher complexity of $\mathcal{F}$ is thus defined by [174]:

$$\hat{\mathfrak{R}}^{\mathfrak{T}}(\mathcal{F}, \mathfrak{T}(S)) = \frac{2}{M} \mathbb{E}_{\sigma} \sum_{j=1}^{p} \omega_j \sup_{c_h\in\mathcal{F}} \sum_{i\in\mathfrak{M}_j} \sigma_i c_h(\boldsymbol{\xi}_i). \qquad (7.55)$$

And the fractional Rademacher complexity is $\mathfrak{R}^{\mathfrak{T}}(\mathcal{F}) = \mathbb{E}_{\mathfrak{T}(S)}\left[\hat{\mathfrak{R}}^{\mathfrak{T}}(\mathcal{F}, \mathfrak{T}(S))\right]$.

We note that the empirical fractional Rademacher complexity is a weighted sum of Rademacher complexities on independent subsets of the transformed set. The fractional term comes from the fractional overlap term found on the transformed examples [88].

With the previous definition and according to (7.52) and (7.54), we have $\forall \delta \in]0, 1]$ the following inequality, which is true with a probability at least equal to $1-\delta$:

$$\forall S, \forall c_h \in \mathcal{F}, \mathcal{L}^{\mathfrak{T}}(c_h) \leqslant \hat{\mathcal{L}}^{\mathfrak{T}}_m(c_h, \mathfrak{T}(S)) + \mathfrak{R}^{\mathfrak{T}}(\ell \circ \mathcal{F}) + \sqrt{\frac{\chi^*(\mathfrak{T}) \ln \frac{1}{\delta}}{2M}}.$$

The function $\mathfrak{T}(S) \mapsto \mathcal{R}^{\mathfrak{T}}_M(\mathcal{F}, \mathfrak{T}(S))$ also satisfies the conditions of Theorem 7.3 with $\forall i, a_i = \frac{2}{M}$. We obtain the following final result, which is a data-dependent bound.

Theorem 7.6 (Data-dependent Generalization Bound for Interdependent Data) *Let $\mathfrak{T} : S \rightarrow (\mathfrak{X} \times \{-1, +1\})^M$ be a transformation function, taking as input a training set S consisting of m independent random variables with values in $\mathcal{X}$. Let $\mathcal{F}$ be a class of functions from $\mathfrak{X}$ to $\{-1, +1\}$. For any $\delta \in]0, 1]$, the following inequality is true with probability at least $1 - \delta$:*

$$\forall c_h \in \mathcal{F}, \mathcal{L}^{\mathfrak{T}}(c_h) \leqslant \hat{\mathcal{L}}^{\mathfrak{T}}_m(c_h, \mathfrak{T}(S)) + \hat{\mathfrak{R}}^{\mathfrak{T}}(\ell \circ \mathcal{F}, \mathfrak{T}(S)) + 3\sqrt{\frac{\chi^*(\mathfrak{T}) \ln \frac{2}{\delta}}{2M}}. \tag{7.56}$$

7.3.3 *Estimation of the Bound for Some Applications*

To better see the application of Theorem 7.6 to the particular case of bipartite ranking presented previously, we will first upper-bound the empirical fractional Rademacher complexity for a class of kernel functions with bounded norm whose supremum can be computed on a training set and, then, we will give the generalization bound for the case of bipartite ranking.

7.3.3.1 Bounding $\mathcal{R}^T_M(\mathcal{F})$ for $\mathcal{F} = \{\xi \mapsto \langle w, \xi \rangle \mid \langle w, w \rangle \leqslant B^2\}$

Suppose that $\|\xi\| \leqslant R, \forall \xi$. The upper bound follows from the equivalent property of Rademacher complexity (Chap. 1, Sect. 2.2.2), as well as from the property (7.50) of exact proper coverings. For the sake of clarity, we will give this calculation. First, note that for a function $c_h \in \mathcal{F}$, we have from the Cauchy-Schwarz inequality

$\forall\, \boldsymbol{\xi} \in \mathfrak{X},\ c_h(\boldsymbol{\xi}) \leqslant B\|\boldsymbol{\xi}\|$. In view of the definition of the empirical fractional Rademacher complexity (7.55), by the linearity of expectation, we have

$$\hat{\mathfrak{R}}^{\mathfrak{T}}(\mathcal{F}, \mathfrak{T}(S)) = \frac{2}{M}\mathbb{E}_\sigma \sum_{j=1}^{p} \omega_j \sup_{\|\boldsymbol{w}\| \leqslant B} \sum_{i \in \mathfrak{M}_j} \langle \boldsymbol{w}, \sigma_i \boldsymbol{\xi}_i \rangle$$

$$\leqslant \frac{2B}{M} \sum_{j=1}^{p} \omega_j \mathbb{E}_\sigma \left\| \sum_{i \in \mathfrak{M}_j} \sigma_i \boldsymbol{\xi}_i \right\| \leqslant \frac{2B}{M} \sum_{j=1}^{p} \omega_j \left(\mathbb{E}_\sigma \left\| \sum_{i \in \mathfrak{M}_j} \sigma_i \boldsymbol{\xi}_i \right\|^2 \right)^{1/2},$$

where the first inequality is by noting $\|\boldsymbol{w}\| \leqslant B$, and applying Cauchy-Schwarz inequality to the dot product, and the second inequality is by Jensen's inequality. As the Rademacher variables are independent, we have $\mathbb{E}[\sigma_k \sigma_i] = \mathbb{E}[\sigma_k]\mathbb{E}[\sigma_i] = 0$ for any distinct k and i. Hence we have

$$\mathbb{E}_\sigma \left\| \sum_{i \in \mathfrak{M}_j} \sigma_i \boldsymbol{\xi}_i \right\|^2 = \mathbb{E}_\sigma \left[\sum_{i,k \in \mathfrak{M}_j} \sigma_i \sigma_k \langle \boldsymbol{\xi}_i, \boldsymbol{\xi}_k \rangle \right] = \sum_{i \in \mathfrak{M}_j} \|\boldsymbol{\xi}_i\|^2,$$

and therefore,

$$\hat{\mathfrak{R}}^{\mathfrak{T}}(\mathcal{F}, \mathfrak{T}(S)) \leqslant \frac{2B}{M} \sum_{j=1}^{p} \omega_j \left(\sum_{i \in \mathfrak{M}_j} \|\boldsymbol{\xi}_i\|^2 \right)^{1/2}.$$

Since we have $\|\boldsymbol{\xi}\| \leqslant R$, then

$$\hat{\mathfrak{R}}^{\mathfrak{T}}(\mathcal{F}, \mathfrak{T}(S)) \leqslant \frac{2BR}{M} \sum_{j=1}^{p} \omega_j \sqrt{|\mathfrak{M}_j|} = \frac{2BR\chi^*(\mathfrak{T})}{M} \sum_{j=1}^{p} \frac{\omega_j}{\chi^*(\mathfrak{T})} \sqrt{|\mathfrak{M}_j|}.$$

By noticing that $\sum_{j=1}^{p} \omega_j / \chi^*(\mathfrak{T}) = 1$, using Jensen's inequality for the square root function yields

$$\mathfrak{R}^{\mathfrak{T}}(\mathcal{F}, \mathfrak{T}(S)) \leqslant \frac{2BR\chi^*(\mathfrak{T})}{M} \sqrt{\sum_{j=1}^{p} \frac{\omega_j}{\chi^*(\mathfrak{T})} |\mathfrak{M}_j|} = \frac{2BR\sqrt{\chi^*(\mathfrak{T})}}{M} \sqrt{\sum_{j=1}^{p} \omega_j |\mathfrak{M}_j|}.$$

Finally, as $\sum_{j=1}^{p} \omega_j |\mathfrak{M}_j| = M$, we get

$$\mathfrak{R}^{\mathfrak{T}}(\mathcal{F}, \mathfrak{T}(S)) \leqslant 2BR\sqrt{\frac{\chi^*(\mathfrak{T})}{M}}.$$

7.3.3.2 Bipartite Ranking

Recall that in this case we want to order the n_+ relevant examples of a training set $S = \left((\mathbf{x}_i, y_i)_{i=1}^m\right)$ in front of the n_- irrelevant examples (assuming $n_+ \leqslant n_-$). The transformation $\mathfrak{T} : \mathcal{X}^m \to \mathfrak{X}^M$ organizes the $M = n_+ n_-$ examples of $\mathfrak{T}(S)$ into pairs that contain a relevant example and an irrelevant example as follows:

$$
\begin{aligned}
\mathfrak{M}_1 &: \left(\mathbf{x}_{\pi(1)}, \mathbf{x}_{\nu(1)}\right), \left(\mathbf{x}_{\pi(2)}, \mathbf{x}_{\nu(2)}\right), \ldots, \left(\mathbf{x}_{\pi(n_+)}, \mathbf{x}_{\nu(n_+)}\right), \\
\mathfrak{M}_2 &: \left(\mathbf{x}_{\pi(1)}, \mathbf{x}_{\nu(2)}\right), \left(\mathbf{x}_{\pi(2)}, \mathbf{x}_{\nu(3)}\right), \ldots, \left(\mathbf{x}_{\pi(n_+)}, \mathbf{x}_{\nu(n_++1)}\right), \\
&\ldots \\
\mathfrak{M}_{n_-} &: \left(\mathbf{x}_{\pi(1)}, \mathbf{x}_{\nu(n_-)}\right), \left(\mathbf{x}_{\pi(2)}, \mathbf{x}_{\nu(n_-+1)}\right), \ldots, \left(\mathbf{x}_{\pi(n_+)}, \mathbf{x}_{\nu(n_-+n_+-1)}\right).
\end{aligned}
$$

where $\pi(i)$ (resp. $\nu(j)$) represents the index of the i-th relevant example (resp. j-th irrelevant example) of $\mathfrak{T}(S)$. This decomposition corresponds to an exact proper covering of $\mathfrak{T}(S)$ illustrated in Fig. 7.3c with $\chi^*(\mathfrak{T}) = \max(n_-, n_+) = n_-$.

Considering the loss function $\Delta(y, z) = \min(1, \max(1 - yz, 0))$ which is 1-Lipshitzian, for a given $\delta \in (0, 1)$, we have according to Theorem 7.6 the following inequality, which holds with a probability of at least $1 - \delta$:

$$
\forall c_h \in \mathcal{F}, \mathcal{L}^{\mathfrak{T}}(c_h) \leqslant \hat{\mathcal{L}}_m^{\mathfrak{T}}(c_h, \mathfrak{T}(S)) + \mathfrak{R}^{\mathfrak{T}}(\Delta \circ \mathcal{F}, \mathfrak{T}(S)) + 3\sqrt{\frac{\chi^*(\mathfrak{T}) \ln \frac{2}{\delta}}{2M}}.
$$

Furthermore, we can show from the properties of Rademacher complexity that $\mathfrak{R}^{\mathfrak{T}}(\Delta \circ \mathcal{F}, \mathfrak{T}(S)) \leqslant \mathfrak{R}^{\mathfrak{T}}(\mathcal{F}, \mathfrak{T}(S))$ (Chap. 1, Sect. 2.2.2) which, from the previous calculation and Theorem 7.6, gives a generalization bound of bipartite ranking functions computable on the training data, which holds with a probability of at least $1 - \delta$:

$$
\begin{aligned}
\forall c_h \in \mathcal{F}, \mathcal{L}^{\mathfrak{T}}(c_h) &\leqslant \hat{\mathcal{L}}_m^{\mathfrak{T}}(c_h, \mathfrak{T}(S)) + \\
&\frac{2B\sqrt{\max(n_-, n_+)}}{n_- n_+} \sqrt{\sum_{i=1}^{n_+} \sum_{j=1}^{n_-} \|\mathbf{x}_{\pi(i)} - \mathbf{x}_{\nu(j)}\|_K^2} + 3\sqrt{\frac{\ln \frac{2}{\delta}}{2\min(n_-, n_+)}}.
\end{aligned}
$$

7.3.3.3 Particular Case: Binary Classification

Binary classification is a special case of the previous framework and corresponds to the transformation function $\mathfrak{T}$, which is identity ($\chi^*(\mathfrak{T}) = 1$). In this case, by taking the cost function $L : z \mapsto \max(1 - z, 0)$, we find the generalization bound of classifiers trained on i.i.d data that we presented in Chap. 2.

To Sum Up

1. *The advancement of information technologies has driven the development of new machine learning frameworks. Learning to Rank is a significant paradigm within this context, focusing on automatically ranking information entities in large datasets.*
2. *Evaluating the performance of Learning to Rank models requires specialized metrics that assess the quality of the ranking. Common metrics include Area Under the ROC Curve (AUC), Mean Average Precision (MAP), and NDCG. These metrics consider the positions of relevant instances in the ranked list and provide a comprehensive evaluation of the model's ranking ability.*
3. *Pointwise approaches treat the ranking problem as a regression or classification task, while pairwise approaches consider pairs of instances and learns to rank by comparing pairs.*
4. *Pairwise approaches reduce the ranking problem to the classification of pairs of examples, where the goal is to determine the relative order within each pair. This reduction transforms the ranking task into a binary classification of the pairs of examples, where each pair is labeled according to the preferred order.*
5. *The effectiveness of Learning to Rank models heavily relies on the features used to represent the instances. Feature engineering involves creating informative features that capture the relevance signals of the instances. This can include textual features, user behavior features, and domain-specific features, all of which contribute to improving the ranking accuracy.*
6. *In the binary classification of the paris of examples, examples are no longer independently distributed due to the inherent dependencies within pairs. This interdependence necessitates the development of new frameworks for learning with interdependent data, ensuring that the ranking model captures the relationships between instances effectively.*
7. *Learning to Rank has a wide range of applications beyond search engines. It is used in recommendation systems to suggest products or content to users, in question-answering systems to rank potential answers, and in document retrieval systems to prioritize relevant documents. Its versatility makes it a valuable technique in many domains where ranking is essential.*
8. *Despite its success, Learning to Rank faces several challenges, such as handling large-scale data, dealing with noisy or incomplete labels, and ensuring fairness and unbiased rankings.*

7.4 Exercises

1. In learning to rank, pointwise and pairwise approaches are two common paradigms.
   (a) Explain the key difference between pointwise and pairwise ranking. Use examples to illustrate how training data is structured for each approach.
   (b) Classify the following methods as pointwise, pairwise, or listwise:
      - Logistic regression predicting relevance scores.
      - SVM comparing document pairs.
      - LambdaMART optimizing NDCG.

   (c) Why might pointwise ranking be more computationally efficient than pairwise ranking?
   (d) What are the limitations of pairwise ranking when dealing with large datasets?

2. Consider a stream of queries, each associated with a set of documents. For each query q, we observe pairs of documents (d_i, d_j) such that d_i should be ranked higher than d_j (i.e., $d_i \succ_q d_j$). Let $\phi(d, q)$ be the feature vector for document d under query q. The goal is to learn a linear ranking function $f(d, q) = \boldsymbol{w}^\top \phi(d, q)$.
   (a) Write the hinge loss for a single pair (d_i, d_j) for query q.
   (b) Assume the current weight vector is $\boldsymbol{w}_t$. Upon observing a new preference pair (d_i, d_j) for query q at time t, derive the on-line update rule for $\boldsymbol{w}_{t+1}$ using stochastic (sub-)gradient descent with learning rate η_t.
   (c) Write the on-line SVM-based ranking algorithm in pseudocode, using the update rule from (b).
   (d) Briefly discuss the advantages and limitations of this on-line approach compared to batch RankSVM.

3. Let $\mathcal{X} \subset \mathbb{R}^d$ be the input space. Consider a training set of m items $x_1, \ldots, x_m \in \mathcal{X}^m$ and a set of pairwise preferences $\mathcal{P} = \{(i, j) : x_i \succ x_j\}$, meaning x_i should be ranked higher than x_j.

   Suppose we seek a linear scoring function $h_{\boldsymbol{w}}(\mathbf{x}) = \boldsymbol{w}^\top \mathbf{x}$ with $\boldsymbol{w} \in \mathbb{R}^d$, and $\mathbf{x} \in \mathcal{X}$ the vector representation of x, such that for all $(i, j) \in \mathcal{P}$, $h_{\boldsymbol{w}}(\mathbf{x}_i) > h_{\boldsymbol{w}}(\mathbf{x}_j)$.
   (a) Describe an adaptation of the Perceptron algorithm for this pairwise ranking setting. Specify the update rule when a pairwise preference is violated.
   (b) Assume that there exists a unit vector $\boldsymbol{w}^*$ and a margin $\rho > 0$ such that for all $(i, j) \in \mathcal{P}$,

$$(\boldsymbol{w}^*)^\top (\mathbf{x}_i - \mathbf{x}_j) \geqslant \rho, \quad \|\mathbf{x}_i - \mathbf{x}_j\| \leqslant R.$$

   Show that the total number of updates (mistakes) made by the algorithm is at most $\frac{R^2}{\rho^2}$.

Hint: Reduce the ranking problem to binary classification on difference vectors $\mathbf{z}_{ij} = \mathbf{x}_i - \mathbf{x}_j$ *with label* $+1$ *for each* $(i, j) \in \mathcal{P}$.

4. Let $\mathcal{X} \subset \mathbb{R}^d$ be the input space. We are given a set of n items $\{x_1, \ldots, x_n\}$ and a set of observed pairwise preferences $\mathcal{P} = \{(i, j) : x_i \succ x_j\}$, meaning item x_i should be ranked higher than x_j.
   We seek a linear scoring function $h_{\boldsymbol{w}}(x) = \boldsymbol{w}^\top \mathbf{x}$ with $\boldsymbol{w} \in \mathbb{R}^d$ and $\mathbf{x}$ the vector representation of x, such that, for all $(i, j) \in \mathcal{P}$, $h_{\boldsymbol{w}}(\mathbf{x}_i) > h_{\boldsymbol{w}}(\mathbf{x}_j)$. To promote robustness, we wish to maximize the minimum margin by which these inequalities are satisfied.
   We consider the following linear program (LP) that finds a weight vector w and a margin $\rho > 0$ such that for all $(i, j) \in \mathcal{P}$, $\boldsymbol{w}^\top(\mathbf{x}_i - \mathbf{x}_j) \geqslant \rho$, and $\|\boldsymbol{w}\|_1 \leqslant 1$:

$$\begin{cases} \max_\rho & \rho \\ \text{s.t.} & \boldsymbol{w}^\top(\mathbf{x}_i - \mathbf{x}_j) \geqslant \rho, \quad \forall (i, j) \in \mathcal{P} \\ & \|\boldsymbol{w}\|_1 \leqslant 1. \end{cases}$$

   (a) Explain how the solution $(\boldsymbol{w}^*, \rho^*)$ to the LP defines a linear ranking function. What does the margin ρ^* represent in terms of the robustness of the ranking? Suppose that the pairwise preferences cannot be perfectly satisfied (i.e., the data is not separable). Consider the modification of the LP formulation by introducing non-negative slack variables $\boldsymbol{\xi}_{ij} \geqslant 0$ for each pair $(i, j) \in \mathcal{P}$ and a regularization parameter $C > 0$, so that the new objective balances margin maximization and total slack:

$$\begin{cases} \max_{\rho, \xi} & \rho - C \sum_{(i,j) \in \mathcal{P}} \xi_{ij} \\ \text{s.t.} & \boldsymbol{w}^\top(\mathbf{x}_i - \mathbf{x}_j) \geqslant \rho - \xi_{ij}, \quad \forall (i, j) \in \mathcal{P} \\ & \xi_{ij} \geqslant 0, \\ & \|\boldsymbol{w}\|_1 \leqslant 1. \end{cases}$$

   (b) Given a new set of items $\{x'_1, \ldots, x'_m\}$, describe how to use the learned $\boldsymbol{w}^*$ to produce a ranking over these items.
   (c) Discuss how the number of constraints in your LP scales with the number of items n and the number of observed pairs $|\mathcal{P}|$. What are practical strategies for solving this LP efficiently when $|\mathcal{P}|$ is large?

5. Let $h : X \to \mathbb{R}$ be a fixed scoring function used to rank the points of X. Suppose you are given a finite sample $T = \{(\mathbf{x}_1, y_1), \ldots, (\mathbf{x}_m, y_m)\}$, where each $y_i \in \{+1, -1\}$. Let p denote the number of positive examples and n the number of negative examples in T.

The empirical AUC of h on T is defined as:

$$\widehat{\mathrm{AUC}}(h;T) = \frac{1}{pn} \sum_{i:y_i=+1} \sum_{j:y_j=-1} \left[\mathbb{1}_{h(\mathbf{x}_i)>h(\mathbf{x}_j)} + \frac{1}{2} \mathbb{1}_{h(\mathbf{x}_i)=h(\mathbf{x}_j)} \right],$$

where $\mathbb{1}_\pi$ is the indicator function, equal to 1 if the predicate π is true and 0 otherwise.
Let $\mathrm{AUC}(h)$ denote the expected AUC of h over independent draws from the data distribution.

(a) Show that $\widehat{\mathrm{AUC}}(h;T)$ can be written as the average of mn independent, bounded random variables. Specify the variables and their bounds.
(b) State Hoeffding's inequality for the average of independent, bounded random variables.
(c) Use Hoeffding's inequality to prove that, for any $\epsilon > 0$,

$$\mathbb{P}\left(\left| \widehat{\mathrm{AUC}}(h;T) - \mathrm{AUC}(h) \right| \geqslant \epsilon \right) \leqslant 2 \exp\left(-2pn\epsilon^2 \right).$$

(d) Interpret the result: What does this bound tell you about the reliability of the empirical AUC as an estimate of the true AUC? How does the sample size affect this guarantee?

Hint: You may assume that the pairs $((\mathbf{x}_i, y_i), (\mathbf{x}_j, y_j))$ are drawn independently for $y_i = +1$, $y_j = -1$, and that the indicator functions are bounded in $[0, 1]$.

6. Let X be a set of items, each labeled with a relevance level $y \in \{1, 2, \ldots, k\}$, where $k \geqslant 2$ and a higher value of y indicates higher relevance.
Let $h : \mathcal{X} \to \mathbb{R}$ be a fixed scoring function used to rank the items, with $\mathcal{X}$ representing the vector space in which the items are embedded.

(a) Define the pairwise k-partite ranking loss of h on a finite sample $S = \{(\mathbf{x}_i, y_i)\}_{i=1}^m$ as

$$\ell_{\mathrm{pair}}(h;S) = \frac{1}{Z} \sum_{i<j} \mathbb{1}_{y_i>y_j} \cdot \mathbb{1}_{h(x_i)\leqslant h(x_j)},$$

where $Z = |\{(i,j) : y_i > y_j\}|$ is the total number of ordered pairs with different relevance levels, and $\mathbb{1}_\pi$ is the indicator function, equal to 1 if the predicate π is true and 0 otherwise.

(i) Explain how this loss generalizes the AUC loss for $k = 2$.
(ii) Give an explicit formula for Z in terms of the counts of each relevance level in S.

(b) Propose a pointwise loss function suitable for k-partite ranking and briefly discuss its advantages and limitations compared to the pairwise loss.

(c) Assume the sample S of size n is drawn i.i.d. from a distribution $\mathcal{D}$ over $\mathcal{X} \times \{1, \ldots, k\}$. Let $\mathcal{L}_{\text{pair}}(h)$ denote the expected pairwise k-partite ranking loss of h under $\mathcal{D}$. Use Hoeffding's inequality to show that, for any $\epsilon > 0$,

$$\mathbb{P}\left(\left|\ell_{\text{pair}}(h; S) - \mathcal{L}_{\text{pair}}(h)\right| \geqslant \epsilon\right) \leqslant 2\exp\left(-2Z\epsilon^2\right).$$

Hint: Argue that the sum in ℓ_{pair} is an average of Z independent, bounded random variables.

7. Consider a dataset $S = \{(\mathbf{x}_i, y_i)\}_{i=1}^m$ where $\mathbf{x}_i \in \mathbb{R}^d$ are feature vectors and $y_i \in \{-1, +1\}$ are binary labels. The goal is to learn a linear scoring function

$$h_{\boldsymbol{w}}(\mathbf{x}) = \boldsymbol{w}^\top \mathbf{x} + w_0,$$

that ranks positive instances higher than negative ones by maximizing the margin between them.

(a) Define the set of positive-negative pairs:

$$\mathcal{P} = \{(i, j) \mid y_i = +1, y_j = -1\}.$$

For each pair $(i, j) \in \mathcal{P}$, enforce the margin constraint with slack variables $\xi_{ij} \geqslant 0$.

(b) Formulate the following linear program to maximize the margin γ while penalizing violations. Use a constant $C > 0$ to balance margin maximization and slack penalty, and the ℓ_1-norm constraint to avoid trivial scaling of $\boldsymbol{w}$.

(c) Show that the dual problem can be written as:

$$\begin{aligned} \min_{\{\alpha_{ij}\}} \quad & \sum_{(i,j)\in\mathcal{P}} \alpha_{ij} \\ \text{s.t.} \quad & \sum_{(i,j)\in\mathcal{P}} \alpha_{ij}(\mathbf{x}_i - \mathbf{x}_j) = \boldsymbol{w}, \\ & 0 \leqslant \alpha_{ij} \leqslant C, \quad \forall (i, j) \in \mathcal{P}. \end{aligned}$$

Interpret the dual variables α_{ij} as weights on pairwise comparisons.

(d) Prove that the primal LP is convex.

(e) Explain that Slater's condition (see Exercise 9, Chap. 5) holds if there exists a $\boldsymbol{w}$ such that for at least one pair (i, j), $\boldsymbol{w}^\top(\mathbf{x}_i - \mathbf{x}_j) > 0$.

(f) Discuss how the parameter C controls the number of support pairs (pairs with $\alpha_{ij} > 0$) and how the ℓ_1-norm constraint promotes sparsity in $\boldsymbol{w}$.

(g) Note that the number of constraints grows as $|\mathcal{P}| = pn$, where p and n are respectively the number of positive and negative instances.

(h) Propose heuristics such as chunking or random sampling of pairs to reduce the computational burden.

(i) Compare the LP formulation to the classical SVM quadratic program:

- Both maximize margins but SVM uses ℓ_2-norm regularization and quadratic programming.
- LP formulation is computationally simpler but may yield less smooth solutions.

8. Consider a ranking scenario where the instance set $\mathcal{X}$ is partitioned into k disjoint subsets:

$$\mathcal{X} = \mathcal{X}_1 \cup \mathcal{X}_2 \cup \cdots \cup \mathcal{X}_k, \quad k \geqslant 1,$$

with $\mathcal{X}_i \cap \mathcal{X}_j = \emptyset$ for $i \neq j$. Each subset $\mathcal{X}_i$ corresponds to a different rank level. The goal is to learn a ranking function $H : \mathcal{X} \to \mathbb{R}$ that respects the ranking:

$$\text{if } x \in \mathcal{X}_i, \quad x' \in \mathcal{X}_j, \quad \text{and } i < j, \quad \text{then } h(x) > h(x').$$

(a) How can the multipartite ranking problem be formulated in terms of k distributions $D_1, D_2, \ldots, D_k$ over each subset $\mathcal{X}_1, \ldots, \mathcal{X}_k$? Write down an expression for the ranking loss that penalizes misordered pairs from different subsets.
(b) Write the pseudocode of the RankBoost algorithm adapted for multipartite ranking. In particular, describe how the distributions over pairs are initialized, updated, and used to train weak rankers at each iteration.
(c) Discuss the computational challenges that arise when extending RankBoost to multipartite ranking. How does the number of pairs grow with k and the sizes of the subsets?
(d) Discuss the computational challenges that arise when extending RankBoost to multipartite ranking. How does the number of pairs grow with k and the sizes of the subsets?
(e) Suggest strategies or heuristics to efficiently implement multipartite RankBoost despite the potentially large number of pairs. What are the advantages and potential drawbacks of using RankBoost in this setting?

9. Consider a multivariate ranking problem where each instance is described by multiple criteria or variables, and the goal is to produce a ranking that accounts for all these variables simultaneously.

(a) Define what is meant by a *multivariate ranking function*. How does it differ from a univariate ranking function?
(b) Suppose you have d variables/features for each instance, and you want to learn a ranking function $h : \mathbb{R}^d \to \mathbb{R}$ that aggregates these variables into a single score for ranking. Propose a linear model for f and write down the form of $h(\mathbf{x})$.

(c) Discuss how pairwise preference constraints can be incorporated into the learning problem to ensure that if instance $\mathbf{x}_i$ is preferred over $\mathbf{x}_j$, then $h(\mathbf{x}_i) > h(\mathbf{x}_j)$.
(d) Formulate an optimization problem (e.g., margin-maximization or hinge-loss minimization) that learns the linear ranking function subject to these pairwise constraints.
(e) Explain how this problem can be extended to handle multivariate ranking where the ranking depends on multiple endpoints or criteria simultaneously, rather than a single scalar target.
(f) Describe how a multivariate ranking loss function might be defined to evaluate the quality of a ranking considering multiple variables jointly.
(g) Discuss possible algorithms or methods to solve the multivariate ranking problem, such as extensions of RankBoost, or other learning-to-rank approaches.

Chapter 8
Semi-Supervised Learning

Building consistent and substantial training sets often requires significant efforts within the framework of international consortia. A significant portion of this work is devoted to data labeling, which can occur at varying levels of precision depending on the task. Due to constraints of time or resources, it is often impractical to carry out the labeling of these training sets. Recognizing that labeled data are expensive while unlabeled data are abundant and contain information relevant to the problem at hand, the learning community has focused, since the late 1990s, on the concept of semi-supervised learning for predictions tasks. This paradigm draws from the two classic theoretical frameworks of learning: supervised and unsupervised learning. In this chapter, we will present the various approaches developed within the framework of semi-supervised learning, highlighting their underlying hypotheses.

8.1 Unsupervised Framework and Central Assumptions

In unsupervised learning, we aim to discover the internal structure of data drawn from an unknown probability distribution $\mathbb{P}(\mathbf{x})$, also known as the data-generating distribution. We do this using a set of observations of size u, for which we lack labeled class information, denoted as $\mathbf{X}_{1:u} = \{\mathbf{x}_1, \ldots, \mathbf{x}_u\}$.[1] The primary goal is to model the probability density of the data $\mathbb{P}(\mathbf{x})$ by making various assumptions about the components that constitute it. Typically, a component of the mixture represents the conditional probability that an observation belongs to a particular cluster (or group), thereby partly explaining the phenomenon under study. These hypotheses are generally based on prior knowledge about the problem or the data. When such knowledge is lacking, we rely on classical and simple density functions from the literature.

[1] In the following, we will also denote this set by $X_{\mathcal{U}}$.

M.-R. Amini, *Advanced Supervised and Semi-supervised Learning*,
Cognitive Technologies, https://doi.org/10.1007/978-3-031-99928-4_8

8.1.1 Mixture Models

The fundamental assumption in this approach is that each observation belongs to exactly one of the clusters $\{G_k\}_{k=1,\dots,K}$. With this assumption and those previously mentioned, the data-generating distribution is modeled as a mixture of densities defined by:

$$\mathbb{P}(\mathbf{x} \mid \Theta) = \sum_{k=1}^{K} \pi_k \mathbb{P}(\mathbf{x} \mid G_k, \theta_k), \tag{8.1}$$

where K is the number of components of the mixture, $\forall k, \pi_k$ the different proportions (or prior probabilities) and $\forall k, \mathbb{P}(\mathbf{x} \mid G_k, \theta_k)$ the conditional densities defined with their parameters θ_k. In the most common case, the analytic forms of the conditional densities $\{\mathbb{P}(\mathbf{x} \mid G_k, \theta_k)\}_{k=1,\dots,K}$ are known and the unknowns are the parameters of the mixture defining these densities, as well as the prior probabilities $\{\pi_k\}_{k=1,\dots,K}$:

$$\Theta = \{\theta_k, \pi_k \mid k \in \{1, \dots, K\}\}.$$

The goal is thus to estimate the parameters Θ for which the densities fit the observations as well as possible. This adjustment is translated by the maximization of the likelihood of the data, which is simply the joint probability of the observations and which, with the fundamental assumption of independence of the observations, is written:

$$\mathbb{P}(\mathbf{X}_{1:u} \mid \Theta) = \mathbb{P}(\mathbf{x}_1, \dots, \mathbf{x}_u \mid \Theta) = \prod_{i=1}^{u} \mathbb{P}(\mathbf{x}_i \mid \Theta) = \prod_{i=1}^{u} \sum_{k=1}^{K} \pi_k \mathbb{P}(\mathbf{x}_i \mid G_k, \theta_k). \tag{8.2}$$

This maximization problem is solved by passing through the logarithm of the likelihood, $\mathcal{L}_M(\Theta) = \ln \mathbb{P}(\mathbf{X}_{1:u} \mid \Theta)$. This passage will not change the value of the maximum; and it just has the effect of transforming in the expression (8.2) the product into a sum, which is simpler to handle. Thus, we seek to maximize the log-likelihood of the mixture (or $\mathcal{L}_M$) with respect to Θ. Duda et al. [49] explain the cases in which it is possible to perform this estimation. They introduce the notion of identifiability; $\mathbb{P}(\mathbf{x} \mid \Theta)$ is said to be identifiable if, for all $\Theta \neq \Theta'$, there exists a $\mathbf{x}$ such that $\mathbb{P}(\mathbf{x} \mid \Theta) \neq \mathbb{P}(\mathbf{x} \mid \Theta')$. If $\mathbb{P}(\mathbf{x} \mid \Theta)$ is identifiable, we can give an estimate of this density. In general, non-identifiability corresponds to discrete distributions. Thus, when there are too many components in the mixture, there may be more unknowns than independent equations and identifiability can be a real problem. For the continuous case, identifiability problems are more solvable.

8.1.2 Estimation of the Parameters of the Mixture

To find the parameters Θ that maximize the log-likelihood function, it is not possible in general to explicitly solve:

$$\frac{\partial \mathcal{L}_M}{\partial \Theta} = 0,$$

since the latter is in the form of a sum of logarithms of a sum and its partial derivatives do not have a simple analytical expression.

8.1.2.1 EM Algorithm

One way to do solve this problem is to use a general class of iterative procedures known as the *Expectation Maximization* (EM) algorithm [45]. This algorithm is a powerful tool in statistical analysis and is the basis for many others such as the CEM algorithm which is the basis for many semi-supervised algorithms. The principle of the EM algorithm is as follows; at each iteration, the values of Θ are re-estimated in order to increase $\mathcal{L}_M$ until a maximum is reached. The main idea of this algorithm is to introduce hidden variables Z so that, if the Z were known, the optimal value of Θ could be found easily. With the introduction of these hidden variables, we have:

$$\mathcal{L}_M(\Theta) = \ln \mathbb{P}(\mathbf{X}_{1:u} \mid \Theta) = \ln \sum_{Z} \mathbb{P}(\mathbf{X}_{1:u}, Z \mid \Theta) \tag{8.3}$$

$$= \ln \sum_{Z} \mathbb{P}(\mathbf{X}_{1:u} \mid Z, \Theta)\mathbb{P}(Z \mid \Theta). \tag{8.4}$$

Denoting the current estimate of the parameters at iteration t, $\Theta^{(t)}$, iteration $t+1$ consists of finding a new value of the parameters Θ that maximizes $\mathcal{L}_M(\Theta) - \mathcal{L}_M(\Theta^{(t)})$:

$$\mathcal{L}_M(\Theta) - \mathcal{L}_M(\Theta^{(t)}) = \ln \frac{\mathbb{P}(\mathbf{X}_{1:u} \mid \Theta)}{\mathbb{P}(\mathbf{X}_{1:u} \mid \Theta^{(t)})},$$

or, according to (8.4):

$$\mathcal{L}_M(\Theta) - \mathcal{L}_M(\Theta^{(t)}) = \ln \frac{\sum_Z \mathbb{P}(\mathbf{X}_{1:u} \mid Z, \Theta)\mathbb{P}(Z \mid \Theta)}{\mathbb{P}(\mathbf{X}_{1:u} \mid \Theta^{(t)})}.$$

By multiplying each term of the sum of the second term by $\frac{\mathbb{P}(Z|\mathbf{X}_{1:u},\Theta^{(t)})}{\mathbb{P}(Z|\mathbf{X}_{1:u},\Theta^{(t)})}$, this gives:

$$\mathcal{L}_M(\Theta) - \mathcal{L}_M(\Theta^{(t)}) = \ln \sum_{Z} \mathbb{P}(Z \mid \mathbf{X}_{1:u}, \Theta^{(t)}) \frac{\mathbb{P}(\mathbf{X}_{1:u} \mid Z, \Theta)\mathbb{P}(Z \mid \Theta)}{\mathbb{P}(Z \mid \mathbf{X}_{1:u}, \Theta^{(t)})\mathbb{P}(\mathbf{X}_{1:u} \mid \Theta^{(t)})}.$$

Using the concavity of the logarithm function, and Jensen's inequality we get:

$$\mathcal{L}_M(\Theta) - \mathcal{L}_M(\Theta^{(t)}) \geqslant \sum_Z \mathbb{P}(Z \mid \mathbf{X}_{1:u}, \Theta^{(t)}) \ln \frac{\mathbb{P}(\mathbf{X}_{1:u} \mid Z, \Theta)\mathbb{P}(Z \mid \Theta)}{\mathbb{P}(\mathbf{X}_{1:u} \mid \Theta^{(t)})\mathbb{P}(Z \mid \mathbf{X}_{1:u}, \Theta^{(t)})}.$$

Which can be rewritten as $\mathcal{L}_M(\Theta) \geqslant Q(\Theta, \Theta^{(t)})$, with:

$$Q(\Theta, \Theta^{(t)}) = \mathcal{L}_M(\Theta^{(t)}) + \sum_Z \mathbb{P}(Z \mid \mathbf{X}_{1:u}, \Theta^{(t)}) \ln \frac{\mathbb{P}(\mathbf{X}_{1:u} \mid Z, \Theta)\mathbb{P}(Z \mid \Theta)}{\mathbb{P}(\mathbf{X}_{1:u} \mid \Theta^{(t)})\mathbb{P}(Z \mid \mathbf{X}_{1:u}, \Theta^{(t)})}.$$

Hence, $\mathcal{L}_M(\Theta)$ is equal to $Q(\Theta, \Theta^{(t)})$ for $\Theta = \Theta^{(t)}$ and it is strictly greater than $Q(\Theta, \Theta^{(t)})$ everywhere else. At iteration $t+1$, we are looking for a new value of Θ which maximizes $Q(\Theta, \Theta^{(t)})$, e.g. $\Theta^{(t+1)} = \text{argmax}_\Theta\, Q(\Theta, \Theta^{(t)})$. By removing terms that do not dependent to Θ, we obtain:

$$\begin{aligned}\Theta^{(t+1)} &= \underset{\Theta}{\text{argmax}} \left[\sum_Z \mathbb{P}(Z \mid \mathbf{X}_{1:u}, \Theta^{(t)}) \ln \mathbb{P}(\mathbf{X}_{1:u}, Z \mid \Theta)\right] \\ &= \underset{\Theta}{\text{argmax}}\, \mathbb{E}_{Z|\mathbf{X}_{1:u}} \left[\ln \mathbb{P}(\mathbf{X}_{1:u}, Z \mid \Theta) \mid \Theta^{(t)}\right].\end{aligned}$$

Therefore, for a given parameter value, the algorithm iteratively estimates a lower bound of the log-likelihood function and then maximizes it. In subsequent iterations, the process begins anew with the updated values, repeating the estimation and maximization steps. This procedure is illustrated in Fig. 8.1 and can be summarized by Algorithm 23. The parameters are often initialized randomly, and it is evident that the algorithm converges to a local minimum of the function $\mathcal{L}_M(\Theta)$.

Input :

- A set of unlabeled examples, $\mathbf{X}_{1:u} = \{\mathbf{x}_1, \cdots, \mathbf{x}_u\}$;
- A precision ϵ;
- A maximum number of iterations, T;

Initialization: Parameters $\Theta^{(0)}$;
repeat
  E step: Estimate the expectation $\mathbb{E}_{Z|\mathbf{X}_{1:u}} \left[\ln \mathbb{P}(\mathbf{X}_{1:u}, Z \mid \Theta) \mid \Theta^{(t)}\right]$;
  M step: Find the parameters $\Theta^{(t+1)}$ which maximize $Q(\Theta, \Theta^{(t)})$;
  $t \leftarrow t+1$;
until $(t > T) \vee (|\mathcal{L}_M(\Theta^{(t+1)}) - \mathcal{L}_M(\Theta^{(t)})| \leqslant \epsilon)$;
output: Model parameters $\Theta^{(t)}$.

Algorithme 23: Expectation Maximization (EM)[45]

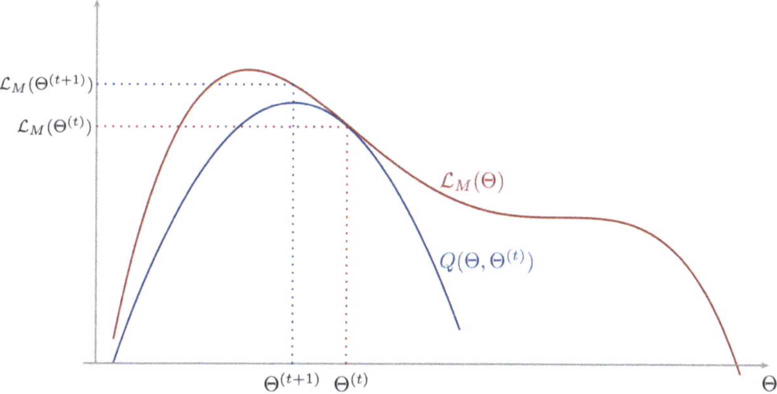

Fig. 8.1 Illustration of the two steps of estimation and maximization of the EM algorithm. The x-axis represents the possible values of the set of parameters Θ and the y-axis gives the logarithm of the likelihood of the data. The EM algorithm computes the function $Q(\Theta, \Theta^{(t)})$ using the current estimate $\Theta^{(t)}$ and gives the new $\Theta^{(t+1)}$ as the maximum point of $Q(\Theta, \Theta^{(t)})$

Indeed, according to the previous formulation:

- $\mathcal{L}_M(\Theta^{(t+1)}) \geqslant Q(\Theta^{(t+1)}, \Theta^{(t)}) \geqslant Q(\Theta^{(t)}, \Theta^{(t)})$,
- $\mathcal{L}_M(\Theta^{(t)}) = Q(\Theta^{(t)}, \Theta^{(t)})$.

Thus, at each iteration we have $\mathcal{L}_M(\Theta^{(t+1)}) \geqslant \mathcal{L}_M(\Theta^{(t)})$ and since the function $\mathcal{L}_M$ is concave, the algorithm is guaranteed to reach a local maximum of the log-likelihood function. One application of this modeling is the clustering task which consists in automatically finding an optimal clustering of the examples into K groups $P = \{G_k; k \in \{1, \dots, K\}\}$, obtained according to the Bayesian decision rule. According to this rule, an example is assigned to the group with which its posterior probability is the largest:

$$\forall \mathbf{x}, \mathbf{x} \in G_k \Leftrightarrow \mathbb{P}(G_k \mid \mathbf{x}, \Theta^*) = \underset{k' \in \{1, \dots, K\}}{\operatorname{argmax}} \ \mathbb{P}(G_{k'} \mid \mathbf{x}, \Theta^*), \tag{8.5}$$

where the posterior probabilities are calculated with Bayes' rule , as well as the estimates of the conditional cluster densities and the mixture parameters, Θ^* [168].

8.1.2.2 CEM Algorithm

Another more direct approach to clusterings follows the maximum complete likelihood technique [162]. In this approach, the hidden variables $(\tilde{\mathbf{z}}_i)_{i=1}^u$ are the group indicator vectors and the goal is to find an optimal cluster of the data

Input :

- A set of unlabeled examples $\mathbf{X}_{1:u} = \{\mathbf{x}_1, \cdots, \mathbf{x}_u\}$;
- A precision, $\epsilon > 0$;

Initialization:

- A random initialization of parameters, $\Theta^{(0)}$;
- An initial clustering of the data $P^{(0)} = \{G_k^{(0)}; k \in \{1, \ldots, K\}\}$;

repeat

E step: With the current parameters $\Theta^{(t)} = \{\pi^{(t)}, \theta^{(t)}\}$, estimate the conditional expectations of the indicator vectors $\mathbb{E}[\tilde{z}_{ik}^{(t)} \mid \mathbf{x}_i; P^{(t)}, \pi^{(t)}, \theta^{(t)}]$. Since these indicator vectors are binary random variables (8.6), for all $i \in \{1, \ldots, u\}$ and for all $k \in \{1, \ldots, K\}$, we have:

$$\mathbb{E}\left[\tilde{z}_{ik}^{(t)} \mid \mathbf{x}_i; P^{(t)}, \Theta^{(t)}\right] = \mathbb{P}(G_k^{(t)} \mid \mathbf{x}_i, \Theta^{(t)}) = \frac{\pi_k^{(t)}\mathbb{P}(\mathbf{x}_i \mid G_k^{(t)}, \theta_k^{(t)})}{\sum_{k=1}^{K} \pi_k^{(t)}\mathbb{P}(\mathbf{x}_i \mid G_k^{(t)}, \theta_k^{(t)})}.$$

C step: Assign to each example $\mathbf{x}_i$ its group; the one whose posterior probability is maximum. Denote $P^{(t+1)} = \{G_k^{(t+1)} \mid k \in \{1, \ldots, K\}\}$ the new partition of the examples;

M step: Find the new $\Theta^{(t+1)}$ that maximize $\mathcal{L}_C(P^{(t+1)}, \Theta^{(t)})$;

$t \leftarrow t + 1$;

until $(t > T) \vee (|\mathcal{L}_C(P^{(t+1)}, \Theta^{(t+1)}) - \mathcal{L}_C(P^{(t)}, \Theta^{(t)})| \leqslant \epsilon)$;

output: Model parameters, $\Theta^{(t)}$, as well as the clustering of the data $P^* = \{G_k^{(t)}; k \in \{1, \ldots, K\}\}$.

Algorithme 24: Classification EM [28]

according to the Bayesian decision rule together with the mixture parameters. As in the case of classification (Chap. 2), these vectors are binary vectors defined as:

$$\forall i, \mathbf{x}_i \text{ belongs to group } G_k \Leftrightarrow \tilde{\mathbf{z}}_i = (\underbrace{\tilde{z}_{i1}, \ldots, \tilde{z}_{ik-1}}_{\text{all equal to } 0}, \underbrace{\tilde{z}_{ik}}_{=1}, \underbrace{\tilde{z}_{ik+1}, \ldots, \tilde{z}_{iK}}_{\text{all equal to } 0}). \tag{8.6}$$

The logarithm of the complete likelihood of the data is:

$$\mathcal{L}_C(P, \Theta) = \ln \prod_{i=1}^{u} \prod_{k=1}^{K} (\mathbb{P}(\mathbf{x}_i, G_k, \Theta))^{\tilde{z}_{ik}} = \sum_{i=1}^{u} \sum_{k=1}^{K} \tilde{z}_{ik} \ln \mathbb{P}(\mathbf{x}_i, G_k, \Theta). \tag{8.7}$$

This criterion is maximized using the classification EM algorithm (CEM) [28]. This algorithm is similar to the EM algorithm except for an additional classification **C** step, in which each data is assigned one and only one component of the mixture.

The convergence of this algorithm is obtained thanks to the two classification steps (**C** step) and maximization steps (**M** step) of Algorithm 24. Indeed:

- At the **C** step, we choose, with the current parameters $\Theta^{(t)}$, a cluster $P^{(t+1)}$ following the optimal Bayes rule:

$$\mathcal{L}_C(P^{(t+1)}, \Theta^{(t)}) \geqslant \mathcal{L}_C(P^{(t)}, \Theta^{(t)}).$$

- At the **M** step, we look for new parameters $\Theta^{(t+1)}$ which maximize $\mathcal{L}_C(P^{(t+1)}, \Theta^{(t)})$:

$$\mathcal{L}_C(P^{(t+1)}, \Theta^{(t+1)}) \geqslant \mathcal{L}_C(P^{(t+1)}, \Theta^{(t)}).$$

Thus, at each iteration t, we have $\mathcal{L}_C(P^{(t+1)}, \Theta^{(t+1)}) \geqslant \mathcal{L}_C(P^{(t)}, \Theta^{(t)})$, and since there are a finite number of clusters on the unlabeled data, the algorithm is guaranteed to converge to a local maximum of the complete likelihood function.

The essential difference between the maximum likelihood and maximum complete likelihood methods is that in the latter case the group indicator vectors of the unlabeled data are part of the model parameters and they are estimated together with the mixture parameters.

8.1.2.3 Particular Case: K-means Algorithm

Let us consider the case where the mixing densities are normal laws with an identity covariance matrix, $\forall k, \boldsymbol{\Sigma}_k = \mathbf{Id}_d$, and where the proportions of the groups are equiprobable:

$$\forall k \in \{1, \dots, K\},\ \mathcal{N}(x|\boldsymbol{\mu}_k, \boldsymbol{\Sigma}_k) = \frac{1}{(2\pi)^{d/2}} \exp^{-\frac{1}{2}||\mathbf{x}-\boldsymbol{\mu}_k||^2},$$

$$\pi_k = \frac{1}{K},$$

where $||.||^2$ represents the square of the Euclidean distance. The parameters of the mixture to be estimated are then $\Theta = \{\boldsymbol{\mu}_k \mid k \in \{1, \dots, K\}\}$. In this case, the logarithm of thecomplete likelihood (8.7) is written as:

$$\begin{aligned}\mathcal{L}_C(P, \Theta) &= -\frac{1}{2}\sum_{i=1}^{u}\sum_{k=1}^{K} \tilde{z}_{ik}||\mathbf{x}_i - \boldsymbol{\mu}_k||^2 + r \\ &= -\frac{1}{2}\sum_{k=1}^{K}\sum_{\mathbf{x}\in G_k} ||\mathbf{x} - \boldsymbol{\mu}_k||^2 + r, \qquad (8.8)\end{aligned}$$

where r is a constant that does not depend on the parameters Θ. Thus, maximizing the logarithm of the classification likelihood with respect to Θ is equivalent to minimizing the sum of the distances:

$$SSR(G_1, \ldots, G_K; \boldsymbol{\mu}_1, \ldots, \boldsymbol{\mu}_K) = \frac{1}{2}\sum_{k=1}^{K}\sum_{\mathbf{x}\in G_k} ||\mathbf{x} - \boldsymbol{\mu}_k||^2. \tag{8.9}$$

The SSR notation stands for *Sum of Squared Residuals*. The gradient of this function with respect to the model parameters $\Theta = (\mu_k)_{1\leqslant k\leqslant K}$ writes:

$$\nabla_\Theta SSR = \left(\sum_{\mathbf{x}\in G_1}(\mathbf{x} - \boldsymbol{\mu}_1), \ldots, \sum_{\mathbf{x}\in G_K}(\mathbf{x} - \boldsymbol{\mu}_K)\right)^\top. \tag{8.10}$$

This criterion thus reaches a local minimum when $\nabla_\Theta SSR = 0$ or, in other words, when the parameters $\{\boldsymbol{\mu}_k \mid k \in \{1, \ldots, K\}\}$ are defined as the centroids of the groups:

$$\forall k \in \{1, \ldots, K\}, \boldsymbol{\mu}_k = \frac{1}{|G_k|}\sum_{\mathbf{x}\in G_k}\mathbf{x}_k, \tag{8.11}$$

where $|G_k|$ represents the number of examples of the group G_k.

The **C** step of Algorithm 24, which also consists of assigning each example of the unlabeled set to the group with which its posterior probability is the greatest, amounts, with the previous hypotheses, to assigning the example to the group when its Euclidean distance to the center of the group is the smallest. The associated standard algorithm then consists, from a set of initial representatives, in iterating the following two operations:

- Reassign each example to the cluster of the closest representative (reassignment step);
- Update the representatives based on the examples present in the group (centroid recalculation step).

These two steps are explained in Algorithm 25, which takes as input a set of unlabeled examples, a number of classes, and a maximum number of iterations. This algorithm corresponds to the standard form of the K-means algorithm. It warrants several remarks.

1. Selection of initial centroids: The selection of initial centroids can be done in different ways, the two most popular being a random draw without replacement of K examples from the collection, or the use of a simple clustering method, such as the one-pass method. In the latter case, it may be necessary to supplement the centroids obtained by other representatives, randomly drawn from the collection, if the number K of groups retained is greater than the number of groups provided

---

Input :

- $X_{\mathcal{U}} = \{\mathbf{x}_1, \cdots, \mathbf{x}_u\}$, a set of unlabeled examples ;
- K, number of clusters;
- T, maximum number of iterations;
- ϵ, precision;

Initialization:

- $\Theta^{(0)} = (\boldsymbol{\mu}_1^{(0)}, \cdots, \boldsymbol{\mu}_K^{(0)})$; // Initial centroids
- An initial partition of the data $P^{(0)}$;
- $t \leftarrow 0$;

repeat

  foreach $k, 1 \leq k \leq K$ **do**

    // Reassignment step (**E** and **C** steps of Algorithm 24)

    $G_k^{(t+1)} \leftarrow \{\mathbf{x} : ||\mathbf{x} - \boldsymbol{\mu}_k^{(t)}||^2 \leq ||\mathbf{x} - \boldsymbol{\mu}_l^{(t)}||^2, \forall l \neq k, 1 \leq l \leq K\}$;

  foreach $k, 1 \leq k \leq K$ **do**

    // Centroid recalculation step (**M** step of Algorithm 24)

    $\boldsymbol{\mu}_k^{(t+1)} \leftarrow \frac{1}{|G_k^{(t+1)}|} \sum_{\mathbf{x} \in G_k^{(t+1)}} \mathbf{x}$; ▷ (8.11)

  $t \leftarrow t + 1$;

until $(t > T) \vee (|\mathcal{SSR}(P^{(t+1)}, \Theta^{(t+1)}) - \mathcal{SSR}(P^{(t)}, \Theta^{(t)})| \leqslant \epsilon)$;

output: $P^* = \{G_1^{(t)}, \ldots, G_K^{(t)}\}$, a clustering of $X_{\mathcal{U}}$ in K groups.

Algorithme 25: K-means

by the one-pass method. It is however important to note that the final solution depends on the initial representatives chosen and that a random choice of these representatives makes the algorithm non-deterministic. To remedy this problem, we in practice run the K-means algorithm several times and choose the partition that minimizes the sum of squares of the residuals (8.9). We must of course ensure that the choice of initial representatives is indeed random; be careful not to systematically use the same seed for the *rand()* function available in the majority of programming languages;

2. Convergence and stopping condition: as we have just seen for the CEM algorithm, the K-means algorithm is guaranteed to converge to a local minimum of the sum of squares of the residuals. However, for algorithmic reasons, it is necessary in the reassignment step and in the event of equality of distance between the current group of an example and another group, not to move the example. The stopping condition concerns both the number of iterations and the stability of the groups obtained. It is sometimes difficult, because of computer rounding problems, to impose that the clustering obtained be stable (stability is generally tested not at the level of the centroids of the groups, but of the examples constituting the group, in order to reduce rounding problems). The condition on the maximum number of iterations to be carried out makes it possible to

overcome this rounding problem. In practice, a few dozen iterations are often sufficient.

3. Complexity: at each iteration, it is necessary to compare the u examples to the K group representatives, which is done in $O(uK)$ (we neglect here the complexity of computing a distance or similarity between two vectors of size d, which is in $O(d)$). Updating the class representatives is dominated by $O(u)$, which finally leads to a total complexity of the order of $O(uKT)$, where T is the total number of iterations.

8.1.3 Central Assumptions

Semi-supervised learning, also called learning with partially labeled data, is a supervised learning where we have both labeled and unlabeled examples. The labeled examples are generally assumed to be too few to obtain a good estimate of the dependencies sought and we want to use the unlabeled examples to obtain a better estimate. To do this, we will assume that there is a set of labeled examples $S = \{(\mathbf{x}_i, y_i) \mid i = 1, \dots, m\}$ from the joint distribution $\mathbb{P}(\mathbf{x}, y)$ (also denoted by $\mathcal{D}$) and a set of unlabeled examples $X_{\mathcal{U}} = \{\mathbf{x}_i \mid i = m + 1, \dots, m + u\}$ assumed to come from the data-generating distribution $\mathbb{P}(\mathbf{x})$. If $X_{\mathcal{U}}$ is empty, we fall back on the problem of supervised learning. If S is empty, we are in the presence of an unsupervised learning problem.

During learning, semi-supervised algorithms estimate labels for the unlabeled examples. We let $\tilde{y}$ and $\tilde{\mathbf{z}}$ be respectively the label and the class indicator vector of the unlabeled example $\mathbf{x} \in X_{\mathcal{U}}$ estimated by these algorithms. The interest of semi-supervised learning arises when $u = |X_{\mathcal{U}}| >> m = |S|$ and the goal is that the knowledge gained on the data-generating distribution, $\mathbb{P}(\mathbf{x})$, through the unlabeled examples can provide useful information in the inference of $\mathbb{P}(y \mid \mathbf{x})$.

If this goal is not achieved, semi-supervised learning will be less efficient than supervised learning and it may even happen that the use of unlabeled data degrades the performance of the learned prediction function [38, 192]. It is then necessary to formulate working hypotheses for taking into account unlabeled data in the supervised learning of a prediction function.

8.1.3.1 Smoothness Assumption

The basic assumption in semi-supervised learning, called the smoothness assumption, states that:

Hypothesis 8.1 *If two examples $\mathbf{x}_1$ and $\mathbf{x}_2$ are close in a high density region, then their class labels y_1 and y_2 should be similar.*

This assumption implies that if two points belong to the same cluster, then their output label is likely to be the same. If, on the other hand, they were separated by a low-density region, then their outputs would be different.

8.1.3.2 Cluster Assumption

Now suppose that examples of the same class form a cluster. Unlabeled data could then help to find the boundary of each cluster more efficiently than if only labeled examples were used. Thus, one way to use unlabeled data would be to find the clusters with a mixture model and then assign class labels to the clusters using the labeled data they contain. The underlying assumption can be stated as:

Hypothesis 8.2 *If two examples* $\mathbf{x}_1$ *and* $\mathbf{x}_2$ *are in the same group, then they are likely to belong to the same class* y.

This assumption could be understood as follows: if there exists a group formed by a dense set of examples, then it is unlikely that they can belong to different classes. This is not equivalent to saying that a class is formed by a single group of examples, but that it is unlikely to find two examples belonging to different classes in the same group. According to the previous smoothness assumption, if we consider clusters of examples as high-density regions, another formulation of the cluster assumption is that the decision boundary passes through low-density regions (Fig. 8.2a). This assumption is the basis of generative (Sect. 8.2) and discriminative (Sect. 8.3) methods for semi-supervised learning.

8.1.3.3 Manifold Assumption

For high-dimensional problems, the previous two assumptions fall short because the search for densities often relies on a notion of distance that loses its relevance in such cases. A third assumption, known as the manifold assumption, on which some semi-supervised models are based, posits that:

Hypothesis 8.3 *For high-dimensional problems, examples are found on locally Euclidean topological spaces (or geometric varieties) of low dimension (Fig. 8.2b).*

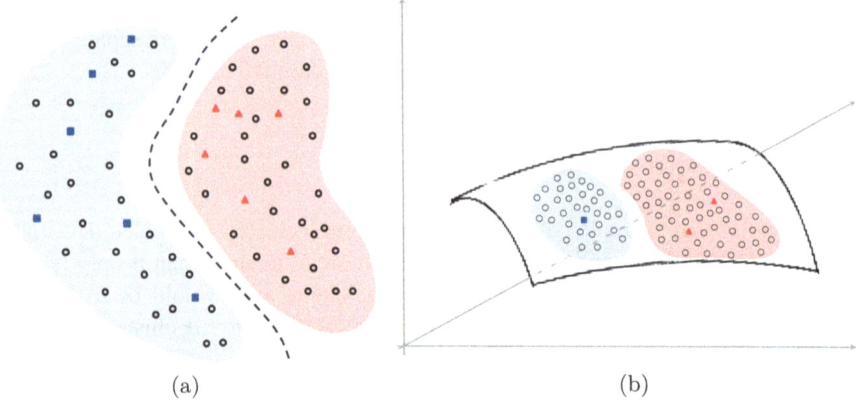

Fig. 8.2 Illustration of the cluster and manifold assumptions. Unlabeled examples are represented by empty circles and labeled examples of different classes by a filled square or triangle. Clusters are considered to be high-density regions and the decision boundary is thus expected to pass through low-density regions (cluster assumption, Figure (**a**)). For a given problem, the examples are assumed to lie on a lower-dimensional geometric manifold (a 2-dimensional manifold, or a curved surface on the given example—manifold assumption, Figure (**b**))

In the following, we will present some classic models of the three families of semi-supervised methods resulting from the previous hypotheses.

8.2 Generative Methods

Semi-supervised prediction with generative models involves estimating the conditional density $\mathbb{P}(\mathbf{x} \mid y, \Theta)$ using, as in the unsupervised case, the EM or CEM algorithm to estimate the parameters Θ of the model. However, the major difference with unsupervised clustering, with for example the EM algorithm, is that the hidden variables associated with the labeled examples are known in advance and correspond to the class labels of these examples.

The basic hypothesis of these models is thus the cluster assumption (Sect. 8.1.3) since, in this case, each group of unlabeled examples corresponds to a class [152]. We can thus interpret semi-supervised learning with generative models (*a*) as a supervised classification where we have additional information on the probability density $\mathbb{P}(\mathbf{x})$ of the data, or (*b*) as a clustering with additional information on the class labels of a subset of examples [14]. In the case where the hypothesis generating the data is known, generative models can become very efficient.

In Sect. 8.2.3, we will present a generative model for document classification in the case where the representation of the documents is based on the occurrence of the terms of a vocabulary in them and therefore when the hypothesis of the multinomial distribution per class holds. In the following, we will first extend the data likelihood criteria to the semi-supervised case (Sect. 8.2.1) and describe the extension of the CEM algorithm to this case (Sect. 8.2.2).

8.2.1 Extension of the Likelihood to Semi-supervised Learning

McLachlan [110, p. 39] extended the complete likelihood to the case where, along with unlabeled examples, one also uses labeled examples to estimate the parameters of a generative model for classification. In this case, the class indicator vectors for the labeled examples, $(\mathbf{z}_i)_{i=1}^m$, are known and remain unchanged during the estimation and classification phases, while the group indicator vectors for the unlabeled examples, $(\tilde{\mathbf{z}}_i)_{i=m+1}^{m+u}$, are estimated as in the unsupervised case. The logarithm of the complete likelihood of the data in the semi-supervised case is written as:

$$\mathcal{L}_{ssC}(P,\Theta) = \sum_{i=1}^{m}\sum_{k=1}^{K} z_{ik} \ln \mathbb{P}(\mathbf{x}_i, y = k, \Theta) + \sum_{i=m+1}^{m+u}\sum_{k=1}^{K} \tilde{z}_{ik} \ln \mathbb{P}(\mathbf{x}_i, \tilde{y} = k, \Theta). \tag{8.12}$$

The first term of this summation concerns the labeled data and the second term concerns the unlabeled examples. Some studies have proposed to modulate the effect of unlabeled examples in learning by adding a real $\lambda \in [0, 1]$ in front of this second term [30, 68, chapter 9]. For comparison, the semi-supervised version of the maximum likelihood criterion is written:

$$\mathcal{L}_{ssM}(\Theta) = \sum_{i=1}^{m} \ln \mathbb{P}(\mathbf{x}_i, y_i, \Theta) + \sum_{i=m+1}^{m+u} \ln\left(\sum_{k=1}^{K} \pi_k \mathbb{P}(\mathbf{x}_i \mid \tilde{y} = k, \theta_k)\right). \tag{8.13}$$

With this criterion, Nigam et al. [127] extended the EM algorithm to take into account labeled data when estimating the parameters of the multinomial Naive Bayes classifier, and they proposed one of the very first semi-supervised algorithms that they applied to the task of document classification. In the following, we will present the extension of the CEM algorithm to the semi-supervised case.

8.2.2 Semi-supervised CEM

The CEM algorithm can be readily adapted to the semi-supervised case by maximizing (8.12) instead of (8.7). In this scenario, groups are treated as classes, and the initial density components $\mathbb{P}(\mathbf{x} \mid \tilde{y}^{(0)} = k, \theta^{(0)}), k \in 1, \dots, K$ are estimated using the labeled data. During the classification step of the algorithm (**C** step), the

Input :

- A labeled training set $S = \{(\mathbf{x}_i, y_i) \mid i \in \{1, \dots, m\}\}$ and an unlabeled training set $X_{\mathcal{U}} = \{\mathbf{x}_i \mid i \in \{m+1, \dots, m+u\}\}$;
- A precision ϵ;
- A maximum number of iterations T;

Initialization:

- Estimate the parameters $\Theta^{(0)}$ over the labeled training set S ;
- $t \leftarrow 0$;

repeat

E step: With the current parameters $\Theta^{(t)} = \{\pi^{(t)}, \theta^{(t)}\}$, estimate the conditional expectations of the group indicator vectors of unlabeled examples. Hence $\forall \mathbf{x}_i \in X_{\mathcal{U}}$

$$\mathbb{E}\left[\tilde{z}_{ik}^{(t)} \mid \mathbf{x}_i; P^{(t)}, \Theta^{(t)}\right] = \mathbb{P}(\tilde{y}^{(t)} = k \mid \mathbf{x}_i, \Theta^{(t)}) = \frac{\pi_k^{(t)} \mathbb{P}(\mathbf{x}_i \mid \tilde{y}^{(t)} = k, \theta_k^{(t)})}{\sum_{k=1}^{K} \pi_k^{(t)} \mathbb{P}(\mathbf{x}_i \mid \tilde{y}^{(t)} = k, \theta_k^{(t)})}.$$

C step: Assign each unlabeled example $\mathbf{x}_i \in X_{\mathcal{U}}$ to the class with which its posterior probability is maximal. Denote $P^{(t+1)}$ this new clustering;

M step: Find the new parameters $\Theta^{(t+1)}$ that maximize the complete log likelihood of the data (8.12);

$t \leftarrow t + 1$;

until $(t > T) \vee (|\mathcal{L}_{ssC}(P^{(t+1)}, \Theta^{(t+1)}) - \mathcal{L}_{ssC}(P^{(t)}, \Theta^{(t)})| \leqslant \epsilon)$;
output: Model parameters Θ^*.

Algorithme 26: Semi-supervised CEM

group (or class) indicator vectors of the unlabeled data $(\tilde{\mathbf{z}}_i)_{i=m+1}^{m+u}$ are estimated, while the class indicator vectors of the labeled examples remain fixed at their known values (as detailed in Algorithm 26). It is evident that the algorithm converges to a local maximum of the criterion (8.12). A key difference from the unsupervised CEM algorithm is the inductive nature of the learned model, which is used to infer class labels for new examples (similar to the supervised models studied so far). In contrast, the goal of clustering is not to derive a general rule but to identify fixed groups of examples.

In the following section, we present the most widely used and known generative model in automatic document classification. This model assumes conditional independence of terms with respect to different classes (Naive Bayes assumption) and, therefore, it is called the Naive Bayes model in the literature.

8.2.3 Application: Semi-supervised Naive-Bayes Model

The semi-supervised Naive Bayes model was initially proposed for document classification [127]. In this model, a document is represented as x_i, , an ordered sequence of terms $x_i = \left\langle w_1, \ldots, w_{l_{x_i}} \right\rangle$ of length l_{x_i}, drawn from a vocabulary $\mathcal{V} = \{w_1, \ldots, w_d\}$ of size d.

Under the *Naive Bayes* assumption, the model posits that the probability of generating a term by a particular class (or component of the mixture) is independent of both the context in which the term appears and its position within the document. This simplifying assumption allows the model to estimate the probability densities of terms in a straightforward manner, despite the potential complexity of the underlying data.

Indeed, the probability densities are given by:

$$\forall k \in \{1, \ldots, K\}, \mathbb{P}(x_i = \left\langle w_1, \ldots, w_{l_{x_i}} \right\rangle \mid y = k) = \prod_{w_j \in x_i} \theta_{j|k}, \tag{8.14}$$

where $\forall j \in \{1, \ldots, d\}, \theta_{j|k} = \mathbb{P}(w_j \mid y = k)$ represents the probability of generation of the term w_j of the vocabulary by the kth component of the mixture. These probabilities verify: $0 \leqslant \theta_{j|k} \leqslant 1$ and $\sum_{j=1}^{d} \theta_{j|k} = 1$.

We then have the following proposition.

Proposition 8.4 *With the Naive Bayes model, we capture the occurrence information of the vocabulary terms in the documents. Thus, if we represent the document x_i as a vector of dimension d, $\mathbf{x}_i = (n_{i,1}, \ldots, n_{i,d})^\top$, where $n_{i,j}, j \in \{1, \ldots, d\}$ represents the number of occurrences of the term with index j of the vocabulary in the document x_i, the probability densities (8.15) become:*

$$\forall k \in \{1, \ldots, K\}, \mathbb{P}(\mathbf{x}_i \mid y = k) = \frac{n_{x_i}\,!}{n_{i,1}! \ \ldots \ n_{i,d}!} \prod_{j=1}^{d} \theta_{j|k}^{n_{i,j}}, \tag{8.15}$$

where $n_{x_i} = n_{i,1} + \ldots + n_{i,d}$ is the total number of occurrences of the vocabulary terms in the document x_i.

Proof The probability that the term with index 1 of the vocabulary appears $n_{i,1}$ times in x_i, given the kth class, is $\binom{n_{x_i}}{n_{i,1}} \theta_{1|k}^{n_{i,1}} \left[1 - \theta_{1|k}\right]^{n_{x_i} - n_{i,1}}$. We remove this term from the set of draws and continue.

The probability that the term with index 2 of the vocabulary appears $n_{i,2}$ times in x_i is this time $\binom{n_{x_i}-n_{i,1}}{n_{i,2}}\left[\frac{\theta_{2|k}}{1-\theta_{1|k}}\right]^{n_{i,2}}\left[1-\frac{\theta_{2|k}}{1-\theta_{1|k}}\right]^{n_{x_i}-n_{i,1}-n_{i,2}}$, (since $n_{i,2}+\ldots+n_{i,d}=n_{x_i}-n_{i,1}$ and $\frac{\theta_{2|k}}{1-\theta_{1|k}}+\ldots+\frac{\theta_{d|k}}{1-\theta_{1|k}}=1$ etc.).

As the draws are independent, we then have:

$$\mathbb{P}(\mathbf{x}_i=(n_{i,1},\ldots,n_{i,d})\mid y=k)=\binom{n_{x_i}}{n_{i,1}}\times\ldots\times\binom{n_{x_i}-n_{i,1}-\ldots-n_{i,d-1}}{n_{i,d}}\times$$

$$\theta_{1|k}^{n_{i,1}}[1-\theta_{1|k}]^{n_{x_i}-n_{i,1}}\times\left[\frac{\theta_{2|k}}{1-\theta_{1|k}}\right]^{n_{i,2}}\left[1-\frac{\theta_{2|k}}{1-\theta_{1|k}}\right]^{n_{x_i}-n_{i,1}-n_{i,2}}\times\ldots$$

Consider the first term of this product:

$$\binom{n_{x_i}}{n_{i,1}}\times\ldots\times\binom{n_{x_i}-n_{i,1}-\ldots-n_{i,d-1}}{n_{i,d}}=$$

$$\frac{n_{x_i}!}{n_{i,1}!\cancel{(n_{x_i}-n_{i,1})!}}\times\frac{\cancel{(n_{x_i}-n_{i,1})!}}{n_{i,2}!\cancel{(n_{x_i}-n_{i,1}-n_{i,2})!}}\times\ldots$$

Simplifying, this gives the multinomial coefficient

$$\binom{n_{x_i}}{n_{i,1}}\times\ldots\times\binom{n_{x_i}-n_{i,1}-\ldots-n_{i,d-1}}{n_{i,d}}=\frac{n_{x_i}!}{n_{i,1}!\ \ldots\ n_{i,d}!}.$$

Simplifying the second term gives the result. □

We then say that the document x_i comes from the multinomial distribution representing the class k, denoted $Mult(n_{x_i};\theta_{1|k},\ldots,\theta_{d|k})$, of support the set of d-tuples of positive or zero integers $\{n_{i,1},\ldots,n_{i,d}\}$ such that $n_{i,1}+n_{i,2}+\ldots+n_{i,d}=n_{x_i}$. The parameters of this generative model are:

$$\Theta=\{\theta_{j|k}:j\in\{1,\ldots,d\},k\in\{1,\ldots,K\};\pi_k:k\in\{1,\ldots,K\}\}.$$

Neglecting the multinomial coefficients that are independent of the parameters Θ; maximizing the logarithm of the complete likelihood (8.12) with the hyperparameter $\lambda\in[0,1]$ counterbalancing the effect of the unlabeled examples, results in solving the following subconstrained optimization problem:

$$\max_{\Theta}\sum_{i=1}^{m}\sum_{k=1}^{K}z_{ik}\left(\ln\pi_k+\sum_{j=1}^{d}n_{i,j}\ln\theta_{j|k}\right)+\lambda\sum_{i=m+1}^{m+u}\sum_{k=1}^{K}\tilde{z}_{ik}\left(\ln\pi_k+\sum_{j=1}^{d}n_{i,j}\ln\theta_{j|k}\right).$$

under the constraints $\forall k\in\{1,\ldots,K\},\ \sum_{j=1}^{d}\theta_{j|k}=1$ and $\sum_{k=1}^{K}\pi_k=1.$

The Lagrangian formulation of this constrained optimization problem using the Lagrange multipliers $\boldsymbol{\beta} = (\beta_1, \ldots, \beta_K)$ and γ, associated with the $K+1$ constraints of the problem, is written as:

$$L_p(\Theta, \boldsymbol{\beta}, \gamma) = \mathcal{L}_{ssC}(P, \Theta) + \sum_{k=1}^{K} \beta_k \sum_{j=1}^{d} \left(1 - \theta_{j|k}\right) + \gamma \left(1 - \sum_{k=1}^{K} \pi_k\right). \tag{8.16}$$

By cancelling the partial derivatives of the Lagrangian with respect to the variables $\theta_{j|k}$; $\forall j, \forall k$ and π_k; $\forall k$, we have:

$$\forall j, \forall k, \frac{\partial L_p}{\partial \theta_{j|k}} = \sum_{i=1}^{m} z_{ik} n_{i,j} \frac{1}{\theta_{j|k}} + \lambda \sum_{i=m+1}^{m+u} \tilde{z}_{ik} n_{i,j} \frac{1}{\theta_{j|k}} - \beta_k = 0$$

$$\forall k, \frac{\partial L_p}{\partial \pi_k} = \sum_{i=1}^{m} z_{ik} \frac{1}{\pi_k} + \lambda \sum_{i=m+1}^{m+u} \tilde{z}_{ik} \frac{1}{\pi_k} - \gamma = 0,$$

and, by reusing the equality constraints, we get:

$$\forall j, \forall k, \theta_{j|k} = \frac{\sum_{i=1}^{m} z_{ik} n_{i,j} + \lambda \sum_{i=m+1}^{m+u} \tilde{z}_{ik} n_{i,j}}{\sum_{i=1}^{m} z_{ik} \sum_{j=1}^{d} n_{i,j} + \lambda \sum_{i=m+1}^{m+u} \tilde{z}_{ik} \sum_{j=1}^{d} n_{i,j}}. \tag{8.17}$$

$$\forall k, \pi_k = \frac{\sum_{i=1}^{m} z_{ik} + \lambda \sum_{i=m+1}^{m+u} \tilde{z}_{ik}}{m + \lambda u}. \tag{8.18}$$

The class indicator vector of each labeled or unlabeled example contains binary components of which only one is equal to 1 (corresponding to the class of the example). We thus have $\forall \mathbf{x}_i \in S, \sum_{k=1}^{K} z_{ik} = 1$ and $\forall \mathbf{x}_i \in X_{\mathcal{U}}, \sum_{k=1}^{K} \tilde{z}_{ik} = 1$. The denominator of (8.18) follows from these equalities.

Furthermore, to practically manage the cases of rare terms that do not appear in any document of a given class and thus lead to zero presence probabilities and calculation errors when switching to the logarithm, the probabilities are smoothed, often using Laplace smoothing which consists of adding 1 to the number of documents of a class containing a term:

$$\forall j, \forall k, \theta_{j|k} = \frac{\sum_{i=1}^{m} z_{ik} n_{i,j} + \lambda \sum_{i=m+1}^{m+u} \tilde{z}_{ik} n_{i,j} + 1}{\sum_{i=1}^{m} z_{ik} \sum_{j=1}^{d} n_{i,j} + \lambda \sum_{i=m+1}^{m+u} \tilde{z}_{ik} \sum_{j=1}^{d} n_{i,j} + d}. \tag{8.19}$$

Finally, steps **E** and **C** can be summarized as computing the logarithm of the joint probabilities for each unlabeled example, with the current model parameters, and assigning it to the class with which this joint probability is largest.

$$\forall \mathbf{x}_i \in X_{\mathcal{U}}, \mathbf{x}_i \text{ belongs to the class } k \Leftrightarrow k = \underset{k' \in \{1,\ldots,K\}}{\operatorname{argmax}} \ln \mathbb{P}(\mathbf{x}_i, \tilde{y} = k', \Theta). \tag{8.20}$$

Neglecting the multinomial coefficients, which, for a given example, are all equal and do not intervene in the decision, we have:

$$\forall k' \ln \mathbb{P}(\mathbf{x}_i, \tilde{y} = k', \Theta) \approx \ln \pi_{k'} + \sum_{j=1}^{d} n_{i,j} \ln \theta_{j|k'}. \tag{8.21}$$

The semi-supervised CEM algorithm, when applied under the multinomial hypothesis for the distribution of examples by class, involves representing and managing examples using a data structure that stores the indices of non-zero attributes and their associated values. This structure is particularly effective for representing sparse vectors, which are common in high-dimensional problems where only a small fraction of each example's attributes are non-zero. An example of this is document classification, where the dimensionality can be in the order of hundreds of thousands, and typically 99% of the document attributes are zero [5, Chapter 2].

8.3 Discriminative Methods

A drawback of generative models is that when their distributional assumptions are no longer valid, their performance tends to decline compared to models trained solely on labeled examples [34]. This observation has inspired numerous efforts to move away from these assumptions. Early works in this direction employed discriminative models to estimate posterior class probabilities for input examples, such as logistic regression (Chap. 4, Sect. 4.2).

These studies demonstrated that maximizing the semi-supervised complete likelihood (8.12) is equivalent to minimizing the empirical risk on both labeled and pseudo-labeled examples. The extension of the semi-supervised CEM algorithm to the discriminative case would in this case be equivalent to the so-called directed decision technique (or self-training algorithm) proposed in the context of adaptive signal processing and which consists in using the outputs predicted by the system for unlabeled examples in order to construct desired outputs [57, 133]. This pseudo-labeling and model learning process is repeated until there is no more change in the labeling of the examples.

In the case where the pseudo-class labels are assigned to the unlabeled examples, by thresholding the outputs of the classifier corresponding to these examples, it can be shown that the self-training algorithm works according to the cluster assumption.

In the following sections, we will present this algorithm for the case of two-class classification which has been mostly studied for semi-supervised learning with these models, as well as other similar techniques that are related to it and proposed in the literature in recent years.

8.3.1 Self-training Algorithm

In the case where we explicitly assume that the output of the discriminative classifier estimates posterior probabilities of classes, we can rewrite the criterion (8.12) so as to highlight these probabilities. Indeed, according to Bayes' rule, we have:

$$\mathcal{L}_{ssC}(P,\Theta) = \sum_{i=1}^{m}\sum_{k=1}^{2} z_{ik} \ln \mathbb{P}(y=k \mid \mathbf{x}_i,\Theta) + \sum_{i=1}^{m} \ln \mathbb{P}(\mathbf{x}_i,\Theta) + \sum_{i=m+1}^{m+u}\sum_{k=1}^{2} \tilde{z}_{ik} \ln \mathbb{P}(\tilde{y}=k \mid \mathbf{x}_i,\Theta) + \sum_{i=1}^{m} \ln \mathbb{P}(\mathbf{x}_i,\Theta).$$

Since no distributional assumptions on the generation of the data are made in the discriminative case (i.e. the $\mathbb{P}(\mathbf{x})$ are not estimated), the optimization of this criterion is equivalent to the optimization of the following criterion [110, p. 261]:

$$\mathcal{L}_{ssD}(P,\Theta) = \sum_{i=1}^{m}\sum_{k=1}^{2} z_{ik} \ln \mathbb{P}(y=k \mid \mathbf{x}_i,\Theta) + \sum_{i=m+1}^{m+u}\sum_{k=1}^{2} \tilde{z}_{ik} \ln \mathbb{P}(\tilde{y}=k \mid \mathbf{x}_i,\Theta). \tag{8.22}$$

In the case where we use logistic discrimination to estimate the posterior probabilities and following the same reasoning as in Chap. 4, Sect. 4.2, it is easy to see that maximizing the criterion (8.22) is equivalent to minimizing the upper bound of the empirical risk with the logistic cost of a linear function $h_{\boldsymbol{w}} : \mathcal{X} \to \mathbb{R}$ on the labeled and pseudo-labeled training sets:

$$\hat{\mathcal{L}}(\boldsymbol{w}) = \frac{1}{m}\sum_{i=1}^{m} \ln(1+e^{-y_i h_{\boldsymbol{w}}(\mathbf{x}_i)}) + \frac{1}{u}\sum_{i=m+1}^{m+u} \ln(1+e^{-\tilde{y}_i h_{\boldsymbol{w}}(\mathbf{x}_i)}). \tag{8.23}$$

The self-training algorithm optimizes the criterion (8.23) through the following steps: A classifier $h_{\boldsymbol{w}}$ is first trained on the labeled dataset S [108, 193]. The algorithm then assigns pseudo-class labels to unlabeled examples $\mathbf{x}_i \in X_{\mathcal{U}}$ based on the classifier's predictions (**C** step).

A new classifier is trained using both the original labeled data and the pseudo-labeled unlabeled data to minimize the empirical risk with a logistic cost ((8.23)—**E** step). The **E** step is straightforward in this context, as posterior probabilities are derived directly from the classifier outputs, making it equivalent to the **M** step. Steps **C** and **M** are repeated until the criterion (8.23) converges.

Optimizing (8.23) is one approach to self-training, but numerous other criteria have been proposed. These criteria aim to better utilize unlabeled data, enhance classifier robustness, or address issues like class imbalance and pseudo-label noise [4]. Each criterion has its own advantages and limitations, and the choice of criterion can significantly affect the performance and convergence of the self-training algorithm.

8.3.1.1 Pseudo-labeling Strategies

Pseudo-labeling is a crucial component of self-training methods, where a portion of unlabeled data is selected for pseudo-labeling to avoid overfitting to the initial classifier. In the following we will review different techniques that have been proposed to determine the subset of examples to pseudo-label, each with its strengths and weaknesses.

Threshold-Based Methods

The underlying hypothesis of threshold-based methods is that the decision boundary passes through low-density regions where classification errors can be made. In this case, the classifier learned at each step is supposed to make the majority of its mistakes on observations close to the decision boundary.

In the case of binary classification, preliminary research suggested to assign pseudo-labels only to unlabeled observations for which the current classifier is the most confident [170]. Hence, considering thresholds θ^- and θ^+ defined for respectively the negative and the positive classes, the pseudo-labeler assigns a pseudo-label $\tilde{y}$ to an instance $\mathbf{x} \in X_{\mathcal{U}}$ such that:

$$\forall \mathbf{x}_i \in X_{\mathcal{U}}, \tilde{y}_i = \begin{cases} 1 & \text{if } h(\mathbf{x}_i) \geqslant \theta^+, \\ -1 & \text{if } h(\mathbf{x}_i) \leqslant \theta^-. \end{cases} \tag{8.24}$$

An unlabeled example $\mathbf{x}$ is pseudo-labeled only if it satisfies specific conditions defined by (8.24). These conditions ensure that the classifier is sufficiently confident in its prediction for the example to be reliably pseudo-labeled. If an unlabeled example does not meet these conditions, it remains unlabeled and is not used in the subsequent training steps.

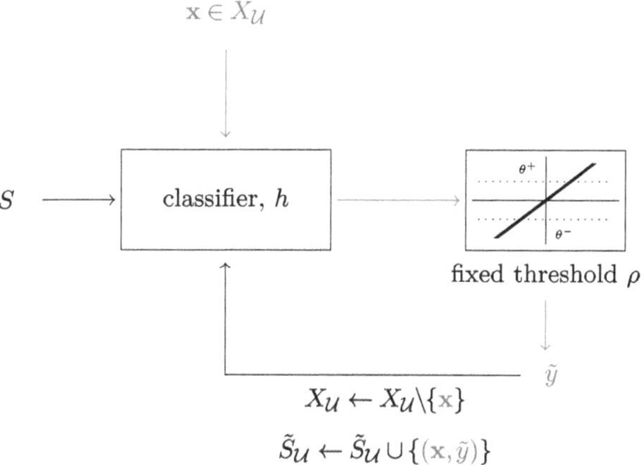

Fig. 8.3 Illustration of the self training algorithm with a margin criterion [170]. A classifier $h : \mathcal{X} \to \mathbb{R}$ is first learned on a training set labeled S. The algorithm then assigns pseudo-labels of classes to the unlabeled examples of the set $X_{\mathcal{U}}$ by thresholding the predictions of the classifier h on these examples (in gray). The pseudo-labeled examples are removed from the set $X_{\mathcal{U}}$ and added to the set $\tilde{S}_{\mathcal{U}}$ and a new classifier is learned using both sets S and $\tilde{S}_{\mathcal{U}}$. These learning and pseudo-labeling processes are repeated until convergence

With this variant of the self-training algorithm, pseudo-class labels are assigned to a portion of the unlabeled examples for which the classifier is most confident. The confidence level is determined by the thresholds θ^+ and θ^- ensuring that only the most reliable predictions are converted into pseudo-labels. This approach helps to mitigate the risk of introducing noise into the training set, which could degrade the classifier's performance.

Once the pseudo-labels are assigned, these pseudo-labeled examples are incorporated into the labeled training set. A new classifier is then trained using this augmented set of labeled data. This iterative process of training and pseudo-labeling continues until there are no more unlabeled examples that can be confidently pseudo-labeled. This process is illustrated in Fig. 8.3.

One effective way to select the thresholds for pseudo-labeling is to set them equal to the average of the positive and negative predictions, as proposed by Tur et al. [170]. This method is straightforward and intuitive, leveraging the classifier's confidence in its predictions to determine which examples should be pseudo-labeled. However, while threshold-based methods are effective in utilizing high-confidence predictions, they also present certain risks.

If the thresholds are not set appropriately, there is a risk of assigning incorrect pseudo-labels to the unlabeled examples. This can introduce noise into the training set and degrade the classifier's performance. Additionally, excessive trust in the confidence measure can be biased, especially when the labeled sample size is small. In such cases, the classifier's confidence may not accurately reflect the true reliability of its predictions, leading to suboptimal pseudo-labeling decisions. In Sect. 8.3.3, we will present an approach to automatically select the threshold for a majority vote classifier by minimizing an upper bound on the classifier's error over the unlabeled training data.

Proportion-Based Methods

Zou et al. [197] adapted the idea of pseudo-labeling with a fixed threshold [170] for multi-class classification by not choosing thresholds but rather fixing a proportion p of the most confident unlabeled data to be pseudo-labeled and then increasing this proportion at each iteration of the algorithm until $p = 0.5$ was reached. Following this idea, Cascante et al. [27] revisited the concept of pseudo-labeling by discussing the iterative process of assigning pseudo-labels to unlabeled data and emphasized the resilience of pseudo-labeling to out-of-distribution samples.

These methods are adaptive to the confidence level of the model and reduce the risk of overfitting to low-confidence predictions. However, they may include incorrect labels if the initial proportion is too high and generally require careful tuning of the proportion parameter [115].

Curriculum Learning-Based Methods

These methods use curriculum learning to pseudo-label easy-to-learn observations before moving on to more complex ones. Zhang et al. [191] proposed an adaptation of curriculum learning to pseudo-labeling, which entails in learning a model using easy-to-learn observations before moving on to more complex ones. The principle is that at the step k of the algorithm, the pseudo-labeler selects unlabeled examples having predictions that are in the $(1 - \alpha_k)$th percentile of the distribution of the maximum probability predictions assumed to follow a Pareto distribution, and where $\alpha_k \in [0, 1]$ is an hyperparameter that varies from 0 to 1 in increments of 0.2. Later on, Dai and Yang [42] proposed a method that combines CLIP's text-image alignment and SAM's mask generation capabilities, along with a multi-source curriculum learning strategy to address noise and excessive focus issues, gradually improving semantic alignment and segmentation precision.

These approaches gradually increase the complexity of pseudo-labeled data and reduce the risk of incorrect labels in early iterations. However they require careful tuning of the hyperparameter α_k and may not be effective if the distribution of predictions is not well-behaved.

8.3.1.2 Theoretical Study and a Variant

The generalization properties of a self-training algorithm have been explored, particularly when the base classifier is a linear classifier [72]. This approach iteratively learns a list of linear classifiers from labeled and unlabeled training data through exploration and pruning phases. During the exploration phase, a classifier is identified by maximizing the unsigned margin among unlabeled examples, and pseudo-labels are assigned to those examples with distances exceeding the current threshold. These pseudo-labels may contain noise.

The training set is then augmented with these noisy pseudo-labeled examples, and a new classifier is trained. This process continues until no unlabeled examples remain for pseudo-labeling. In the pruning phase, pseudo-labeled examples with distances to the last hyperplane greater than the associated unsigned margin are discarded.

The authors demonstrate that the misclassification error of the resulting sequence of classifiers is bounded, ensuring that the semi-supervised approach does not degrade performance compared to a classifier trained solely on the initial labeled training set.

A variant of the self-training algorithm involves performing the pseudo-labeling and classifier training steps just once on both labeled and pseudo-labeled examples. This variant is known as the one-pass self-training algorithm. The consistency of this algorithm has been studied when the pseudo-labeling step is carried out using the 1-nearest neighbor algorithm, which assigns the same class label to the unlabeled example that is closest, in terms of Euclidean distance, to a labeled example in the training set [172].

8.3.2 *Transductive Support Vector Machines*

Transductive learning, introduced by Vapnik [177], aims to produce a prediction function for only a fixed number of unlabeled examples given $X_{\mathcal{U}} = \{\mathbf{x}_i \mid i = m{+}1, \dots, m{+}u\}$. This framework is motivated by the fact that for some applications it is not necessary to learn a general rule as in the inductive case [36, 120], but just to accurately predict the outputs of examples from an unlabeled or test base.[2] The transductive error defined for this type of learning is thus the average number of predicted outputs $\{\tilde{y}_i \mid \mathbf{x}_i \in X_{\mathcal{U}}, i \in \{m+1, \dots, m+u\}\}$ different from the true outputs of the unlabeled examples $\{y_i \mid i \in \{m+1, \dots, m+u\}\}$:

$$R_u(\mathfrak{A}_T) = \frac{1}{u} \sum_{i=m+1}^{m+u} \mathbb{1}_{y_i \neq \tilde{y}_i}, \tag{8.25}$$

where $\mathfrak{A}_T : \mathcal{X} \to \mathcal{Y}$ is the transductive algorithm that predicts the output $\tilde{y}_i \in \mathcal{Y}$ for the example $\mathbf{x}_i \in X_{\mathcal{U}}$ of the unlabeled basis. Since the size of this set is finite, the class of functions considered, $\mathcal{F} = \{-1, +1\}^{m+u}$, to find the transductive prediction function $\mathfrak{A}_T$ is also finite [47]. According to the principle of structural

[2] Since these unlabeled examples are known a priori and are observed by the learning algorithm during the training phase, transductive learning is a special case of semi-supervised learning.

risk minimization (Chap. 1, Sect. 1.3.3), $\mathcal{F}$ can be defined according to a nested structure:

$$\mathcal{F}_1 \subset \mathcal{F}_2 \subset \ldots \subset \mathcal{F} = \{-1, +1\}^{m+u}. \tag{8.26}$$

This structure normally reflects a priori knowledge about the learning problem to be addressed and it must be constructed in such a way that with a high probability, the correct prediction of the class labels of the labeled and unlabeled training examples is contained in a class of functions $\mathcal{F}_k$ of small size. In particular, [47] show that, for a $\delta \in (0, 1)$, we have:

$$\mathbb{P}\left(R_u(\mathfrak{A}_T) \leqslant R_m(\mathfrak{A}_T) + \Gamma(m, u, |\mathcal{F}_k|, \delta)\right) \geqslant 1 - \delta, \tag{8.27}$$

where $R_m(\mathfrak{A}_T)$ is the transductive error (or empirical classification error) on the training set and $\Gamma(m, u, |\mathcal{F}_k|, \delta)$ is a complexity term depending on the number of labeled examples m, the number of unlabeled examples u, and the size $|\mathcal{F}_k|$ of the function class $\mathcal{F}_k$. To find the best function class among the existing ones, transductive algorithms generally use the distribution of the unsigned margins of the unlabeled examples to guide the search for the prediction function by forcing it to respect the clustering of the data according to the assumptions of semi-supervised learning (Sect. 8.1.3). According to the SRM principle, the learning algorithm should choose the labeling of the unlabeled examples $\{\tilde{y}_i \mid \mathbf{x}_i \in X_{\mathcal{U}}\} \in \mathcal{F}_k$ for which the empirical classification error $R_m(\mathfrak{A}_T)$ and the size of the function class $|\mathcal{F}_k|$ minimize the bound (8.27). In the special case where the empirical error is zero, minimizing the bound would be equivalent to finding the labeling of the unlabeled examples with the largest margin.

Transductive support vector machines (TSVM) arise from this paradigm and find the best hyperplane in the feature space, in the sense of the margin, that best separates the labeled examples and does not cross high-density regions. To do this, TSVM build a structure on the class of functions $\mathcal{F}$ by ordering all the outputs of the unlabeled examples with respect to their margins. The first optimization problem associated with this paradigm, known as hard-margin TSVM, is the following [177]:

$$\min_{\bar{\boldsymbol{w}}, w_0, \tilde{y}_{m+1}, \ldots, \tilde{y}_{m+u}} \frac{1}{2}||\bar{\boldsymbol{w}}||^2 \tag{8.28}$$

$$\begin{aligned} \text{subject to } & \forall i \in \{1, \ldots, m\},\ y_i\left(\langle \bar{\boldsymbol{w}}, \mathbf{x}_i\rangle + w_0\right) - 1 \geqslant 0 \\ & \forall i \in \{m+1, \ldots, m+u\},\ \tilde{y}_i\left(\langle \bar{\boldsymbol{w}}, \mathbf{x}_i\rangle + w_0\right) - 1 \geqslant 0 \\ & \forall i \in \{m+1, \ldots, m+u\},\ \tilde{y}_i \in \{-1, +1\}. \end{aligned} \tag{8.29}$$

The solutions to this problem are the labels $\tilde{y}_{m+1}, \ldots, \tilde{y}_{m+u}$ of the unlabeled examples for which the hyperplane with parameters $\{\bar{\boldsymbol{w}}, w_0\}$ separates the examples of the two labeled and unlabeled training sets with the largest margin. Figure 8.4 illustrates the solutions found by the SVM and TSVM algorithms on a toy problem.

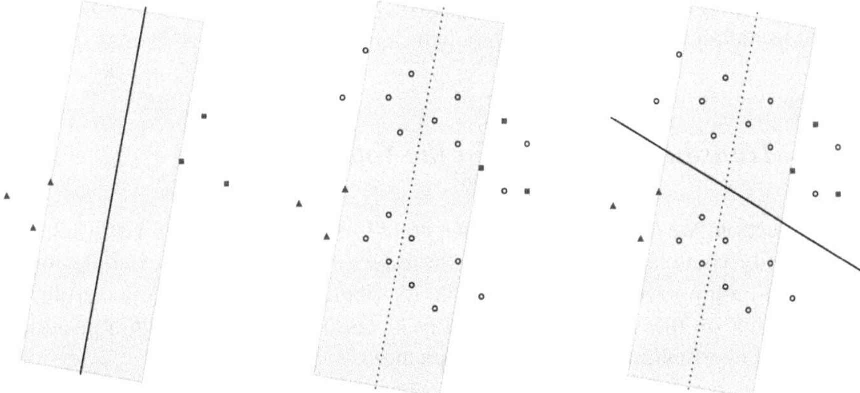

Fig. 8.4 Illustration of the hyperplanes found by the SVM (left) and TSVM (right) algorithms, for a binary classification problem, labeled examples are represented by circles and squares and unlabeled examples by asterisks. The SVM algorithm finds the separating hyperplane, ignoring the unlabeled examples, while the TSVM finds the solution that separates the two classes by not crossing through the dense regions

The fundamental difference between these two algorithms is that the SVM also finds a hyperplane with a larger margin, but which just separates the examples of the labeled training set without taking into account the presence of the unlabeled examples. By introducing slack variables associated with the labeled and unlabeled examples, [91] extended the optimization problem (8.28) to the non-separable case:

$$\min_{\bar{\boldsymbol{w}}, w_0, \tilde{y}_{m+1}, \ldots, \tilde{y}_{m+u}, \xi_1, \ldots, \xi_m, \tilde{\xi}_1, \ldots, \tilde{\xi}_u} \frac{1}{2}||\bar{\boldsymbol{w}}||^2 + C\sum_{i=1}^{m} \xi_i + \tilde{C}\sum_{i=1}^{u} \tilde{\xi}_i \tag{8.30}$$

$$\begin{aligned} \text{s.t. } & \forall i \in \{1, \ldots, m\}, y_i\left(\langle \bar{\boldsymbol{w}}, \mathbf{x}_i \rangle + w_0\right) \geqslant 1 - \xi_i \\ & \forall i \in \{m+1, \ldots, m+u\}, \tilde{y}_i\left(\langle \bar{\boldsymbol{w}}, \mathbf{x}_i \rangle + w_0\right) \geqslant 1 - \tilde{\xi}_{i-m} \\ & \forall i \in \{m+1, \ldots, m+u\}, \tilde{y}_i \in \{-1, +1\} \qquad (8.31) \\ & \forall i \in \{1, \ldots, m\}, \xi_i \geqslant 0 \\ & \forall i \in \{1, \ldots, u\}, \tilde{\xi}_i \geqslant 0. \end{aligned}$$

For both problems (8.28) and (8.30), the pseudo-labels of the unlabeled examples are taken as integers and according to the constraints (8.29) and (8.31), both problems are written as optimization problems with quadratic objective functions and linear constraints. Currently, there is no efficient algorithm to find a global optimal solution to these problems. However, an approximate optimization of these problems is proposed by [89, 91] with a coordinate-by-coordinate descent optimization method [109], which consists in alternately choosing a coordinate at each iteration and minimizing the objective function along this direction with a simple technique such as line search (Chap. 3, Sect. 3.3).

8.3.3 Transductive Bounds for the Voted Classifier

In this section, we will present another bound on the transductive risk (8.25) of the majority vote classifier also called the Bayes classifier, on the examples of an unlabeled training set by considering the distribution of the unsigned margins of this classifier on these examples. The Bayes classifier is defined with a posteriori probability distribution Q on a class of functions $\mathcal{F}$ by:

$$\begin{aligned} B_Q : \mathcal{X} &\to \{-1, +1\} \\ \mathbf{x} &\mapsto \text{sign}\left[\mathbb{E}_{f\sim Q} f(\mathbf{x})\right]. \end{aligned} \tag{8.32}$$

B_Q thus translates the majority vote of the classifiers of $\mathcal{F}$, with their associated weights, on the class label of an example. The voting classifier with normalized weights from the AdaBoost algorithm (presented in Chap. 6, Sect. 6.1) is an example.

The transductive risk of the Bayes classifier on all the examples of an unlabeled training base $X_{\mathcal{U}}$ (8.25) is thus defined by:

$$R_u(B_Q) = \frac{1}{u} \sum_{\mathbf{x} \in X_{\mathcal{U}}} \mathbb{1}_{B_Q(\mathbf{x}) \neq y}. \tag{8.33}$$

Similarly, we define the associated Gibbs classifier, G_Q, as the prediction function obtained with a random draw from the set $\mathcal{F}$ following the distribution Q and its transductive risk is also:

$$R_u(G_Q) = \frac{1}{u} \sum_{\mathbf{x} \in X_{\mathcal{U}}} \mathbb{E}_{f\sim Q}\left[\mathbb{1}_{f(\mathbf{x}) \neq y}\right]. \tag{8.34}$$

Considering the distribution of unsigned margins of unlabeled examples:

$$\forall \mathbf{x} \in \mathcal{X}, m_Q(\mathbf{x}) = \left|\mathbb{E}_{f\sim Q} f(\mathbf{x})\right|. \tag{8.35}$$

We can bound the transductive risk of the Bayes classifier committed on examples having an unsigned margin above a given threshold ρ:

$$R_{u\wedge\rho}(B_Q) = \frac{1}{u} \sum_{\mathbf{x} \in X_{\mathcal{U}}} \mathbb{1}_{B_Q(\mathbf{x}) \neq y \wedge m_Q(\mathbf{x}) > \rho}. \tag{8.36}$$

Theorem 8.5 (Transductive Risk Bound for Bayes Classifier with Margin Threshold [3]) *Let the Bayes classifier B_Q be defined as in (8.32). For all Q, all $\rho \geqslant 0$ and all $\delta \in (0, 1]$ and with a probability at least equal to $1-\delta$, we have:*

$$R_{u\wedge\rho}(B_Q) \leqslant \inf_{\gamma\in(\rho,1]} \left\{ \mathbb{P}_u(\rho < m_Q(\mathbf{x}) < \gamma) + \frac{1}{\gamma} \left\lfloor K_u^\delta(Q) + M_Q^{\leqslant}(\rho) - M_Q^{<}(\gamma) \right\rfloor_+ \right\}, \tag{8.37}$$

where $\mathbb{P}_u(C)$ is the fraction of unlabeled examples satisfying condition C, $K_u^\delta(Q) = R_u^\delta(G_Q) + \frac{1}{2}(\mathbb{E}_u\left[m_Q(\mathbf{x})\right] - 1)$, $M_Q^{\lhd}(z) = \mathbb{E}_u\left[m_Q(\mathbf{x})\mathbb{1}_{m_Q(\mathbf{x})\lhd z}\right]$ with $\lhd$ being $<$ or $\leqslant$ and $\lfloor z \rfloor_+ = \mathbb{1}_{z>0}z$.

Proof Let be the set of unsigned margins of unlabeled data of size N:

$$\{\gamma_i, i = 1, \ldots, N\} = \{m_Q(\mathbf{x}) \mid \mathbf{x} \in X_{\mathcal{U}} \wedge m_Q(\mathbf{x}) > 0\}.$$

Let $k = \max\{i \mid \gamma_i \leqslant \rho\}$, the largest index of margin smaller than ρ, and $b_i = \mathbb{P}_u(B_Q(\mathbf{x}) \neq y \wedge m_Q(\mathbf{x}) = \gamma_i), \forall i \in \{1, \ldots, N\}$, the fraction of unlabeled examples misclassified by the Bayes classifier and with unsigned margins equal to γ_i. The transductive risk of the Bayes classifier committed on examples with an unsigned margin above a certain threshold ρ (8.36) is then expressed as:

$$\forall \rho \in [0, 1], R_{u\wedge\rho}(B_Q) = \sum_{i=k+1}^{N} b_i. \tag{8.38}$$

By the definition of the unsigned margin and the fact that the functions $f \in \mathcal{F}$ have values in $\{-1, +1\}$, we have:

$$\forall \mathbf{x} \in X_{\mathcal{U}}, m_Q(\mathbf{x}) = |\mathbb{E}_{f\sim Q} f(\mathbf{x})| = |1 - 2\mathbb{E}_{f\sim Q}[\mathbb{1}_{f(\mathbf{x})\neq y}]|. \tag{8.39}$$

Moreover, the Bayes classifier misclassifies an unlabeled example $\mathbf{x} \in X_{\mathcal{U}}$ if and only if $\mathbb{E}_{f\sim Q}[\mathbb{1}_{f(\mathbf{x})\neq y}] > \frac{1}{2}$, that is, according to (8.39):

$$\mathbb{E}_{f\sim Q}[\mathbb{1}_{f(\mathbf{x})\neq y}] = \frac{1}{2}(1 + m_Q(\mathbf{x}))\mathbb{1}_{B_Q(\mathbf{x})\neq y} + \frac{1}{2}(1 - m_Q(\mathbf{x}))\mathbb{1}_{B_Q(\mathbf{x})=y}. \tag{8.40}$$

By averaging the previous equation over all unlabeled examples, the risk of the Gibbs classifier (8.34) can be expressed with respect to b_i and γ_i:

$$
\begin{aligned}
R_u(G_Q) &= \frac{1}{u}\sum_{\mathbf{x}\in X_{\mathcal{U}}} m_Q(\mathbf{x})\mathbb{1}_{B_Q(\mathbf{x})\neq y} + \frac{1}{2}(1-\mathbb{E}_u[m_Q(\mathbf{x})]) \\
&= \sum_{i=1}^{N} b_i\gamma_i + \frac{1}{2}(1-\mathbb{E}_u[m_Q(\mathbf{x})]).
\end{aligned} \tag{8.41}
$$

Suppose we have a bound $R_u^{\delta}(G_Q)$ on the error of the Gibbs classifier $R_u(G_Q)$ that holds with probability at least $1-\delta$. From Eqs. (8.38) and (8.41), we can establish, with a worst-case consideration, a bound on the joint risk:

$$
R_{u\wedge\rho}(B_Q) \leqslant \max_{b_1,\dots,b_N} \sum_{i=k+1}^{N} b_i \tag{8.42}
$$

$$
\text{s.t. } \forall i,\ 0\leqslant b_i\leqslant \mathbb{P}_u(m_Q(\mathbf{x})=\gamma_i) \text{ and } \sum_{i=1}^{N} b_i\gamma_i \leqslant \underbrace{R_u^{\delta}(G_Q)-\frac{1}{2}(1-\mathbb{E}_u[m_Q(\mathbf{x})])}_{K_u^{\delta}(Q)}.
$$

This is a linear optimization problem (LP) [43] and the solution of the maximum of the linear function $\sum_{i=k+1}^{N} b_i$ on a convex polyhedron, as detailed in Exercise 5. It can be expressed as follows.

$$
b_i = \begin{cases} 0, & \text{if } i\leqslant k \\ \min\left(\mathbb{P}_u(m_Q(\mathbf{x})=\gamma_i), \left\lfloor \frac{K_u^{\delta}(Q)-\sum_{k<j<i}\gamma_j\mathbb{P}_u(m_Q(\mathbf{x})=\gamma_j)}{\gamma_i}\right\rfloor_+\right) & \text{,otherwise.} \end{cases}
$$

Taking $I=\max\left\{i \mid K_u^{\delta}(Q)-\sum_{k<j<i}\gamma_j\mathbb{P}_u(m_Q(\mathbf{x})=\gamma_j)>0\right\}$, we get:

- If $i<I, b_i=\mathbb{P}_u(m_Q(\mathbf{x})=\gamma_i)$.
- If $i=I, b_I=\frac{K_u^{\delta}(Q)-\sum_{k<j<I}\gamma_j\mathbb{P}_u(m_Q(\mathbf{x})=\gamma_j)}{\gamma_I}$.
- If $i>I, b_i=0$.

Hence:

$$
R_{u\wedge\rho}(B_Q) \leqslant \mathbb{P}_u(\rho<m_Q(\mathbf{x})<\gamma_I) + \frac{K_u^{\delta}(Q)+M_Q^{\leqslant}(\rho)-M_Q^{<}(\gamma_I)}{\gamma_I}, \tag{8.43}
$$

where $M_Q^{<}(\gamma_i) = \sum_{j=1}^{j<i} \gamma_j \mathbb{P}_u(m_Q(\mathbf{x}) = \gamma_j)$. To complete the demonstration, we note that the following function is decreasing for $\gamma < \gamma_I$:

$$\Psi : \gamma \mapsto \mathbb{P}_u(\theta < m_Q(\mathbf{x}) < \gamma) + \frac{K_u^{\delta}(Q) + M_Q^{\leqslant}(\rho) - M_Q^{<}(\gamma)}{\gamma}.$$

□

In the case where $\rho = 0$ and as $M_Q^{\leqslant}(0) = 0$ and that:

$$R_u(B_Q) = R_{u\wedge 0}(B_Q) + \mathbb{P}_u(B_Q(\mathbf{x}) \neq y \wedge m_Q(\mathbf{x}) = 0) \leqslant R_{u\wedge 0}(B_Q) + \mathbb{P}_u(m_Q(\mathbf{x}) = 0),$$

we have the following corollary which gives a bound on the transductive error of the Bayes classifier (8.33).

Corollary 8.6 *Let the Bayes classifier B_Q be defined as in (8.32). For any Q, any $\delta \in (0, 1]$ and with probability at least equal to $1 - \delta$, we have:*

$$R_u(B_Q) \leqslant \inf_{\gamma \in (\rho, 1]} \left\{ \mathbb{P}_u(m_Q(\mathbf{x}) < \gamma) + \frac{1}{\gamma} \left\lfloor K_u^{\delta}(Q) - M_Q^{<}(\gamma) \right\rfloor_+ \right\}. \tag{8.44}$$

$K_u^{\delta}(Q) = R_u^{\delta}(G_Q) + \frac{1}{2}(\mathbb{E}_u\left[m_Q(\mathbf{x})\right] - 1)$, $M_Q^{<}(z) = \mathbb{E}_u\left[m_Q(\mathbf{x})\mathbb{1}_{m_Q(\mathbf{x})<z}\right]$ *and* $\lfloor z \rfloor_+ = \mathbb{1}_{z>0} z$.

The interesting result of this bound is that, in the case where we consider the margin as a confidence indicator, it is then possible to have a good estimate of the transductive error of the Bayes classifier following the cluster assumption mentioned in Sect. 8.1.3. This result is stated in the following proposition.

Proposition 8.7 *Suppose that $\forall \mathbf{x} \in X_{\mathcal{U}}, m_Q(\mathbf{x}) > 0$, $\delta \in (0, 1]$ and that $\exists L \in (0, 1]$ such that $\forall \gamma > 0$:*

$$\mathbb{P}_u(B_Q(\mathbf{x}) \neq y \wedge m_Q(\mathbf{x}) = \gamma) \neq 0 \Rightarrow \mathbb{P}_u(B_Q(\mathbf{x}) \neq y \wedge m_Q(\mathbf{x}) < \gamma) \geqslant L \mathbb{P}_u(m_Q(\mathbf{x}) < \gamma).$$

We then have with a probability at least equal to $1 - \delta$:

$$F_u^{\delta}(Q) - R_u(B_Q) \leqslant \frac{1 - L}{L} R_u(B_Q) + \frac{R_u^{\delta}(G_Q) - R_u(G_Q)}{\gamma^*}, \tag{8.45}$$

(continued)

Proposition 8.7 (continued)
where $\gamma^* = \sup\{\gamma \mid \mathbb{P}_u(B_Q(\mathbf{x}) \neq y \wedge m_Q(\mathbf{x}) = \gamma) \neq 0\}$ *and* $F_u^\delta(Q) = \inf_{\gamma \in (\rho,1]} \left\{\mathbb{P}_u(m_Q(\mathbf{x}) < \gamma) + \frac{1}{\gamma}\left\lfloor K_u^\delta(Q) - M_Q^<(\gamma)\right\rfloor_+\right\}$ *is the bound on the transductive error of the Bayes classifier (8.44).*

Proof Suppose that:

$$R_u(B_Q) \geqslant \mathbb{P}_u(B_Q(\mathbf{x}) \neq y \wedge m_Q(\mathbf{x}) < \gamma^*) + \frac{1}{\gamma^*}\left\lfloor K_u(Q) - M_Q^<(\gamma^*)\right\rfloor_+, \tag{8.46}$$

where γ^* is the largest margin on which the Bayes classifier makes an error ; $\gamma^* = \sup\{\gamma \mid \mathbb{P}_u(B_Q(\mathbf{x}) \neq y \wedge m_Q(\mathbf{x}) = \gamma) \neq 0\}$ and $K_u(Q) = R_u(G_Q) + \frac{1}{2}(\mathbb{E}_u\left[m_Q(\mathbf{x})\right] - 1)$. Using the hypothesis of the proposition:

$$R_u(B_Q) \geqslant L\mathbb{P}_u(B_Q(\mathbf{x}) < \gamma^*) + \frac{1}{\gamma^*}\left\lfloor K_u(Q) - M_Q^<(\gamma^*)\right\rfloor_+. \tag{8.47}$$

For $\delta \in (0,1]$, we have by definition the following inequality $F_u^\delta(Q) \leqslant \mathbb{P}_u(B_Q(\mathbf{x}) < \gamma^*) + \frac{1}{\gamma^*}\left\lfloor K_u(Q) - M_Q^<(\gamma^*)\right\rfloor_+$ which holds with probability at least $1-\delta$. So, with the same probability, we have:

$$F_u^\delta(Q) - R_u(B_Q) \leqslant (1-L)\mathbb{P}_u(m_Q(\mathbf{x}) < \gamma^*) + \frac{R_u^\delta(G_Q) - R_u(G_Q)}{\gamma^*}. \tag{8.48}$$

This is because

$$\left\lfloor K_u^\delta(Q) - M_Q^<(\gamma^*)\right\rfloor_+ - \left\lfloor K_u(Q) - M_Q^<(\gamma^*)\right\rfloor_+ \leqslant R_u^\delta(G_Q) - R_u(G_Q).$$

From the inequality (8.47), we also have $\mathbb{P}_u(B_Q(\mathbf{x}) < \gamma^*) \leqslant \frac{1}{L}R_u(B_Q)$, which, replacing it in (8.47), finally gives the result. □

The interpretation of the bound (8.45) is that, following the cluster assumption, the hyperplane induced by the Bayes classifier does not cross high-density regions and the classifier thus makes most of its errors in low-margin regions. In this case, the constant B will be close to 1 and we can derive a very good estimate of the error of the Bayes classifier (8.32). Feofanov et al. [53, 53, 54] extended the transductive bound (Theorem 8.5) to the multiclass classification setting.

8.3.4 Multi-view Learning Based on Pseudo-labeling

For applications where data are produced by multiple information sources, for example images described by multiple visual or textual descriptors, or documents written in multiple languages, pseudo-labeling of examples could be done based on the predictions provided by the classifiers associated with each of the sources. Each representation of an example corresponding to an information source characterizes a view of the example and the paradigm that is based on the exploitation of these views to learn the classifiers is called multi-view learning.

The three families of methods developed following this concept use the redundancy in the different data representations and attempt to reduce the disagreement between the class predictions of the view-specific classifiers, to improve their performances. These algorithms achieve this goal either by projecting the view-specific representations into a common canonical space [10], or by constraining the classifiers to have similar outputs by adding a term estimating the disagreement in their objective functions [156], or by pseudo-labeling the unlabeled examples based on the output of each of these classifiers [16].

This latter approach was initiated in the seminal work of Blum et Mitchell [16], who proposed the co-training algorithm for multi-view two-modality problems. This model assumes that each representation is rich enough to learn the parameters of both classifiers in the case where enough labeled examples are available. The two classifiers are first trained separately on the labeled data. We then randomly draw a portion of the examples from the unlabeled base $X_{\mathcal{U}}$, which are labeled by each of the two classifiers, the output estimated by the first classifier serving as the desired output for the second classifier and vice versa. The implementation that is proposed is quite simple and is similar to the directed decision algorithm, with the difference that we have here two classifiers, each alternately providing the labels corresponding to the desired outputs of the other classifier. Blum et Mitchell [16] used this algorithm to classify web pages into two relevant/irrelevant classes with respect to a given task. We start with two different representations for each page, namely the bag of words contained in it and the bag of words contained in the hyperlink pointing to the page, and the classifiers used in the algorithm are Naive Bayes models that rely on each of these representations.

The effectiveness of this algorithm has been proven in this context of web page classification, but the algorithm has a number of limitations. First, the fact of not pseudo-labeling all the unlabeled examples, but just a part, forces the two classifiers to, slowly, agree on the output labels that each of them should have. However, in later studies, it has been shown that the agreement between the outputs of the different classifiers associated with the views is the basis of the multi-view learning paradigm [105, 156]. In particular, Leskes [105] proved that by constraining the classifiers associated with the different views to have similar outputs on the unlabeled examples, it is possible to restrict the search spaces in the classes of associated functions and thus to have a better estimate of the generalization bound

Input : A learning algorithm: $\mathfrak{A}$;
Labeled S and unlabeled $X_{\mathcal{U}}$ training sets;
Initialization: $t \leftarrow 0$; $\tilde{S}_U \leftarrow \emptyset$; For each view v, train a classifier $h_v^{(0)}$ on S with Algorithm $\mathfrak{A}$.
repeat
  $U \leftarrow \emptyset$;
  Estimate the parameters of the margin $(\rho_v^{(t)})_{v=1}^V$;
  for $\mathbf{x} = (x^1, \dots, x^V) \in X_{\mathcal{U}}$ **do**
    $\forall v, \tilde{y}_v^{(t)} \leftarrow \begin{cases} \text{sign}(h_v^{(t)}) \text{ if } h_v^{(t)}(x^v) > \rho_v^{(t)} \\ 0 \text{ else} \end{cases}$
    if $\sum_{v=1}^{V} \tilde{y}_v^{(t)} > \frac{V}{2}$ **then**
      $\tilde{S}_U \leftarrow \tilde{S}_U \cup \{(\mathbf{x}, +1)\}$;
      $U \leftarrow U \cup \{\mathbf{x}\}$;
    else
      if $\sum_{v=1}^{V} \tilde{y}_v^{(t)} < -\frac{V}{2}$ **then**
        $\tilde{S}_U \leftarrow \tilde{S}_U \cup \{(\mathbf{x}, -1)\}$;
        $U \leftarrow U \cup \{\mathbf{x}\}$;
  $X_{\mathcal{U}} \leftarrow X_{\mathcal{U}} \setminus U$;
  $t \leftarrow t + 1$;
  For each view v, train the classifier $h_v^{(t)}$ over $S \cup \tilde{S}_U$ with Algorithm $\mathfrak{A}$;
until $U = \emptyset \vee X_{\mathcal{U}} = \emptyset$;
output : The classifiers $(h_v^{(t)})_{v=1}^V$.

Algorithme 27: Multi-view Self-training Algorithm

for each of these functions by decreasing the class complexity terms involved in these bounds.

Algorithm 27 is a simple extension of the directed decision algorithm with automatic threshold selection to the multiview case that implements this idea. In this case, the classifiers associated with each view are first trained on the labeled examples of the training set and pseudo-labeling thresholds are automatically estimated based on their unsigned margin distributions computed on the unlabeled examples. Each classifier then assigns pseudo-class labels to the unlabeled examples based on its threshold computed in the previous step and pseudo-labels are assigned to these examples based on the majority vote. New classifiers are then trained based on the labeled and pseudo-labeled examples and this pseudo-labeling and training process is repeated until there are no more examples to pseudo-label as in the directed decision algorithm.

8.4 Graphical Models

We have seen that the generative and discriminative methods proposed in semi-supervised learning exploit the geometry of the data through density estimation techniques or based on the unsigned margin of the examples. The last family of semi-supervised methods uses an empirical graph $G = (V, E)$ built on the labeled and unlabeled examples to express their geometry.

The nodes $V = 1, \ldots, m+u$ of this graph represent the training examples and the edges E translate the similarities between the examples. These similarities are usually given by a positive symmetric matrix $\boldsymbol{W} = [W_{ij}]_{i,j}$, where $\forall (i, j) \in \{1, \ldots, m+u\}^2$ the weight W_{ij} is non-zero if and only if the examples of indices i and j are connected, or if $(i, j) \in E \times E$ is an edge of the graph G. The two examples of similarity matrices commonly used in the literature are:

- The binary matrix of k-nearest neighbors:

$$\forall (i, j) \in \{1, \ldots, m+u\}^2;\ W_{ij} = 1 \text{ if and only if the example } \mathbf{x}_i \text{ is among the } k\text{-nearest neighbors of the example } \mathbf{x}_j.$$

- The Gaussian similarity matrix with parameter σ:

$$\forall (i, j) \in \{1, \ldots, m+u\}^2;\ W_{ij} = e^{-\frac{\|\mathbf{x}_i - \mathbf{x}_j\|^2}{2\sigma^2}}. \tag{8.49}$$

By convention, $W_{ii} = 0$. In the remainder of this section, we will present a set of semi-supervised techniques based on label propagation in graphs.

8.4.1 Label Propagation

A simple idea to take advantage of G built on examples is to propagate the labels of the examples across the graph. Nodes $1, \ldots, m$ associated with labeled examples are assigned class labels, $+1$ or -1, of these examples; and nodes associated with unlabeled examples $m+1, \ldots, m+u$ associated with unlabeled examples are assigned the label 0. The algorithms proposed following this framework, called *label propagation* algorithms, are quite similar and they propagate the label of each node of the graph to its neighbors [194–196]. The goal of these algorithms is that the found labels, $\tilde{Y} = (\tilde{Y}_m, \tilde{Y}_u)$, are consistent with the class labels of the labeled examples, $Y_m = (y_1, \ldots, y_m)$, and also with the geometry of the data induced by the structure of the graph G and expressed by the matrix $\boldsymbol{W}$.

The consistency between the initial labels of the labeled examples, Y_m, and the estimated labels for these examples, $\tilde{Y}_m$, is measured by:

$$\sum_{i=1}^{m}(\tilde{y}_i - y_i)^2 = \|\tilde{Y}_m - Y_m\|^2$$
$$= \|\boldsymbol{S}\tilde{Y} - \boldsymbol{S}Y\|^2, \tag{8.50}$$

where $\boldsymbol{S}$ is the diagonal block matrix with its first m diagonal elements equal to 1 and its other elements all zero.

The consistency with the geometry of the examples, on the other hand, follows the continuity (or variety) hypothesis. It consists in penalizing rapid changes in $\tilde{Y}$ between close examples, given the matrix $\boldsymbol{W}$. The latter is thus measured by:

$$\frac{1}{2}\sum_{i=1}^{m+u}\sum_{j=1}^{m+u} W_{ij}(\tilde{y}_i - \tilde{y}_j)^2 = \frac{1}{2}\left(2\sum_{i=1}^{m+u}\tilde{y}_i^2\sum_{j=1}^{m+u} W_{ij} - 2\sum_{i,j=1}^{m+u} W_{ij}\tilde{y}_i\tilde{y}_j\right)$$
$$= \tilde{Y}(\boldsymbol{D}\ominus\boldsymbol{W})\tilde{Y}, \tag{8.51}$$

where $\boldsymbol{D} = [D_{ij}]$ is the diagonal matrix defined by $D_{ii} = \sum_{j=1}^{m+u} W_{ij}$, $\ominus$ represents the term-by-term matrix subtraction and $(\boldsymbol{D}\ominus\boldsymbol{W})$ is called the unnormalized Laplacian matrix.

The objective function considered thus expresses a compromise between these two terms (8.50) and (8.51):

$$\Delta(\tilde{Y}) = \|\boldsymbol{S}\tilde{Y} - \boldsymbol{S}Y\|^2 + \lambda\tilde{Y}(\boldsymbol{D}\ominus\boldsymbol{W})\tilde{Y}, \tag{8.52}$$

where $\lambda \in (0, 1)$ modulates this trade-off. The derivative of the objective function is thus:

$$\frac{\partial\Delta(\tilde{Y})}{\partial\tilde{Y}} = 2\left[\boldsymbol{S}(\tilde{Y} - Y) + \lambda(\boldsymbol{D}\ominus\boldsymbol{W})\tilde{Y}\right]$$
$$= 2\left[(\boldsymbol{S}\oplus\lambda(\boldsymbol{D}\ominus\boldsymbol{W}))\,\tilde{Y} - \boldsymbol{S}Y\right],$$

where $\oplus$ is the term-by-term matrix addition. Furthermore, the Hessian matrix of the objective function:

$$\frac{\partial^2\Delta(\tilde{Y})}{\partial\tilde{Y}\partial\tilde{Y}^\top} = 2\left(\boldsymbol{S}\oplus\lambda(\boldsymbol{D}\ominus\boldsymbol{W})\right),$$

Input : Labeled S and unlabeled $X_{\mathcal{U}}$ training sets;
Initialization:

- $t \leftarrow 0$;
- Estimate the similarity matrix $\boldsymbol{W}$ (8.49) (for $i \neq j$, $W_{ii} \leftarrow 0$);
- Construct the matrix $\boldsymbol{N} \leftarrow \boldsymbol{D}^{-1/2}\boldsymbol{W}\boldsymbol{D}^{-1/2}$ where $\boldsymbol{D}$ is the diagonal matrix defined by $D_{ii} \leftarrow \sum_j W_{ij}$;
- Set $\tilde{Y}^{(0)} \leftarrow (y_1, \dots, y_m, 0, 0, \dots, 0)$;
- Choose the parameter $\alpha \in (0, 1)$;

repeat
  $\tilde{Y}^{(t+1)} \leftarrow \alpha \boldsymbol{N} \tilde{Y}^{(t)} + (1-\alpha)\tilde{Y}^{(0)}$;
  $t \leftarrow t + 1$;
until *convergence*;
output : The class labels after convergence $\tilde{Y}^* = (y_1^*, \dots, y_{m+u}^*)$. Assign the class labels to the each example $\mathbf{x}_i \in X_{\mathcal{U}}$ using the sign of y_i^*.

Algorithme 28: Label propagation for semi-supervised learning

is a positive definite matrix, which ensures that the minimum of $\Delta(\tilde{Y})$ is reached when its derivative vanishes, that is:

$$\tilde{Y}^* = (\boldsymbol{S} \oplus \lambda(\boldsymbol{D} \ominus \boldsymbol{W}))^{-1}\boldsymbol{S}Y. \tag{8.53}$$

We note that the pseudo-labels of the unlabeled examples are thus obtained by a simple matrix inversion and that this matrix depends only on the unnormalized Laplacian matrix. Other techniques based on label propagation are a variant of the previous formulation. We can notably cite the one proposed by Zhou et al. [194] which is an iterative approach (Algorithm 28) where, at each iteration, a node i of the graph receives a contribution from its neighbor j (in the form of a normalized weighting of the weight of the edge (i, j)), with in addition a small contribution from its initial label. Let the update rule be:

$$\tilde{Y}^{(t+1)} = \alpha \boldsymbol{N} \tilde{Y}^{(t)} + (1-\alpha)\tilde{Y}^{(0)}, \tag{8.54}$$

where $\tilde{Y}^{(0)} = (\underbrace{y_1, \dots, y_m}_{=Y_m}, \underbrace{0, 0, \dots, 0}_{=Y_u})$ is the vector of initial labels, $\boldsymbol{D}$ is the diagonal matrix defined by $D_{ii} = \sum_j W_{ij}$, $\boldsymbol{N} = \boldsymbol{D}^{-1/2}\boldsymbol{W}\boldsymbol{D}^{-1/2}$ is the normalized weight matrix and α is a real value in $(0, 1)$. The proof of convergence of Algorithm 28 follows the update rule (8.54). Indeed, after t iterations, we have:

$$\tilde{Y}^{(t+1)} = (\alpha \boldsymbol{N})^t \tilde{Y}^{(0)} + (1-\alpha)\sum_{l=0}^{t}(\alpha \boldsymbol{N})^l \tilde{Y}^{(0)}. \tag{8.55}$$

By definition of the diagonal matrix $\boldsymbol{D}$, the square matrix $\boldsymbol{N}$ is the normalized Laplacian matrix whose elements are positive real numbers between 0 and 1 and the sum of the elements of each of its rows is 1. The matrix $\boldsymbol{N}$ is thus by definition a stochastic matrix (or a Markov matrix) and its eigenvalues are less than or equal to 1 [100, Chapter 2]. Since α is a positive real number strictly less than 1, the eigenvalues of the matrix $\alpha\boldsymbol{N}$ are all strictly less than 1 and we have $\lim_{t\to\infty}(\alpha\boldsymbol{N})^t = 0$. Similarly, $(\alpha\boldsymbol{N})^l; l \in \mathbb{N}$ is a geometric matrix sequence with common ratio $\alpha\boldsymbol{N}$ and therefore $\lim_{t\to\infty}\sum_{l=0}^{t}(\alpha\boldsymbol{N})^l = (\boldsymbol{I} \ominus \alpha\boldsymbol{N})^{-1}$, where $\boldsymbol{I}$ is the identity matrix. According to these results, the vector of pseudo-labels of the examples $\tilde{Y}^{(t)}$ then converges to:

$$\lim_{t\to\infty}\tilde{Y}^{(t+1)} = \tilde{Y}^* = (1-\alpha)(\boldsymbol{I} \ominus \alpha\boldsymbol{N})^{-1}\tilde{Y}^{(0)}. \tag{8.56}$$

The objective function corresponding to this problem is as follows:

$$\begin{aligned}\Delta_n(\tilde{Y}) &= \|\tilde{Y} - \boldsymbol{S}Y\|^2 + \frac{\lambda}{2}\sum_{i=1}^{m+u}\sum_{j=1}^{m+u} W_{ij}\left(\frac{\tilde{y}_i}{\sqrt{D_{ii}}} - \frac{\tilde{y}_j}{\sqrt{D_{jj}}}\right)^2 \\ &= \|\tilde{Y}_m - Y_m\|^2 + \|\tilde{Y}_u\|^2 + \lambda\tilde{Y}^\top(\boldsymbol{I} \ominus \boldsymbol{N})\tilde{Y} \\ &= \|\tilde{Y}_m - Y_m\|^2 + \|\tilde{Y}_u\|^2 + \lambda\left(\boldsymbol{D}^{-1/2}\tilde{Y}\right)^\top(\boldsymbol{D} \ominus \boldsymbol{W})\left(\boldsymbol{D}^{-1/2}\tilde{Y}\right).\end{aligned} \tag{8.57}$$

Indeed, the derivative of this function with respect to the pseudo-labels $\tilde{Y}$ is:

$$\frac{\partial\Delta_n(\tilde{Y})}{\partial\tilde{Y}} = 2\left[\tilde{Y} - \boldsymbol{S}Y + \lambda\left(\tilde{Y} - \boldsymbol{N}\tilde{Y}\right)\right],$$

and it is canceled for:

$$\tilde{Y} = ((1+\lambda)\boldsymbol{I} \ominus \lambda\boldsymbol{N})^{-1}\boldsymbol{S}Y,$$

which for $\lambda = \alpha/(1-\alpha)$ is the same solution as the one found by Algorithm 28 after convergence (8.56). Compared to the criterion (8.52), the two major differences are (a) the normalized Laplacian matrix and (b) the term $\|\tilde{Y}_m - Y_m\|^2 + \|\tilde{Y}_u\|^2$ which not only constrains the pseudo-labels to match the labels of the labeled examples, but also the pseudo-labels of the unlabeled examples not to reach large values.

8.4.2 Markov Random Walks

A variant of the label propagation Algorithm 28 introduced by [163] considers a Markovian random walk[3] on the graph G defined with transition probabilities between nodes i and j estimated on the similarity matrix by:

$$\forall (i, j), p_{ij} = \frac{W_{ij}}{\sum_l W_{il}}. \tag{8.58}$$

The similarity W_{ij} is defined with a Gaussian kernel (8.49) for the neighbors of nodes i and j and is set to 0 everywhere else. The proposed algorithm first initializes the probabilities of belonging to the class $+1$ for all nodes of the graph, $\mathbb{P}(y = 1 \mid i), i \in V$ using the EM algorithm and estimates, for each example $\mathbf{x}_j$, the probability $\mathbb{P}^{(t)}(y_s = 1 \mid j)$ of starting from an example of class $y_s = 1$ to arrive at the example $\mathbf{x}_j$ after t random walks, defined as:

$$\mathbb{P}^{(t)}(y_s = 1 \mid j) = \sum_{i=1}^{m+u} \mathbb{P}(y = 1 \mid i)\mathbb{P}_{1\to t}(i \mid j). \tag{8.59}$$

where $\mathbb{P}_{1\to t}(i \mid j)$ is the probability of starting from node i and arriving at node j after t random walks. Node j is then assigned the pseudo-class label $+1$, in the case where $\mathbb{P}^{(t)}(y_s = 1 \mid j) > \frac{1}{2}$, and -1 otherwise. In practice, the choice of the value of t has a large impact on the performance of the random walk algorithm and is not easy to do. An alternative, proposed in [195, 196], is to assign a pseudo-class label to node i according to the probability of arriving at a node with label $+1$ from this node i to arrive at the labeled node, $\mathbb{P}(y_e = 1 \mid i)$, by performing a random walk. In the case where the example $\mathbf{x}_i$ is labeled, we have:

$$\mathbb{P}(y_e = 1 \mid i) = \begin{cases} 1 & \text{if } y_i = 1, \\ 0 & \text{otherwise.} \end{cases}$$

If $\mathbf{x}_i$ is unlabeled, we have the following relation:

$$\mathbb{P}(y_e = 1 \mid i) = \sum_{j=1}^{m+u} \mathbb{P}(y_e = 1 \mid j)p_{ij}, \tag{8.60}$$

where p_{ij} is the transition probability defined in (8.58).

---

[3] A random walk models systems with discrete dynamics composed of a succession of random steps [121]. The Markovian character of the process reflects the complete decorrelation between the random steps.

By setting $\forall\, i,\ \tilde{z}_i = \mathbb{P}(y_e = 1 \mid i)$ and $\tilde{Z} = (\tilde{Z}_m\ \tilde{Z}_u)$ the corresponding vector divided into two labeled and unlabeled parts, and by also subdividing the matrices $\boldsymbol{D}$ and $\boldsymbol{W}$ into four parts:

$$\boldsymbol{D} = \begin{pmatrix} \boldsymbol{D}_{mm} & 0 \\ 0 & \boldsymbol{D}_{uu} \end{pmatrix}, \qquad \boldsymbol{W} = \begin{pmatrix} \boldsymbol{W}_{mm} & \boldsymbol{W}_{mu} \\ \boldsymbol{W}_{um} & \boldsymbol{W}_{uu}. \end{pmatrix}$$

Equation (8.60) can be written as:

$$\begin{aligned} \tilde{Z}_u &= \left(\boldsymbol{D}_{uu}^{-1}\boldsymbol{W}_{um}\ \ \boldsymbol{D}_{uu}^{-1}\boldsymbol{W}_{uu}\right)\begin{pmatrix} \tilde{Z}_m \\ \tilde{Z}_u \end{pmatrix} \\ &= \boldsymbol{D}_{uu}^{-1}\left(\boldsymbol{W}_{um}\tilde{Z}_m + \boldsymbol{W}_{uu}\tilde{Z}_u\right). \end{aligned} \tag{8.61}$$

Equation (8.61) leads to the following linear system:

$$(\boldsymbol{D} \ominus \boldsymbol{W})_{uu}\tilde{Z}_u = \boldsymbol{W}_{um}\tilde{Z}_m. \tag{8.62}$$

We note that if $(\tilde{Z}_m \tilde{Z}_u)$ is a solution to the previous equation, then $(\tilde{Y}_m \tilde{Y}_u)$ is defined as:

$$\begin{aligned} \tilde{Y}_m &= 2\tilde{Z}_m - \mathbf{1}_m = Y_m \\ \tilde{Y}_u &= 2\tilde{Z}_u - \mathbf{1}_u, \end{aligned}$$

where $\mathbf{1}_m$ and $\mathbf{1}_u$ are respectively vectors of dimension m and u whose elements are all equal to 1, is also a solution. This last equality allows to express the linear system with respect to the vectors of the labels and pseudo-labels of the examples, that is:

$$\tilde{Y}_u = (\boldsymbol{D} \ominus \boldsymbol{W})_{uu}^{-1}\boldsymbol{W}_{um}Y_m.$$

The Markov Random Walk approach to label propagation offers a robust framework for semi-supervised learning, leveraging the graph structure to propagate labels effectively. By considering the geometric relationships within the data, this method enhances the accuracy of pseudo-labeling, particularly in scenarios with limited labeled data. The iterative nature of the algorithm, combined with its probabilistic foundation, ensures that the model adapts dynamically to the underlying data distribution, making it a powerful tool in various applications.

To Sum Up

1. *The Expectation-Maximization (EM) algorithm is an iterative method for finding maximum likelihood estimates in probabilistic models, especially when data is incomplete or has hidden variables. It is widely used in unsupervised learning for tasks like clustering and density estimation.*
2. *Clustering algorithms group similar data points together based on various distance metrics. These methods help in discovering the inherent structure within the data without relying on labeled examples.*
3. *The Classification Expectation-Maximization (CEM) algorithm extends the EM algorithm by explicitly defining the hidden variables as those corresponding to the clusters and incorporating a classification step.*
4. *The K-means algorithm is a special case of the CEM algorithm in which the density functions are multivariate Gaussian with uniform cluster proportions and covariance matrices that are identity matrices.*
5. *Semi-supervised learning is particularly beneficial in domains where acquiring labeled data is expensive or time-consuming, such as medical imaging [134] and natural language processing [193].*
6. *Semi-supervised learning techniques exploit the geometry of the data to learn a prediction function. These methods leverage both labeled and unlabeled data to improve performance, especially when labeled data is scarce.*
7. *Generative semi-supervised learning uses generative models to learn the underlying data distribution from both labeled and unlabeled data, enhancing the ability to classify or predict outcomes by leveraging the structure learned from the unlabeled data.*
8. *Generative approaches may perform poorly if the data do not follow the distributional assumptions on which these approaches are based.*
9. *Discriminative semi-supervised learning utilizes the geometry of the data by focusing on the unsigned margins of the examples.*
10. *Self-training is a discriminative semi-supervised technique where a model is initially trained on labeled data and then used to pseudo-label unlabeled data. This iterative process can enhance the model's performance by gradually incorporating more labeled examples.*
11. *Graph-based semi-supervised learning relies on the empirical graph constructed from the data to capture its geometric properties.*
12. *Graph-based methods extend the use of graph structures to propagate labels from a small set of labeled data points to a larger set of unlabeled points. This approach assumes that nearby nodes in the graph are likely to share the same label.*

8.5 Exercises

1. Consider a mixture of densities with K components:

$$\mathbb{P}(\mathbf{x}|\Theta) = \sum_{k=1}^{K} \pi_k \mathcal{N}(\mathbf{x}|\boldsymbol{\mu}_k, \boldsymbol{\Sigma}_k),$$

where, $\mathcal{N}$ is the normal distribution with parameters $\Theta = \{\pi_k, \boldsymbol{\mu}_k, \boldsymbol{\Sigma}_k \mid k \in \{1, \ldots, K\}\}$.

For each data point $\mathbf{x}_i$, define $\gamma_k^{(t)}(\mathbf{x}_i)$ as the posterior probability that $\mathbf{x}_i$ was generated by component k, given the current parameters $\Theta^{(t)}$:

$$\gamma_k^{(t)}(\mathbf{x}_i) = \frac{\pi_k^{(t)} \mathcal{N}(\mathbf{x}_i|\boldsymbol{\mu}_k^{(t)}, \boldsymbol{\Sigma}_k^{(t)})}{\sum_{j=1}^{K} \pi_j^{(t)} \mathcal{N}(\mathbf{x}_i|\boldsymbol{\mu}_j^{(t)}, \boldsymbol{\Sigma}_j^{(t)})}.$$

(a) Derive the **E** step (computation of posterior probabilities $\gamma_k^{(t)}(\mathbf{x}_i)$) and the **M** step updates for $\pi_k, \boldsymbol{\mu}_k, \boldsymbol{\Sigma}_k$.
(b) Explain why the **M** step for π_k involves normalizing the sum of responsibilities.
(c) Suppose $\boldsymbol{\Sigma}_k$ becomes singular during iterations (e.g., due to a cluster with a single point).

- How does this affect the likelihood computation?
- Propose a regularization strategy to avoid numerical instability.

(d) Prove that for a mixture of densities with two identical initial clusters, the EM algorithm will merge them into a single cluster.
Hint: Analyze the updates of the posteriori probabilities when $\boldsymbol{\mu}_1^{(0)} = \boldsymbol{\mu}_2^{(0)}$.

2. Assume $\mathbb{P}(X, Z \mid \Theta)$ belongs to an exponential family:

$$\mathbb{P}(X, Z \mid \Theta) = h(X, Z) \exp(\eta(\Theta)^\top T(X, Z) - A(\eta(\Theta))),$$

where $h(x)$ is the base measure, $T(X, Z)$ is the sufficient statistic, $A(\eta)$ is the log-partition (normalizer) function, and $\eta(\Theta)$ is called the natural parameter (or canonical parameter), which is a function of the original parameter Θ and appears linearly in the exponent.

(a) Show that the **E** step of the EM algorithm reduces to computing $\mathbb{E}[T(X, Z) \mid X, \Theta^{(t)}]$.
(b) Prove that the **M** step has a closed-form solution.
(c) For a Bernoulli mixture model, identify the sufficient statistics $T(X, Z)$ and derive the corresponding **M** step.

3. EM Failure Modes

(a) Construct an example where EM converges to a suboptimal local maximum (e.g., poorly separated clusters in a mixture of densities).
(b) Consider a mixture of densities where $\boldsymbol{\mu}_1 = \boldsymbol{\mu}_2$ and $\pi_1 = \pi_2$. Show that the log-likelihood is invariant to swapping component labels, leading to multiple global maxima.
(c) Assume the true data-generating distribution is a uniform mixture of three Gaussians, but the EM algorithm is applied with $K = 2$.

   - Characterize the asymptotic behavior of the estimated parameters as the number of samples tends to infinity.
   - Derive the bias in $\boldsymbol{\mu}_1$ and $\boldsymbol{\mu}_2$.

4. The evidence lower bound (ELBO) is defined as:

$$\mathcal{L}(q, \Theta) = \mathbb{E}_q[\log \mathbb{P}(X, Z \mid \Theta)] - \mathbb{E}_q[\log q(Z)].$$

   (a) Show that EM maximizes $\mathcal{L}(q, \Theta)$ by alternating updates to $q(Z)$ (**E** step) and Θ (**M** step).
   (b) Relate the ELBO to the KL divergence $KL(q(Z)\|\mathbb{P}(X, Z \mid \Theta))$
   (c) Suppose the **M** step only partially maximizes $Q(\Theta, \Theta^{(t)})$. Prove that the ELBO still increases monotonically.

5. Consider the linear programming problem of (8.42):

$$\begin{aligned} \max_{b_1,\ldots,b_N} \quad & \sum_{i=k+1}^{N} b_i \\ \text{s.t.} \quad & 0 \leqslant b_i \leqslant \frac{\gamma_i}{N} \quad \forall i = 1, \ldots, N, \\ & \sum_{i=1}^{N} b_i \gamma_i \leqslant K_u^\delta(Q). \end{aligned}$$

   Let us introduce Lagrange multipliers:

   - $\lambda_i \geqslant 0$ for the constraints $b_i \geqslant 0$,
   - $\mu_i \geqslant 0$ for the constraints $b_i \leqslant \frac{\gamma_i}{N}$,
   - $\nu \geqslant 0$ for the budget constraint $\sum_{i=1}^{N} b_i \gamma_i \leqslant K_u^\delta(Q)$.

   (a) Write the Lagrangian function for this problem.
   (b) Write the Karush-Kuhn-Tucker (KKT) conditions, including:

      - Stationarity,
      - Primal feasibility,
      - Dual feasibility,
      - Complementary slackness.

(c) Compute the partial derivatives of the Lagrangian with respect to b_i and write the stationarity conditions explicitly for $i \leqslant k$ and $i \geqslant k+1$.

(d) Suppose that there exists $I \in \{k+1, N-1\}$ such that $\forall j \geqslant I+1, K_u^\delta(Q) \leqslant 0$. Explain why, in an optimal solution, the variables $b_1, \ldots, b_k$ and $b_{I+1}, \ldots, b_N$ should be set to 0 in order to maximize the objective.

(e) For $k+1 \leqslant i \leqslant I$, describe a greedy algorithm to allocate the b_i's, given the budget constraint, that maximizes the objective. Specifically:

- Sort indices $i \geqslant k+1$ by increasing γ_i.
- Allocate $b_i = \frac{\gamma_i}{N}$ for as many i as possible until the budget is exhausted.
- If the budget is insufficient for the last term, adjust b_i so that the budget constraint is met with equality.

(f) For each constraint, explain the economic interpretation of the complementary slackness condition in this context.

(g) Discuss the interpretation of the dual variable ν in the context of this problem.

6. Laplacian Regularized Least Squares (LapRLS) is a semi-supervised learning algorithm that incorporates both labeled and unlabeled data by encouraging the learned function to be smooth with respect to the geometry of all the data. This is achieved by adding a regularization term based on the graph Laplacian constructed from both labeled and unlabeled samples.

Suppose you are given m labeled data points $\{(\mathbf{x}_i, y_i)\}_{i=1}^m$ and u unlabeled data points $\{\mathbf{x}_j\}_{j=m+1}^{m+u}$. Let $f : \mathbb{R}^d \to \mathbb{R}$ be the prediction function. The LapRLS optimization problem is:

$$\min_f \frac{1}{m} \sum_{i=1}^{m} (f(\mathbf{x}_i) - y_i)^2 + \lambda \|f\|_{\mathcal{H}}^2 + \gamma \sum_{j,k=1}^{m+u} W_{jk}(f(\mathbf{x}_j) - f(\mathbf{x}_k))^2,$$

where $\|f\|_{\mathcal{H}}^2$ is a norm in a reproducing kernel Hilbert space (RKHS) that regularizes model complexity, W_{jk} is the weight of the edge between x_j and x_k in the similarity graph. The third term is the **manifold regularization** term, which can be written as $f^\top L f$ where L is the graph Laplacian matrix: $\boldsymbol{L} = \boldsymbol{D} \ominus \boldsymbol{W}$, with $\boldsymbol{D}$ the diagonal degree matrix ($D_{jj} = \sum_k W_{jk}$).

(a) Explain how the unlabeled data contributes to the regularization term. Why does minimizing the Laplacian term $\sum_{j,k} W_{jk}(f(\mathbf{x}_j) - f(\mathbf{x}_k))^2$ encourage "smoothness" of f along the data manifold?

(b) What is the intuition behind this approach in the context of semi-supervised learning?

7. Let $\{\mathbf{x}_1, \ldots, \mathbf{x}_{m+u}\}$ be a dataset with m labeled points $\{(\mathbf{x}_i, y_i)\}_{i=1}^m$ and u unlabeled points $\{\mathbf{x}_j\}_{j=m+1}^{m+u}$. Consider a graph $G = (V, E)$ constructed by connecting all data points, with edge weights W_{ij} encoding similarity between x_i and x_j.

(a) Describe two methods for constructing the similarity graph G used in graph-based semi-supervised learning. For each method, specify how the edge weights W_{ij} are defined.
(b) Explain the principle of the label propagation algorithm for semi-supervised classification.
(c) Write the update equation for the label scores f_i at each node, and describe how the labels of labeled nodes are handled during propagation.
(d) The label propagation solution can be interpreted as finding a harmonic function on the graph. Define what it means for a function f on the nodes to be harmonic with respect to the graph.
(e) Show that the label propagation algorithm converges to the unique harmonic function that agrees with the labeled data.
(f) Define the (unnormalized) graph Laplacian matrix $\boldsymbol{L}$.
(g) Write the objective function for graph Laplacian regularized semi-supervised learning and explain the role of the unlabeled data in this objective.

Appendix A
Reminders on Probabilities

There are several substantial works on probability theory, notably those of [11] and [52]. In this appendix, we will briefly recall the common notions in probability, as well as the important results which arise from them and which we have used throughout this book.

A.1 Probability Measure

The probability measure of an event is a positive real number between 0 and 1 reflecting the degree of likelihood that we attribute to the occurrence of this event during a random experiment.

A.1.1 Measurable Space

A random experiment is a test in which all possible outcomes are known, but the outcome cannot be predicted with certainty. The set of all possible outcomes, associated with the random experiment and denoted Ω, is called the fundamental set or universe of possibilities.

We generally distinguish three types of random experiments depending on whether the associated fundamental set is finite, countably infinite, or continuously infinite.

An event associated with a random trial is a logical proposition taking the value true or false at the end of the trial and it is represented by a part of the fundamental set.

In the case where the fundamental set Ω is not countable, we generally restrict ourselves to a family of parts $\mathcal{P}$ of Ω stable by passage to the complement and

M.-R. Amini, *Advanced Supervised and Semi-supervised Learning*,
Cognitive Technologies, https://doi.org/10.1007/978-3-031-99928-4

by countable union called σ-algebra (or σ-field), and verifying the following three properties:

(a) $\Omega \in \mathcal{P}$
(b) if $\forall n \in \mathbb{N}^*$, $E_n \in \mathcal{P}$ then $\bigcup_{n\in\mathbb{N}^*} E_n \in \mathcal{P}$
(c) if $E \in \mathcal{P}$ then its complementary part $\bar{E}$ is also in $\mathcal{P}$; $\bar{E} \in \mathcal{P}$

We then call the pair $(\Omega, \mathcal{P})$ a measurable space and we define a probability measure, the application $\mathbb{P} : \mathcal{P} \to [0, 1]$ verifying the following Kolmogorov axioms:

1. $\mathbb{P}(\Omega) = 1$
2. For any countably finite or infinite sequence of events, $(E_n)_{n\in\mathbb{S}}$ of $\mathcal{P}$ pairwise incompatible ($\forall i \neq j$, $E_i \cap E_j = \emptyset$), with $\mathbb{S} \subseteq \mathbb{N}$:

$$\mathbb{P}\left(\bigcup_{i\in\mathbb{S}} E_i\right) = \sum_{i\in\mathbb{S}} \mathbb{P}(E_i) \tag{A.1}$$

The second axiom is called σ-additivity and, in the case where $\mathbb{S} = \mathbb{N}$, we admit the existence of the following limit:

$$\mathbb{P}\left(\bigcup_{i\in\mathbb{N}} E_i\right) = \lim_{n\to\infty} \sum_{i=1}^{n} \mathbb{P}(E_i) \in [0, 1]$$

Moreover, in the case of n events $(E_i)_{i=1}^n$ which are two by two incompatible and whose union is the certain event, $\bigcup_{i=1}^{n} E_i = \Omega$, also called the complete system of events, we have according to Kolmogorov's second axiom the following result, known as the law of total probabilities:

$$\forall B \in \mathcal{P}, \mathbb{P}(B) = \sum_{i=1}^{n} \mathbb{P}(B \cap E_i) \tag{A.2}$$

Indeed, any event B can be written as a union of incompatible events two by two, based on the complete system of events $(E_i)_{i=1}^n$; $B = \bigcup_{i=1}^{n} (E_i \cap B)$.

A.1.2 Probability Space

We call the probability space, $(\Omega, \mathcal{P}, \mathbb{P})$, the measurable space $(\Omega, \mathcal{P})$ equipped with its probability measure $\mathbb{P} : \mathcal{P} \to [0, 1]$. An event $E \in \mathcal{P}$ is said to be feasible if $\mathbb{P}(E) > 0$. We can thus list the fundamental results that follow from the preceding axioms:

1. $\mathbb{P}(\Omega) = \mathbb{P}(\Omega \cup \emptyset) = \mathbb{P}(\Omega) + \mathbb{P}(\emptyset)$, thus $\mathbb{P}(\emptyset) = 0$
2. $\forall E \in \mathcal{P}, \mathbb{P}(E) + \mathbb{P}(\bar{E}) = \mathbb{P}(E \cup \bar{E}) = \mathbb{P}(\Omega) = 1$, E and $\bar{E}$ are incompatible
3. $\forall (E_i, E_j) \in \mathcal{P} \times \mathcal{P}$, if $E_i \subset E_j$ we have $E_j = E_i \cup (E_j \setminus E_i)$ and
$$\mathbb{P}(E_j) = \mathbb{P}(E_i) + \mathbb{P}(E_j \setminus E_i) \geqslant \mathbb{P}(E_i)$$
4. $\forall (E_i, E_j) \in \mathcal{P} \times \mathcal{P}, \mathbb{P}(E_i \setminus E_j) = \mathbb{P}(E_i) - \mathbb{P}(E_i \cap E_j)$ as
   $E_i = (E_i \cap E_j) \cup (E_i \cap \bar{E}_j) = (E_i \cap E_j) \cup (E_i \setminus E_j)$
5. $\forall (E_i, E_j) \in \mathcal{P} \times \mathcal{P}, \mathbb{P}(E_i \cup E_j) = \mathbb{P}(E_i) + \mathbb{P}(E_j) - \mathbb{P}(E_i \cap E_j)$ as
   $E_i \cup E_j = (E_i \cap \bar{E}_j) \cup (E_i \cap E_j) \cup (\bar{E}_i \cap E_j) = (E_i \setminus E_j) \cup (E_i \cap E_j) \cup (E_j \setminus E_i)$
   and $(E_i \setminus E_j)$, $(E_i \cap E_j)$, $(E_j \setminus E_i)$ are two by two incompatible.

The last equality leads to the Boole's inequality or the union bound used in several important results in learning, which in the general case is written:

$$\forall n \in \mathbb{N}, \mathbb{P}\left(\bigcup_{i=1}^{n} E_i\right) \leqslant \sum_{i=1}^{n} \mathbb{P}(E_i) \tag{A.3}$$

A.2 Conditional Probability

In a forecast, the knowledge of additional information can affect the outcome of the forecast; this concept is quantified through the notion of conditional probability.

Formally, let $(\Omega, \mathcal{P}, \mathbb{P})$ be a probability space and E be a feasible event ($\mathbb{P}(E) > 0$). We call conditional probability measure, assuming the event E to have occurred, the application $\mathbb{P} : \mathcal{P} \to [0, 1]$ which to any event $A \in \mathcal{P}$ associates with it:

$$\mathbb{P}(A \mid E) = \frac{\mathbb{P}(A \cap E)}{\mathbb{P}(E)} \tag{A.4}$$

With this definition, it is easy to see that conditional probability verifies Kolmogorov's axioms. The Eq. (A.4) leads to the product rule of probability which, for two feasible events E and A of a probability space $(\Omega, \mathcal{P}, \mathbb{P})$, is stated as follows:

$$\mathbb{P}(A \cap E) = \mathbb{P}(A \mid E) \times \mathbb{P}(E) = \mathbb{P}(E \mid A) \times \mathbb{P}(A) \tag{A.5}$$

A.2.1 Bayes Rule

This definition naturally leads to the Bayes rule [15] which is stated as follows: $(\Omega, \mathcal{P}, \mathbb{P})$ a probability space and E a feasible event of $\mathcal{P}$, let $(A_i)_{i\in\mathbb{S}\subset\mathbb{N}}$ be a family of events of $\mathcal{P}$ pairwise incompatible verifying $\bigcup_{i\in\mathbb{S}} A_i = \Omega$, we then have:

$$\forall i \in \mathbb{S}, \mathbb{P}(A_i \mid E) = \frac{\mathbb{P}(E \mid A_i) \times \mathbb{P}(A_i)}{\sum_{j\in\mathbb{S}} \mathbb{P}(E \mid A_j) \times \mathbb{P}(A_j)} \tag{A.6}$$

The proof of this formula follows directly from Eqs. (A.4) and (A.5). Indeed, according to these equations, we have for any event E and A_i of $\mathcal{P}$:

$$\mathbb{P}(A_i \mid E) = \frac{\mathbb{P}(E \mid A_i) \times \mathbb{P}(A_i)}{\mathbb{P}(E)}$$

Furthermore, since the events $(A_i)_{i\in\mathbb{S}\subset\mathbb{N}}$ are two by two incompatible, we have according to the law of total probabilities (Eq. (A.2)):

$$\mathbb{P}(E) = \sum_{j\in\mathbb{S}} \mathbb{P}(E \cap A_j) = \sum_{j\in\mathbb{S}} \mathbb{P}(E \mid A_j) \times \mathbb{P}(A_j)$$

Equation A.6 was independently found by Thomas Bayes [15] and Pierre-Simon Laplace [99][1] and is sometimes called the formula for the probability of causes; in some cases, the events A_j can be seen as all the incompatible causes of a consequence. Indeed, if in $\mathcal{P}$ we know the probabilities of the events A_j (or the prior probabilities of the causes) and also the conditional probabilities of the consequence E knowing the causes A_j having occurred, we can then estimate with Bayes rule (Eq. (A.6)) the posterior probabilities that each of the causes A_j is responsible for the consequence E, once the latter is observed.

Example A.1 Consider the Monty Hall problem inspired by the TV game show *Let's Make a Deal*. In this problem, a host is pitted against a player who is placed in front of three closed doors. Behind one of the doors is a valuable prize and behind each of the other two is an unimportant prize. The host asks the player to point to one of the doors where the player thinks the valuable prize is. He then opens the door that is not pointed to by the player and where one of the unimportant prizes is located. The contestant then has the choice of opening the door he had initially pointed to or the third door. The question is: what is the probability that the valuable prize is behind one of these two doors, given the door opened by the host?

Let: $A_i, i \in \{1, 2, 3\}$, the event: *the valuable prize is behind door number i.* Suppose that the player designates door $i = 2$ and that the presenter chooses to

[1] http://gallica.bnf.fr/ark:/12148/bpt6k77596b.image.f32.langFR.

open, among doors 1 and 3, door 3. Let E be this last event. We have in this case, i.e. *E: the presenter opens door 3 knowing that the door designated by the player is 2*:

- The a priori probabilities of the causes $\mathbb{P}(A_i) = \frac{1}{3}$
- The conditional probabilities of the consequence E knowing these causes: $\mathbb{P}(E \mid A_3) = 0, \mathbb{P}(E \mid A_2) = \frac{1}{2}$ and $\mathbb{P}(E \mid A_1) = 1$.

  According to the formula for total probabilities (Eq. (A.2)):

$$\begin{aligned}\mathbb{P}(E) &= \mathbb{P}(E \mid A_1) \times \mathbb{P}(A_1) + \mathbb{P}(E \mid A_2) \times \mathbb{P}(A_2) + \mathbb{P}(E \mid A_3) \times \mathbb{P}(A_3)\\ &= 1 \times \frac{1}{3} + \frac{1}{2} \times \frac{1}{3} + 0 \times \frac{1}{3} = \frac{1}{2}\end{aligned}$$

According to the Bayes rule, we deduce the posterior probabilities of each of the causes leading to consequence E (i.e. the presenter opened the door that he knew was not the one where the valuable prize is):

$$\mathbb{P}(A_1 \mid E) = \frac{\mathbb{P}(E \mid A_1) \times \mathbb{P}(A_1)}{\mathbb{P}(E)} = \frac{2}{3}, \mathbb{P}(A_2 \mid E) = \frac{\mathbb{P}(E \mid A_2) \times \mathbb{P}(A_2)}{\mathbb{P}(E)} = \frac{1}{3},$$

A.2.2 Independence in Probability

In the case where two events in a probability space are independent, we see that knowledge of one will not affect the prediction of the other. This notion is quantified as follows: two events A and E of a probability space $(\Omega, \mathcal{P}, \mathbb{P})$ are said to be independent with respect to the probability measure $\mathbb{P}$, or again $\mathbb{P}$-independent, if and only if:

$$\mathbb{P}(A \cap E) = \mathbb{P}(A) \times \mathbb{P}(E). \tag{A.7}$$

A.3 Real-valued Random Variables

A real-valued random variable X, often denoted r.v., is an application of the fundamental set Ω to the set of real numbers which, to any element of Ω associates a real number:

$$\begin{aligned} X : \Omega &\to \mathbb{R}\\ \omega &\mapsto X(\omega)\end{aligned}$$

The idea underlying this definition is that in practice, we are more interested in the value that we attribute to a random event, than in the outcome of the event itself. This r.v. is thus defined on a probability space $(\Omega, \mathcal{P}, \mathbb{P})$, which implies being able to probabilize events on the arrival space $\mathbb{R}$. To do this, we construct a probabilized space on $\mathbb{R}$ equipped with its tribe $\mathcal{B}(\mathbb{R})$, generated by the intervals of the form $]-\infty, b[$, and a probability measure $\mathbb{P} : \mathcal{B}(\mathbb{R}) \rightarrow [0, 1]$ defined by:

$$\forall B \in \mathcal{B}(\mathbb{R}), \mathbb{P}(B) = \mathbb{P}(X^{-1}(B)) = \mathbb{P}\left(\{\omega_i \in \Omega \mid X(\omega_i) \in B\}\right)$$

So, $\mathbb{P}(B)$ is the measure by $\mathbb{P}$ of the reciprocal image of B by X. The condition for the existence of $\mathbb{P}$ is that the reciprocal image $X^{-1}(B)$ is an element of $\mathcal{P}$ on which we can apply the probability measure $\mathbb{P}$. In this case, we say that the real random variable is a measurable application of any probabilizable space in $(\mathbb{R}, \mathcal{B}(\mathbb{R}))$ and we will call the probability data $\mathbb{P}$ on $(\mathbb{R}, \mathcal{B}(\mathbb{R}))$ distribution of X.

A.3.1 Distribution Function

The Borel tribe $\mathcal{B}(\mathbb{R})$ is generated by intervals of the form $]-\infty, b[$ and any event of $\mathcal{B}(\mathbb{R})$ can then be defined from these intervals and set operations. Knowledge of the probabilities $\mathbb{P}(]-\infty, b[) = \mathbb{P}(X < b)$ is thus sufficient to probabilize $(\mathbb{R}, \mathcal{B}(\mathbb{R}))$ by $\mathbb{P}$. In this case, for a r.v., the application $F : \mathbb{R} \rightarrow [0, 1]$ which to any real x associates $F(x) = \mathbb{P}(X < x)$ is called the distribution function of X and any random variable X whose distribution function F is continuous and differentiable except at a finite number of points, is called a random variable with density. We can clearly see that the definition of the law of X is equivalent to that of its distribution function. Thus, to define the law of X it will be necessary to know how to calculate $\mathbb{P}(B)$. Moreover, as $\mathbb{P}$ is a set function, it is possible to calculate the probability of an interval via the distribution function which depends on a single variable. A direct consequence of the previous definition is the following result:

$$\mathbb{P}(a \leqslant X < b) = F(b) - F(a) \tag{A.8}$$

since $\{a \leqslant X < b\} = \{X < b\} \setminus \{X < a\}$ and according to property iii) stated in Sect. A.1.2 we have:

$$\mathbb{P}(a \leqslant X < b) = \mathbb{P}(\{X < b\} \setminus \{X < a\}) = \mathbb{P}(\{X < b\}) - \mathbb{P}(\{X < a\})$$

Finally, the density of a random variable, p, is defined as the derivative of its distribution function F, or:

$$\forall x \in \mathbb{R}, F(x) = \int_{-\infty}^{x} p(t)dt \tag{A.9}$$

Any random variable whose distribution function is a step function is referred to as a discrete variable. The definition of a discrete variable involves specifying a set $\mathfrak{X}$ of possible values, which is at most countable, and probabilities $\mathbb{P}(X = x) = p(x)$ that satisfy the following conditions:

$$\forall x \in \mathfrak{X},\, p(x) \geqslant 0,\, \sum_{x\in\mathfrak{X}} p(x) = 1$$

The data of the family $(p(x))_{x\in\mathfrak{X}}$ is called the law of X. We say that a discrete r.v. follows a Bernoulli law of parameter q in the special case where $\mathfrak{X} = \{0, 1\}$ with $q = p(1)$.

A.3.2 Expectation and Variance of a Random Variable

The mathematical expectation of a r.v., defined on a domain $\mathfrak{D} \subseteq \mathbb{R}$ of $\mathbb{R}$, is given by:

$$\mathbb{E}(X) = \int_{\mathfrak{D}} xp(x)dx$$

According to the additivity property of the integral, we have:

$$\forall (a, b) \in \mathbb{R} \times \mathbb{R},\, \mathbb{E}(aX + b) = a\mathbb{E}(X) + b \tag{A.10}$$

By analogy, the expectation of a discrete r.v. X is:

$$\mathbb{E}(X) = \sum_{x\in\mathfrak{X}} xp(x)$$

Another important property that follows from Eq. (A.7) concerns the expectation of a product of n independent random variables $X_1, \ldots, X_n$ two by two:

$$\mathbb{E}[X_1 X_2 \ldots X_n] = \prod_{i=1}^{n} \mathbb{E}[X_i]$$

We define the variance of the r.v. in the same way:

$$\mathbb{V}(X) = \mathbb{E}[(X - \mathbb{E}(X))]^2 \tag{A.11}$$

According to the additivity property of expectation, the variance of X can be written as $\mathbb{V}(X) = \mathbb{E}(X^2) - [\mathbb{E}(X)]^2$. Indeed:

$$\begin{aligned}\mathbb{V}(X) &= \mathbb{E}[(X - \mathbb{E}(X))]^2 = \mathbb{E}[(X^2 - 2X\mathbb{E}(X) + \mathbb{E}^2(X))] \\ &= \mathbb{E}[X^2] - 2\mathbb{E}^2(X) + \mathbb{E}^2(X) \\ &= \mathbb{E}[X^2] - [\mathbb{E}(X)]^2\end{aligned}$$

For a discrete r.v. following a Bernoulli distribution of parameter q, we have:

$$\begin{aligned}\mathbb{E}[X] &= 1 \times q + 0 \times (1 - q) = q \\ \mathbb{V}(X) &= \mathbb{E}[X^2] - \mathbb{E}^2[X] = q - q^2 = q(1 - q)\end{aligned}$$

The variance of a discrete r.v. following the Bernoulli distribution is thus less than $\frac{1}{4}$.

$$\begin{aligned}\forall (a, b) \in \mathbb{R} \times \mathbb{R}, \mathbb{V}(aX + b) &= \mathbb{E}[(aX + b - a\mathbb{E}(X) - b)]^2 = \mathbb{E}[a^2(X - \mathbb{E}(X))]^2 \\ &= a^2\mathbb{E}[(X - \mathbb{E}(X))]^2 = a^2\mathbb{V}(X)\end{aligned}$$

Finally, we call the standard deviation of the r.v. X the following quantity, if it exists:

$$\sigma(X) = \sqrt{\mathbb{V}(X)}$$

A.3.3 Concentration Inequalities

Concentration inequalities estimate the concentration of the distribution of a r.v. X around its expectation, $\mathbb{E}[X]$, and they are used to establish important probability and learning results such as those presented in [22]. In this section, we will present the two inequalities most often used for the derivation of the generalization bounds presented in Chap. 1, namely the Chebychev inequality [166] and the Hoeffding inequality [81]. These two inequalities are based on the following lemma:

Lemma A.1 *Let $I \subseteq \mathbb{R}$ be a real interval and $g : I \to \mathbb{R}_+$ be a strictly positive function. For a real $\epsilon \in I$, let $b \in \mathbb{R}_+^*$ verify $\forall x \in I, x \geqslant \epsilon$. Then $g(x) \geqslant b$. In this case, for any r.v. X taking its value in I, we have:*

$$\mathbb{P}(X \geqslant \epsilon) \leqslant \frac{\mathbb{E}[g(X)]}{b} \tag{A.12}$$

Proof Let $\mathfrak{D}_1 = \{x \in I \mid x \geqslant \epsilon\}$. We then have $\mathfrak{D}_1 \subseteq I$. Since g is a strictly positive function, it follows

$$\begin{aligned}\mathbb{E}[g(X)] = \int_I g(x)p(x)dx &\geqslant \int_{\mathfrak{D}_1} g(x)p(x)dx \geqslant b\int_{\mathfrak{D}_1} p(x)dx \\ &\geqslant b\mathbb{P}(X \geqslant \epsilon)\end{aligned}$$

□

Suppose that the function g is strictly increasing i.e. $\forall x \geqslant \epsilon, g(x) \geqslant g(\epsilon)$. Taking $b = g(\epsilon)$ in the inequality (A.12), we get:

$$\mathbb{P}(X \geqslant \epsilon) \leqslant \frac{\mathbb{E}[g(X)]}{g(\epsilon)} \tag{A.13}$$

A.3.3.1 Markov Inequality

In the case where X is a non-negative r.v., we can bound the probability that X takes values greater than a positive multiple of its expectation, called the Markov inequality. This inequality follows directly from the previous result (Eq. (A.13)) by considering the identity function, $g : z \mapsto z$, and an interval of the set $\mathbb{R}_+$.

Theorem A.1 (Markov Inequality) *Let X be a non-negative r.v., with expectation $\mathbb{E}(X) > 0$. We then have for any strictly positive real $\epsilon > 0$:*

$$\mathbb{P}(X \geqslant \epsilon) \leqslant \frac{\mathbb{E}[X]}{\epsilon}. \tag{A.14}$$

A.3.3.2 Chebyshev Inequality

From Markov's inequality we can deduce another stronger inequality, which bounds the probability of deviation of the difference in values between a r.v. and its expectation, called the Chebyshev inequality:

Theorem A.2 (Chebyshev Inequality) *Let X be a r.v. of expectation $\mathbb{E}(X)$ and variance $\mathbb{V}$. We then have for any strictly positive real $\epsilon > 0$:*

$$\mathbb{P}\left(|X - \mathbb{E}[X]| \geqslant \epsilon\right) \leqslant \frac{\mathbb{V}(X)}{\epsilon^2}. \tag{A.15}$$

Proof We first note that the events $\{|X - \mathbb{E}[X]| \geqslant \epsilon\}$ and $\{(X-\mathbb{E}[X])^2 \geqslant \epsilon^2\}$ are equivalent. Then applying Markov's inequality (Eq. (A.15)) to the r.v. $Y = (X - \mathbb{E}[X])^2 \geqslant 0$, with expectation $\mathbb{E}[Y] = \mathbb{V}(X)$, we obtain:

$$\begin{aligned}\mathbb{P}\left(|X - \mathbb{E}[X]| \geqslant \epsilon\right) &= \mathbb{P}\left((X - \mathbb{E}[X])^2 \geqslant \epsilon^2\right)\\ &= \mathbb{P}\left(Y \geqslant \epsilon^2\right)\\ &\leqslant \frac{\mathbb{E}[Y]}{\epsilon^2} = \frac{\mathbb{V}(X)}{\epsilon^2}.\end{aligned}$$

□

A.3.3.3 Chernoff Inequality

Another important result is the Chernoff inequality, also follows from the inequality (A.13) by considering the exponential function defined by $\forall s > 0, g : z \mapsto e^{sz}$, and an interval of $\mathbb{R}$, that is:

Theorem A.3 (Chernoff Inequality) *Let X be a r.v. Then for all real numbers $s > 0$ and $\epsilon > 0$ we have:*

$$\mathbb{P}(X \geqslant \epsilon) \leqslant e^{-s\epsilon}\mathbb{E}[e^{sX}]. \tag{A.16}$$

The choice of the non-negative function $\forall s > 0, g : z \mapsto e^{sz}$ removes the condition of non-negativity of the r.v. X imposed in Markov's inequality, but adds the expectation of the r.v. e^{sX} which must be bounded in turn.

In the case where the r.v. X is centered ($\mathbb{E}[X] = 0$) and bounded in an interval $[a, b] \subset \mathbb{R}$, we can bound this expectation:

Lemma A.2 (Hoeffding's Lemma [31]) *Let a and b be two real numbers such that $a < b$ and X be a r.v. bounded in the interval $[a, b]$ and centered (i.e. $\mathbb{E}[X] = 0$). Then for any strictly positive real number, $s > 0$, we have the following inequality:*

$$\mathbb{E}[e^{sX}] \leqslant e^{\frac{s^2(b-a)^2}{8}}. \tag{A.17}$$

Proof We can write any real number $x \in [a, b]$ as $x = \frac{b-x}{b-a}a + \frac{x-a}{b-a}b$. By convexity of the function $x \mapsto e^{sx}$ and Jensen's inequality, we have:

$$e^{sX} \leqslant \frac{b-X}{b-a}e^{sa} + \frac{X-a}{b-a}e^{sb} \tag{A.18}$$

By passing to the expectation and using its linearity property and the fact that the r.v. is centered, we deduce:

$$\begin{aligned}\mathbb{E}[e^{sX}] &\leqslant \frac{b-\mathbb{E}[X]}{b-a}e^{sa} + \frac{\mathbb{E}[X]-a}{b-a}e^{sb} \\ &\leqslant \underbrace{\frac{b}{b-a}e^{sa} - \frac{a}{b-a}e^{sb}}_{=e^{G(s)}}\end{aligned}$$

Let $G(s) = \ln\left(\frac{b}{b-a}e^{sa} + \frac{-a}{b-a}e^{sb}\right)$, $(b-a)s = z, \theta = \frac{-a}{b-a}$, and therefore $1 - \theta = \frac{b}{b-a}$. It follows that:

$$\begin{aligned}G(s) &= \ln\left(\frac{b}{b-a}e^{sa} + \frac{-a}{b-a}e^{sb}\right) \\ &= sa + \ln\left(\frac{b}{b-a} + \frac{-a}{b-a}e^{s(b-a)}\right).\end{aligned}$$

Consider the function $z \mapsto H(z)$, with:

$$H(z) = -\theta z + \ln(1 - \theta + \theta e^z)$$

The first and the second derivatives of H are:

$$\begin{aligned}H'(z) &= -\theta + \frac{\theta e^z}{1-\theta+\theta e^z} \\ H''(z) &= \frac{(1-\theta)\theta e^z}{(1-\theta+\theta e^z)^2}\end{aligned}$$

H'' is in the form $\frac{xy}{(x+y)^2}$, and since:

$$\begin{aligned}(x-y)^2 = (x+y)^2 - 4xy \geqslant 0 \\ (x+y)^2 \geqslant 4xy,\end{aligned}$$

we have $\frac{xy}{(x+y)^2} \leqslant \frac{1}{4}$, i.e. $\forall z > 0, H''(z) \leqslant \frac{1}{4}$. We also note that $H(0) = H'(0) = 0$; according to the Taylor-Lagrange formula, there exists a $t \in [0, z]$ such that:

$$H(z) = H(0) + H'(0)z + H''(t)\frac{z^2}{2} \leqslant \frac{z^2}{8},$$

which completes the proof. □

A.3.3.4 Hoeffding's Inequality

Hoeffding's inequality concerns the sums of independent random variables, and as we discussed in the Chap. 1, it is one of the most widely used tools in proofs of learning theory results.

Theorem A.4 (Hoeffding Inequality [81])

$$\mathbb{P}(S_m - \mathbb{E}[S_m] \geqslant \epsilon) \leqslant \exp\left(\frac{-2\epsilon^2}{\sum_{i=1}^m (b_i - a_i)^2}\right) \tag{A.19}$$

$$\mathbb{P}(S_m - \mathbb{E}[S_m] \leqslant -\epsilon) \leqslant \exp\left(\frac{-2\epsilon^2}{\sum_{i=1}^m (b_i - a_i)^2}\right) \tag{A.20}$$

Proof According to Chernoff's inequality (A.16), we have for all $\epsilon > 0$ and $s > 0$:

$$\mathbb{P}(S_m - \mathbb{E}[S_m] \geqslant \epsilon) \leqslant e^{-s\epsilon}\mathbb{E}[e^{S_m - \mathbb{E}[S_m]}]$$

Let $\forall i \in \{1, \dots, m\}, Y_i = X_i - \mathbb{E}[X_i]$, i.e. $\forall i, \mathbb{E}[Y_i] = 0$ and $S_m - \mathbb{E}[S_m] = \sum_{i=1}^m Y_i$. According to the independence of the r.v., we obtain:

$$\begin{aligned}\mathbb{P}(S_m - \mathbb{E}[S_m] \geqslant \epsilon) &\leqslant e^{-s\epsilon}\mathbb{E}[e^{Y_1 + \dots + Y_m}] \\ &= e^{-s\epsilon}\prod_{i=1}^m \mathbb{E}[e^{sY_i}]\end{aligned}$$

Like each of the r.v. is bounded in the interval $[a_i - \mathbb{E}[X_i], b_i - \mathbb{E}[X_i]]$, we are able to apply Chernoff's lemma to each term of the previous product and, according to the algebraic property of the exponential function, we have:

$$\mathbb{P}(S_m - \mathbb{E}[S_m] \geqslant \epsilon) \leqslant \exp\left(-s\epsilon + \frac{s^2\sum_{i=1}^m (b_i - a_i)^2}{8}\right) \tag{A.21}$$

The previous inequality is true for all $s > 0$. In particular, $s = \frac{4\epsilon}{\sum_{i=1}^{m}(b_i - a_i)^2}$ which realizes the minimum of the right bound and completes the proof of the first inequality. The second inequality follows the previous one by setting $\forall i \in \{1, \dots, m\}, Z_i = \mathbb{E}[X_i] - X_i$. □

References

1. S. Agarwal et al. “Generalization Bounds for the Area Under the ROC Curve”. In: *Journal of Machine Learning Research* 6 (2005), pp. 393–425. ISSN: 1533–7928.
2. E. L. Allwin, R. E. Schapire, and Y. Singer. “Reducing Multiclass to Binary: A Unifying Approach for Margin Classifiers”. In: *Journal of Machine Learning Research* 1 (2000), pp. 113–141.
3. M.-R. Amini, N. Usunier, and F. Laviolette. “A Transductive Bound for the Voted Classifier with an Application to Semi-supervised Learning”. In: *Advances in Neural Information Processing Systems (NeurIPS 21)*. 2009, pp. 65–72.
4. M.-R. Amini et al. “Self-training: A survey”. In: *Neurocomputing* 616 (2025), p. 128904.
5. Massih-Reza Amini and Éric Gaussier. *Recherche d’Information - applications, modèles et algorithmes*. Eyrolles, 2013, pp. I–XIX, 1–233. ISBN: 978-2-212-13532-9.
6. J. A. Anderson. “Logistic Discrimination”. In: *Handbook of Statistics* 2 (1982), pp. 169–191.
7. J. A. Anderson and E. Rosenfeld. *Neurocomputing: Foundations of Research*. MIT Press, 1988.
8. A. Antos et al. “Data-dependent Margin-based Generalization Bounds for Classification”. In: *Journal of Machine Learning Research* 3 (2003), pp. 73–98.
9. K. A. Atkinson. *An introduction to numerical analysis*. John Wiley and Sons, 1988.
10. F. Bach, R.G. Lanckriet, and M.I. Jordan. “Multiple Kernel Learning, conic duality, and the SMO algorithm”. In: *Proceedings of the Twenty-first International Conference on Machine Learning*. 2004.
11. P. Barbé and M. Ledoux. *Probabilité*. EDP Sciences, 2007.
12. P. L. Bartlett, M. I. Jordan, and J. D. McAuliffe. “Convexity, Classification, and Risk Bounds”. In: *Journal of the American Statistical Association* 101.473 (2006), pp. 138–156.
13. P. L. Bartlett and S. Mendelson. “Rademacher and Gaussian complexities: risk bounds and structural results”. In: *Journal of Machine Learning Research* 3 (2003), pp. 463–482.
14. S. Basu, A. Banerjee, and R. J. Mooney. “Semi-supervised Clustering by Seeding”. In: *Proceedings of the Nineteenth International Conference on Machine Learning*. 2002, pp. 27–34.
15. T. Bayes. “An Essay towards solving a Problem in the Doctrine of Chances”. In: *Philosophical Transactions of the Royal Society of London* 53 (1763), pp. 370–418.
16. A. Blum and T.M. Mitchell. “Combining labeled and unlabeled data with co-training”. In: *Proceedings of the 11th Annual Conference on Learning Theory*. 1998, pp. 92–100.
17. A. Blumer et al. “Learnability and the Vapnik-Chervonenkis dimension”. In: *Journal of the ACM* 36 (1989), pp. 929–965.

M.-R. Amini, *Advanced Supervised and Semi-supervised Learning*,
Cognitive Technologies, https://doi.org/10.1007/978-3-031-99928-4

18. J. F. Bonnans et al. *Numerical optimization, theoretical and numerical aspects*. Springer Verlag, 2006.
19. L. Bottou. “Online Algorithms and Stochastic Approximations”. In: *Online Learning and Neural Networks*. Ed. by David Saad. revised, oct 2012. Cambridge, UK: Cambridge University Press, 1998. URL: http://leon.bottou.org/papers/bottou-98x.
20. L. Bottou. “Une Approche théorique de l’Apprentissage Connexionniste: Applications à la Reconnaissance de la Parole”. PhD thesis. Or-say, France: Université de Paris XI, 1991. URL: http://leon.bottou.org/papers/bottou-91a.
21. S. Boucheron, O. Bousquet, and G. Lugosi. “Theory of classification : a survey of some recent advances”. In: *ESAIM: Probability and Statistics* (2005), pp. 323–375.
22. S. Boucheron, G. Lugosi, and P. Massart. *Concentration Inequalities: A Nonasymptotic Theory of Independence*. Oxford University Press, 2013.
23. O. Bousquet, S. Boucheron, and G. Lugosi. “Introduction to Statistical Learning Theory”. In: *Advanced Lectures on Machine Learning*. 2003, pp. 169–207.
24. S. Boyd and L. Vandenberghe. *Convex Optimization*. New York, NY, USA: Cambridge University Press, 2004. ISBN: 0521833787.
25. H. Brönnimann and M. T. Goodrich. “Almost Optimal Set Covers in Finite VC-Dimension.” In: *Discrete and Computational Geometry* 14.4 (1995), pp. 463–479.
26. C. Calauzènes, N. Usunier, and P. Gallinari. “On the (Non-)existence of Convex, Calibrated Surrogate Losses for Ranking”. In: *Advances in Neural Information Processing Systems (NeurIPS 25)*. 2012, pp. 197–205.
27. P. Cascante-Bonilla et al. “Curriculum labeling: Revisiting pseudo-labeling for semi-supervised learning”. In: *AAAI Conference on Artificial Intelligence*. 2021, pp. 6912–6920.
28. G. Celeux and G. Govaert. “A Classification EM Algorithm for Clustering and Two Stochastic Versions”. In: *Computational Statistics and Data Analysis* 14.3 (1992), pp. 315–332.
29. N. Cesa-Bianchi and D. Haussler. “A graph-theoretic generalization of the Sauer-Shelah lemma”. In: *Discrete Applied Mathematics* 86 (1998), pp. 27–35.
30. O. Chapelle, B. Schölkopf, and A. Zien, eds. *Semi-Supervised Learning*. Cambridge, MA: MIT Press, 2006.
31. H. Chernoff. “A Measure of Asymptotic Efficiency for Tests of a Hypothesis Based on the sum of Observations”. In: *Annals of Mathematical Statistics* 23.4 (1952), pp. 493–507.
32. W. Chu and S. S. Keerthi. “New approaches to support vector ordinal regression”. In: *22th International Conference on Machine Learning, ICML 2005*. 2005, pp. 145–152.
33. S. Clinchant and E. Gaussier. “Information-based models for *ad* hoc IR”. In: *SIGIR’10, conference on Research and development in information retrieval*. Geneva, Switzerland, 2010. ISBN: 978-1-4503-0153-4.
34. I. Cohen et al. “Semisupervised Learning of Classifiers: Theory, Algorithms, and Their Application to Human-Computer Interaction”. In: *IEEE Transactions on Pattern Analysis and Machine Intelligence* 26.12 (2004), pp. 1553–1567.
35. W. W. Cohen, R. E. Schapire, and Y. Singer. “Learning to Order Things”. In: *Advances in Neural Information Processing Systems (NeurIPS 10)*. 1998, pp. 451–457.
36. P. Colombo et al. “Transductive Learning for Textual Few-Shot Classification in API-based Embedding Models”. In: *Proceedings of the Conference on Empirical Methods in Natural Language Processing EMNLP*. 2023, pp. 4214–4231.
37. C. Cortes and M. Mohri. “AUC Optimization vs. Error Rate Minimization”. In: *Advances in Neural Information Processing Systems (NeurIPS 16)*. 2004, pp. 313–320.
38. F. G. Cozman and I. Cohen. “Unlabeled Data Can Degrade Classification Performance of Generative Classifiers”. In: *Fifteenth International Florida Artificial Intelligence Society Conference*. 2002, pp. 327–331.
39. K. Crammer and Y. Singer. “On the algorithmic implementation of multi class kernel-based vector machines”. In: *Journal of Machine Learning Research* 2 (2001), pp. 265–292.
40. K. Crammer and Y. Singer. “Pranking with ranking”. In: *Advances in Neural Information Processing Systems (NeurIPS 14)*. 2002, pp. 641–647.

41. G. Cybenko. In: *Mathematics of Control, Signals, and Systems (MCSS)* 2.4 (1989), pp. 303–314.
42. Q. Dai and S. Yang. "Curriculum Point Prompting for Weakly-Supervised Referring Image Segmentation". In: *Proceedings of the IEEE/CVF Conference on Computer Vision and Pattern Recognition (CVPR)*. 2024, pp. 13711–13722.
43. G. B. Dantzig. "Maximization of a linear function of variables subject to linear inequalities". In: *Activity Analysis of Production and Alloca- tion*. Ed. by T. C. Koopmans. Wiley, New York, 1951, pp. 339–347.
44. W. C. Davidon. "Variable metric method for minimization". In: *Journal of the Society for Industrial and Applied Mathematics on Optimization (SIOPT)* 1.1 (1991), pp. 1–17.
45. A. P. Dempster, N. M. Laird, and D. B. Rubin. "Maximum Likelihood from Incomplete Data via the EM Algorithm". In: *Journal of the Royal Statistical Society. Series B (Methodological)* 39.1 (1977), pp. 1–38.
46. J. E. Dennis Jr. and R. B. Schnabel. *Numerical Methods for Unconstrained Optimization and Nonlinear Equations (Classics in Applied Mathematics, 16)*. Soc for Industrial & Applied Math, 1996.
47. P. Derbeko, E. El-Yaniv, and R. Meir. "Error bounds for transductive learning via compression and clustering". In: *Advances in Neural Information Processing Systems (NeurIPS 15)*. 2003, pp. 1085–1092.
48. P. Deuflhard. *Newton Methods for Nonlinear Problems: Affine Invariance and Adaptive Algorithms*. Springer Verlag, 2004.
49. R. O. Duda, P. E. Hart, and D. G. Stork. *Pattern Classification*. Wiley, 2001.
50. A. Ehrenfeucht et al. "A general lower bound on the number of examples needed for learning". In: *Information and Computation* 82 (1989), pp. 247–261.
51. R. E. Fan et al. "LIBLINEAR: A Library for Large Linear Classification". In: *Journal of Machine Learning Research* 9 (2008), pp. 1871–1874.
52. W. Feller. *An Introduction to Probability Theory and Its Applications*. Wiley, 1968.
53. V. Feofanov, E. Devijver, and M.-R. Amini. "Multi-class Probabilistic Bounds for Majority Vote Classifiers with Partially Labeled Data". In: *Journal of Machine Learning Research* 25.104 (2024), pp. 1–47.
54. V. Feofanov, E. Devijver, and M.-R. Amini. "Transductive Bounds for the Multi-Class Majority Vote Classifier". In: *AAAI Conference on Artificial Intelligence*. 2019, pp. 3566–3573.
55. R. Fletcher. *Practical methods of optimization*. New York, USA: John Wiley & Sons, 1987.
56. R. Fletcher and C. M. Reeves. "Function minimization by conjugate gradients". In: *The Computer Journal* 7.2 (1964), pp. 149–154.
57. S. C. Fralick. "Learning to Recognize Patterns without a Teacher". In: *IEEE Transactions on Information Theory* 13.1 (1967), pp. 57–64.
58. Y. Freund and R. E. Schapire. "A Decision-theoretic Generalization of On-line Learning and an Application to Boosting". In: *Journal of Computer and System Sciences* 55.1 (1997), pp. 119–139.
59. Y. Freund and R. E. Schapire. "Large margin classification using the perceptron algorithm". In: *Machine Learning Journal* 37 (1999), pp. 277–296.
60. Y. Freund et al. "An Efficient Boosting Algorithm for Combining Preferences". In: *Journal of Machine Learning Research* 4 (2003), pp. 933–969.
61. K. Fukunaga. *Introduction to Statistical Pattern Recognition*. New York, USA: Academic Press, 1972.
62. K. Fukushima. "Neocognitron: A Self-organizing Neural Network Model for a Mechanism of Pattern Recognition Unaffected by Shift in Position". In: *Biological Cybernetics* 36.4 (1980).
63. M. R. Genesereth and N. J. Nilsson. *Logical Foundations of Artificial Intelligence*. San Francisco, CA, USA: Morgan Kaufmann Publishers Inc., 1987.
64. P. E. Gill and M. W. Leonard. "Reduced-Hessian Quasi-Newton Methods for Unconstrained Optimization". In: *Journal of the Society for Industrial and Applied Mathematics on Optimization (SIOPT)* 12.1 (2001).

65. E. Giné. “Empirical processes and applications: an overview”. In: *Bernoulli* 2.1 (1996), pp. 1–28.
66. I. Goodfellow, Y. Bengio, and A. Courville. *Deep Learning*. MIT Press, 2016.
67. I. Goodfellow et al. “Generative adversarial networks”. In: *Communications of the ACM* 63.11 (2020), pp. 139–144.
68. Y. Grandvalet and Y. Bengio. “Semi-supervised Learning by Entropy Minimization”. In: *Advances in Neural Information Processing Systems (NeurIPS 17)*. MIT Press, 2005, pp. 529–536.
69. I. Guefassa and Y. Chaib. “A modified conjugate gradient method for unconstrained optimization with application in regression function”. In: *RAIRO Operations Research* 59.1 (2025), pp. 1–22.
70. Y. Guermeur. “Sample complexity of classifiers taking values in R^Q, application to multi-class SVMs”. In: *Communications in Statistics - Theory and Methods* 39.3 (2010), pp. 543–557.
71. Y. Guermeur. “SVM Multiclasses, Théorie et Applications”. Habilitation à diriger des recherches. Université Nancy 1, 2007.
72. L. Hadjadj, M.-R. Amini, and S. Louhichi. “Self-Training of Half-spaces with Generalization Guarantees under Massart Mislabeling Noise Model”. In: *International Joint Conference on Artificial Intelligence - IJCAI*. 2023, pp. 3777–3785.
73. R. W. Hamming. “Error detecting and error correcting codes”. In: *Bell System Technical Journal* 29.2 (1950), pp. 147–160.
74. T. Hastie, R. Tibshirani, and J. Friedman. *The Elements of Statistical Learning*. Springer, 2001.
75. D. Hebb. *The Organization of Behavior*. John Wiley, 1949.
76. R. Herbrich et al. “Learning preference relations for information retrieval”. In: *Proceedings of the AAAI Workshop Text Categorization and Machine Learning, Madison, USA*. 1998.
77. M. R. Hestenes and E. Stiefel. “Methods of Conjugate Gradients for Solving Linear Systems”. In: *Journal of Research of the National Bureau of Standards* 49 (1952), pp. 409–436.
78. S. Hill et al. “Average Precision and the Problem of Generalisation”. In: *SIGIR Workshop on Mathematical and Formal Methods in Information Retrieval*. 2002.
79. G. E. Hinton et al. “Improving neural networks by preventing co-adaptation of feature detectors”. In: *CoRR* abs/1207.0580 (2012).
80. S. Hochreiter and J. Schmidhuber. “Long Short-Term Memory”. In: *Neural Computation* 9.8 (1997), pp. 1735–1780.
81. W. Hoeffding. “Probability inequalities for sums of bounded random variables”. In: *Journal of the American Statistical Association* 58 (1963), pp. 13–30.
82. A. E. Hoerl and R. W. Kennard. “Ridge Regression: Applications to Nonorthogonal Problems”. In: *Technometrics* 12 (1970), pp. 69–82.
83. B. Hofmann et al. “Ill-posed problems and the conjugate gradient method”. In: *Journal of Inverse and Ill-posed Problems* 30.5 (2022), pp. 659–677.
84. J. J. Hopfield. “Neurons with graded response have collective computational properties like those of two-state neurons”. In: *Proceedings of the National Academy of Sciences USA*. 1984, pp. 3088–3092.
85. K. Hornik, M. Stinchcombe, and H. White. “Multilayer feedforward networks are universal approximators”. In: *Neural Networks* 2.5 (1989), pp. 359–366.
86. D. H. Hubel and T. N. Wiesel. “Receptive fields and functional architecture of monkey striate cortex”. In: *Journal of Physiology* 195.1 (1968), pp. 215–243.
87. S. Ioffe and C. Szegedy. “Batch Normalization: Accelerating Deep Network Training by Reducing Internal Covariate Shift”. In: *Proceedings of the 32nd International Conference on Machine Learning*. 2015, pp. 448–456.
88. S. Janson. “Large deviations for sums of partly dependent random variables”. In: *Random Structures and Algorithms* 24.3 (2004), pp. 234–248. ISSN: 1042-9832.
89. T. Joachims. *Learning to Classify Text Using Support Vector Machines: Methods, Theory and Algorithms*. Norwell, MA, USA: Kluwer Academic Publishers, 2002.

90. T. Joachims. “Making large-Scale SVM Learning Practical”. In: *Advances in Kernel Methods - Support Vector Learning*. Ed. by B. Schölkopf, C. Burges, and A. Smola. Cambridge, MA: MIT Press, 1999. Chap. 11, pp. 169–184.
91. T. Joachims. “Transductive inference for text classification using support vector machines”. In: *Proceedings of the* 16th *International Conference on Machine Learning*. 1999, pp. 200–209.
92. D. Kearns and L. G. Valiant. *Learning Boolean formulae or finite automata is as hard as factoring*. Tech. rep. TR-14-88. Harvard University Aiken Computation Laboratory, 1988.
93. N. Keskar et al. “On Large-Batch Training for Deep Learning: Generalization Gap and Sharp Minima”. In: *5th International Conference on Learning Representations, ICLR*. 2017.
94. R. Kohavi. “A study of cross-validation and bootstrap for accuracy estimation and model selection”. In: *Proceedings of the* 14th *International Joint Conference on Artificial Intelligence (IJCNN)*. 1995, pp. 1137–1143.
95. V. I. Koltchinskii. “Rademacher penalties and structural risk minimization”. In: *IEEE Transactions on Information Theory* 47.5 (2001), pp. 1902–1914.
96. V. I. Koltchinskii and D. Panchenko. “Rademacher Processes and bounding the risk of function learning”. In: *High Dimensional Probability II*. Ed. by E. Giné, D. Mason, and J. Wellner. 2000, pp. 443–459.
97. A. Krizhevsky, I. Sutskever, and G. E. Hinton. “ImageNet Classification with Deep Convolutional Neural Networks”. In: *Advances in Neural Information Processing Systems 25*. 2012, pp. 1097–1105.
98. J. Langford. “Tutorial on Practical Prediction Theory for Classification”. In: *Journal of Machine Learning Research* 6 (2005), pp. 273–306. ISSN: 1532-4435.
99. P. S. Laplace. “Mémoire sur la Probabilité des Causes par les Événements”. In: *Académie Royale des sciences de Paris (Savants étrangers)* 6 (1771), pp. 621–656.
100. G. Latouche and V. Ramaswami. *Introduction to matrix analytic methods in stochastic modeling*. ASA-SIAM Series on Statistics and Applied Probability. Philadelphia, Pa. SIAM, Society for Industrial and Applied Mathematics Alexandria, Va. ASA, American Statistical Association, 1999.
101. Duc M. Le et al. “Accelerated Gradient Approach For Deep Neural Network-Based Adaptive Control of Unknown Nonlinear Systems”. In: *IEEE Transactions on Neural Networks and Learning Systems* 36.4 (2025), pp. 6299–6312. DOI: https://doi.org/10.1109/TNNLS.2024.3361643.
102. Y. LeCun, L. Bottou, and Y. Bengio. “Reading Checks with Multilayer Graph Transformer Networks”. In: *Proceedings of the 1997 IEEE International Conference on Acoustics, Speech, and Signal Processing (ICASSP ’97)*. 1997.
103. Y. LeCun et al. “Backpropagation applied to handwritten zip code recognition”. In: *Neural Computation* 1.4 (1989), pp. 541–551.
104. Y. Lee, Y. Lin, and G. Wahba. “Multicategory support vector machines: Theory and application to the classification of microarray data and satellite radiance data”. In: *Journal of the American Statistical Association* 99.465 (2004), pp. 67–81.
105. B. Leskes. “The value of agreement, a new boosting algorithm”. In: *Proceedings of Conference on Learning Theory (COLT)*. 2005, pp. 95–110.
106. Ming Li et al. “Convergence property of Nesterovaccelerated adaptive moment assessment algorithm for stochastic optimization”. In: *Mathematical Methods in the Applied Sciences* 47.5 (2024), pp. 8311–8327. DOI: https://doi.org/10.1002/mma.10174.
107. Y. Li et al. “Multi-Innovation Nesterov accelerated gradient parameter identification for ARX model”. In: *Journal of Vibration and Control* 30.3-4 (2024), pp. 494–507. DOI: https://doi.org/10.1177/10775463231207117.
108. H. Luo. “Self-Training for Natural Language Processing”. PhD thesis. Massachusetts Institute of Technology, USA, 2022.
109. Z. Luo and P. Tseng. “On the convergence of the coordinate descent method for convex differentiable minimization”. In: *Journal of Optimization Theory and Applications* 72.1 (1992), pp. 7–35.

110. G. J. Machlachlan. *Discriminant Analysis and Statistical Pattern Recognition*. Wiley Interscience, 1992.
111. P. Massart. "Some applications of concentration inequalities to statistics". In: *Annales de la faculté des sciences de Toulouse* 9.2 (2000), pp. 245–303.
112. P. McCullagh. "Regression models for ordinal data". In: *Journal of the Royal Statistical Society. Series B (Methodological)* 42.2 (1980), pp. 109–142.
113. W. S. McCulloch and W. Pitts. "A logical calculus of the ideas immanent in nervous activity". In: *Bulletin of Mathematical Biophysics* 5 (1943), pp. 115–133.
114. C. McDiarmid. "On the method of bounded differences". In: *Surveys in combinatorics* 141 (1989), pp. 148–188.
115. Y. Meng et al. "Text Classification Using Label Names Only: A Language Model Self-Training Approach". In: *Proceedings of the 2020 Conference on Empirical Methods in Natural Language Processing (EMNLP)*. Association for Computational Linguistics, 2020, pp. 9006–9017.
116. J. Mercer. "Functions of positive and negative type and their connection with the theory of integral equations". In: *Philosophical Transactions of the Royal Society* 209 (1909), pp. 415–446.
117. M. Minsky. "A Neural-Analogue Calculator Based upon a Probability Model of Reinforcement". In: *Harvard University Psychological Laboratories, Cambridge, Massachusetts* (1952).
118. M. Minsky and S. Papert. *Perceptrons: An Introduction to Computational Geometry*. MIT Press, 1969.
119. M. Mohri, A. Rostamizadeh, and A. Talwalkar. *Foundations of Machine Learning*. MIT Press, 2012. ISBN: 9780262018258.
120. O. Montasser, S. Hanneke, and N. Srebro. "Transductive Robust Learning Guarantees". In: *Proceedings of The 25th International Conference on Artificial Intelligence and Statistics - AISTATS*. 2022, pp. 11461–11471.
121. E. W. Montroll. "Random Walks in Multidimensional Spaces, Especially on Periodic Lattices". In: *Journal of the Society for Industrial and Applied Mathematics (SIAM)* 4.4 (1956), pp. 241–260.
122. A. Nemirovski et al. "Robust Stochastic Approximation Approach to Stochastic Programming". In: *Journal of the Society for Industrial and Applied Mathematics on Optimization (SIOPT)* 19.4 (2009), pp. 1574–1609.
123. A. S. Nemirovski and D. B. Yudin. *Problem complexity and method efficiency in optimization*. Wiley-Interscience, 1983.
124. Y. Nesterov. "A method of solving a convex programming problem with convergence rate $O(1/k^2)$". In: *Soviet Mathematics Doklady* 27.2 (1983), pp. 372–376.
125. Y. Nesterov. "A method of solving a convex programming problem with convergence rate $O(1/k^2)$". In: *Soviet Mathematics Doklady* 27 (1983), pp. 372–376.
126. Y. Nesterov. *Introductory Lectures on Convex Optimization: A Basic Course*. Kluwer Academic Publishers, 2004.
127. K. Nigam et al. "Text Classification from Labeled and Unlabeled Documents Using EM". In: *Machine Learning Journal* 39.2–3 (2000), pp. 103–134.
128. N. J. Nilsson. *Learning machines; foundations of trainable pattern-classifying systems*. McGraw-Hill, 1965.
129. J. Nocedal and S. J. Wright. *Numerical Optimization*. Springer, 2006.
130. Y. Notay. "Flexible Conjugate Gradients". In: *SIAM Journal on Scientific Computing* 22.4 (2000), pp. 1444–1460.
131. A. B. Novikoff. "On convergence proofs on perceptrons". In: *Symposium on the Mathematical Theory of Automata* 12 (1962), pp. 615–622.
132. D. Parker. *Learning logic*. Tech. rep. TR-87. Cambridge, MA: Center for Computational Research in Economics and Management Science, MIT, 1985.
133. E. A. Patrick, J. P. Costello, and F. C. Monds. "Decision-Directed Estimation of a Two-Class Decision Boundary". In: *IEEE Transactions on Information Theory* 9.3 (1970), pp. 197–205.

134. Z. Peng et al. “FaxMatch: Multi-Curriculum Pseudo-Labeling for semi-supervised medical image classification”. In: *Medical Physics* 50.5 (2023), pp. 3210–3222.
135. E. Polak. *Computational methods in optimization*. Academic press, 1971.
136. E. Polak and G. Ribiere. “Note sur la convergence de méthodes de directions conjuguées”. In: *ESAIM: Mathematical Modelling and Numerical Analysis - Modélisation Mathématique et Analyse Numérique* 3.R1 (1969), pp. 35–43.
137. B. T. Polyak. “Some methods of speeding up the convergence of iteration methods”. In: *USSR Computational Mathematics and Mathematical Physics* 4.5 (1964), pp. 1–17.
138. T. Qin, T.-Y. Liu, and H. Li. *A General Approximation Framework for Direct Optimization of Information Retrieval Measures*. Tech. rep. MSR-TR-2008-164. Microsoft Research, 2008.
139. A. Rakotomamonjy. “Optimizing Area Under Roc Curve with SVMs.” In: *1st International workshop on ROC Analysis in Artificial Intelligence*. 2004, pp. 71–80.
140. M. D. Richard and R. P. Lippman. “Neural Network Classifiers Estimate Bayesian a posteriori Probabilities”. In: *Neural Computation* 3.4 (1991), pp. 461–483.
141. S. E. Robertson and S. Walker. “Some Simple Effective Approximations to the 2-Poisson Model for Probabilistic Weighted Retrieval”. In: *SIGIR’94, conference on Research and development in information retrieval*. 1994, pp. 232–241.
142. A. J. Robinson and F. Fallside. *The Utility Driven Dynamic Error Propagation Network*. Tech. rep. Cambridge, UK: Engineering Department, Cambridge University, 1987.
143. F. Rosenblatt. “The Perceptron: A Probabilistic Model for Information Storage and Organization in the Brain”. In: *Psychological Review* 65 (1958), pp. 386–408.
144. C. Rudin et al. “Margin-Based Ranking Meets Boosting in the Middle”. In: *Conference On Learning Theory (COLT)*. 2005.
145. D. E. Rumelhart, G. E. Hinton, and R. J. Williams. “Learning internal representations by error propagation”. In: *Parallel Distributed Processing: Explorations in the Microstructure of Cognition* I (1986).
146. G. Salton. “A Vector Space Model for Automatic Indexing”. In: *Communications of the ACM* 18.11 (1975), pp. 613–620.
147. N. Sauer. “On the density of families of sets”. In: *Journal of Combinatorial Theory* 13.1 (1972), pp. 145–147.
148. R. E. Schapire. “The Strength of Weak Learnability”. In: *Machine Learning* 5.2 (1990), pp. 197–227.
149. R. E. Schapire. “Theoretical Views of Boosting and Applications”. In: *Proceedings of the 10th International Conference on Algorithmic Learning Theory*. 1999, pp. 13–25.
150. R. E. Schapire et al. “Boosting the margin: a new explanation for the effectiveness of voting methods”. In: *The Annals of Statistics* 26.5 (1998), pp. 1651–1680.
151. B. Schölkopf and A. J. Smola. *Learning with kernels: support vector machines, regularization, optimization, and beyond*. MIT Press, 2002.
152. M. Seeger. *Learning with Labeled and Unlabeled Data*. Tech. rep. 2001.
153. S. Shalev-Shwartz et al. “Pegasos: primal estimated sub-gradient solver for SVM”. In: *Mathematical Programming* 127.1 (2011), pp. 3–30.
154. A. Shashua and A. Levin. “Ranking with Large Margin Principle: Two Approaches”. In: *Advances in Neural Information Processing Systems (NeurIPS 15)*. 2003, pp. 961–968.
155. S. Shelah. “A combinatorial problem: Stability and order for models and theories in infinity languages”. In: *Pacific Journal of Mathematics* 41 (1972), pp. 247–261.
156. V. Sindhwani, P. Niyogi, and M. Belkin. “A Co-regularization Approach to Semi-supervised Learning with Multiple Views”. In: *ICML- 05 Workshop on Learning with Multiple Views*. 2005, pp. 74–79.
157. N. Srivastava et al. “Dropout: A Simple Way to Prevent Neural Networks from Overfitting”. In: *Journal of Machine Learning Research* 15 (2014), pp. 1929–1958.
158. N. Srivastava et al. “Dropout: a simple way to prevent neural networks from overfitting”. In: *Journal of Machine Learning Research* 15.1 (2014), pp. 1929–1958.
159. M. H. Stone. “The Generalized Weierstrass Approximation Theorem”. In: *Mathematics Magazine* 21.4 (1948), pp. 167–184.

160. C. Sun et al. "Revisiting Unreasonable Effectiveness of Data in Deep Learning Era". In: *2017 IEEE International Conference on Computer Vision (ICCV)*. 2017, pp. 843–852.
161. I. Sutskever et al. "On the importance of initialization and momentum in deep learning". In: *Proceedings of the 30th International Conference on Machine Learning*. 2013, pp. 1139–1147.
162. M. J. Symons. "Clustering criteria and multivariate normal mixtures". In: *Biometrics* 37.1 (1981), pp. 35–43.
163. M. Szummer and T. Jaakkola. "Partially labeled classification with Markov random walks". In: *Advances in Neural Information Processing Systems (NeurIPS 14)*. 2002, pp. 945–952.
164. J. S. Taylor and N. Cristianini. *Kernel Methods for Pattern Analysis*. New York, USA: Cambridge Press University, 2004.
165. M. Taylor et al. "SoftRank: Optimising Non-Smooth Rank Metrics". In: *WSDM 2008*. 2008.
166. P. L. Tchebychev. "Des valeurs moyennes". In: *Journal de mathématiques pures et appliquées* 2.12 (1867), pp. 177–184.
167. R. Tibshirani. "Regression Shrinkage and Selection via the Lasso". In: *Journal of the Royal Statistical Society (Series B)* 58 (1996), pp. 267–288.
168. D. M. Titterington, A. F. M. Smith, and U. E. Smith. *Statistical Analysis of Finite Mixture Distributions*. Wiley, New York, 1985.
169. J. Truett, J. Cornfield, and W. Kannel. "A multivariate analysis of the risk of coronary heart disease in Framingham". In: *Journal of Chronic Diseases* 20.7 (1967), pp. 511–524.
170. G. Tür, D. Z. Hakkani-Tür, and R. E. Schapire. "Combining active and semi-supervised learning for spoken language understanding". In: *Speech Communication* 45.2 (2005), pp. 171–186.
171. A. M. Turing. "Computing machinery and intelligence". In: *Mind* 59 (1950), pp. 433–460.
172. R. Urner, S. Shalev-Shwartz, and S. Ben-David. "Access to Unlabeled Data can Speed up Prediction Time". In: *28th International Conference on Machine Learning, ICML 2011*. 2011, pp. 641–648.
173. N. Usunier. "Apprentissage de fonctions dordonnancement: une étude théorique de la réduction à la classification et deux applications à la Recherche d'Information". PhD thesis. Université Pierre & Marie Curie, 2006.
174. N. Usunier, M.-R. Amini, and P. Gallinari. "Generalization error bounds for classifiers trained with interdependent data". In: *Ad- vances in Neural Information Processing Systems (NeurIPS 18)*. 2006, pp. 1369–1376.
175. L. J. Valiant. "The Theory of the Learnable". In: *Communications of the ACM* 27.11 (1984), pp. 1134–1142.
176. V. N. Vapnik. *The nature of statistical learning theory*. New York, NY, USA, 1995.
177. V. N. Vapnik. *The nature of statistical learning theory (second edition)*. Springer-Verlag, 1999.
178. V. N. Vapnik and A. J. Chervonenkis. "On the uniform convergence of relative frequencies of events to their probabilities". In: *Theory of Probability and its Applications* 16 (1971), pp. 264–280.
179. A. Vaswani et al. "Attention is All you Need". In: *Advances in Neural Information Processing Systems*. Vol. 30. 2017.
180. A. Waibel. "Phoneme Recognition Using Time-Delay Neural Net- works". In: *Meeting of the Institute of Electrical, Information and Communication Engineers*. 1987.
181. P. J. Werbos. "Beyond Regression: New Tools for Prediction and Analysis in the Behavioral Sciences". PhD thesis. Harvard University, 1974.
182. J. Weston and C. Watkins. "Support vector machines for multi-class pattern recognition". In: *European Symposium on Artificial Neural Netwroks (ESANN)*. 1999, pp. 219–224.
183. G. Widrow and M. Hoff. "Adaptive switching circuits". In: *Institute of Radio Engineers, Western Electronic Show and Convention, Con- vention Record* 4 (1960), pp. 96–104.
184. R. J. Williams and D. Zipser. "Gradient-based learning algorithms for recurrent networks and their computational complexity". In: *Back- propagation: theory, architectures, and applications*. Ed. by B. Schölkopf, C. Burges, and A. Smola. 1995. Chap. 4, pp. 433–486.

185. P. Wolfe. "Convergence Conditions for Ascent Methods". In: *SIAM Review* 11.2 (1966), pp. 226–235.
186. S. K. M. Wong and Y. Y. Yao. "Linear Structure in Information Retrieval". In: *SIGIR'88, conference on Research and Development in Information Retrieval*. 1988, pp. 219–232.
187. S. J. Wright and B. Recht. *Optimization for Data Analysis*. Cambridge University Press, 2022.
188. J. Xu and H. Li. "AdaRank: A Boosting Algorithm for Information Retrieval". In: *SIGIR '07, conference on Research and Development in Information Retrieval*. 2007, pp. 391–398.
189. J. Xu et al. "Directly Optimizing Evaluation Measures in Learning to Rank". In: *SIGIR '08, conference on Research and Development in Information Retrieval*. 2008, pp. 107–114.
190. Y. Yue et al. "A Support Vector Method for Optimizing Average Precision". In: *SIGIR '07, conference on Research and Development in Information Retrieval*. 2007, pp. 271–278.
191. B. Zhang et al. "FlexMatch: Boosting Semi-Supervised Learning with Curriculum Pseudo Labeling". In: *Advances in Neural Information Processing Systems - NeurIPS*. 2021, pp. 18408–18419.
192. T. Zhang and F. J. Oles. "A probability analysis on the value of unlabeled data for classification problems". In: *17th International Conference on Machine Learning*. 2000.
193. Z. Zhang, E. Strubell, and E. H. Hovy. "Data-efficient Active Learning for Structured Prediction with Partial Annotation and Self-Training". In: *Findings of the Association for Computational Linguistics: EMNLP*. Association for Computational Linguistics, 2023, pp. 12991–13008.
194. D. Zhou et al. "Learning with local and global consistency". In: *Advances in Neural Information Processing Systems (NeurIPS 16)*. MIT Press, 2004, pp. 321–328.
195. X. Zhu and Z. Ghahramani. *Learning from labeled and unlabeled data with label propagation*. Tech. rep. CMU-CALD-02-107. Carnegie Mellon University, 2002.
196. X. Zhu, Z. Ghahramani, and J. Lafferty. "Semi-supervised learning using Gaussian fields and harmonic functions". In: 20th *International Conference on Machine Learning*. 2003, pp. 912–919.
197. Y. Zou et al. "Unsupervised domain adaptation for semantic segmentation via class-balanced self-training". In: *European conference on computer vision - ECCV*. 2018, pp. 289–305.
198. G. Zoutendijk. "Some Recent Development in Nonlinear Programming". In: *5th Conference on Optimization Techniques, Part 1*. 1973, pp. 407–417.
199. X. Zuo, R. Xiong, and X. Li. "NALA: A Nesterov Accelerated Look-Ahead Optimizer for Deep Learning". In: *PeerJ Computer Science* 9 (2023).

Index

M.-R. Amini, *Advanced Supervised and Semi-supervised Learning*,
Cognitive Technologies, https://doi.org/10.1007/978-3-031-99928-4